Integrated Systems: Data Driven Engineering

Mohammad-Reza Alam · Madjid Fathi
Editors

Integrated Systems: Data Driven Engineering

Springer

Editors
Mohammad-Reza Alam
Department of Mechanical Engineering
University of California
Berkeley, CA, USA

Madjid Fathi
Institute for Knowledge Based Systems
and Knowledge Management
Research Center for Knowledge
Management and Intelligent Systems
University of Siegen
Siegen, Germany

ISBN 978-3-031-53651-9 ISBN 978-3-031-53652-6 (eBook)
https://doi.org/10.1007/978-3-031-53652-6

This Springer imprint is published by the registered company Springer Nature Switzerland AG
The registered company address is: Gewerbestrasse 11, 6330 Cham, Switzerland

Preface

The term "integrated system" refers to a seamless collaborative operation of a large number of (potentially unrelated) subsystems to achieve a specific goal. It involves combining multiple subsystems or components, such as hardware, software, networks, and workflows, into one larger system that works as a whole. Integrated systems, widely employed across various scientific and technological domains today, aim to enhance coherence, efficiency, and the overall quality of the system functionality.

A familiar example of an integrated system is a smartphone, which, to deliver the expected smooth functionality, must have a consistent and flawless integration of various subsystems such as electronics, wireless technology, camera, software, user interface, energy management, heat control, smart materials and composite design, ergonomics, and more. Back in the old days, creation of most goods relied mainly on the skills of a single, or small group of artisans, with one (usually specific) expertise. A carpenter, for example, could craft most of the household items. In today's world, however, a lot more is expected from a product than a mere functionality. Take a simple chair, for example. Modern chairs are expected to be ergonomic, necessitating extensive research into people's habits and the health implications of factors like shape, height, and back angle adjustments. Furthermore, factors such as stability, quality, reliability, safety, production methodology, economic considerations, market competition, environmental impacts, and more must also be taken into account for a successful venture.

The concept of "integrated systems" extends beyond consumer products and applies to most modern challenges in science, technology, and society. Astrophysics, for instance, relies on precise measurements primarily accomplished through highly accurate space telescopes fabricated using advanced manufacturing methods. Once constructed the entire telescope must be transported into orbit using efficient and powerful rockets. In the realm of medicine, surgical procedures increasingly depend on a variety of precision machines, ultrasonic devices, X-rays, MRIs, and modern nuclear technologies, bringing together various fields of science and engineering.

As evident from the examples above, the swift pace of technological advancement has heightened the demand for multidisciplinary approaches to tackle intricate and complex challenges of the modern world. In response, a successful strategy is the integration of systems and technologies that involves combining elements like human resources, technology, and environmental factors to attain a shared objective. Integrated systems and technologies yield numerous advantages, including increased efficiency, enhanced accuracy, higher flexibility, learning capability, adaptability of digital system, superior decision making, and cost savings. Through process streamlining, waste reduction, and overall efficiency improvement, integrated systems and technologies offer a more comprehensive and effective approach to problem-solving.

Despite these advantages, there are still challenges and gaps in our understanding of how to effectively integrate various systems. These challenges include complexity, compatibility, security, cost, resistance to change, customized maintenance, and the need for regular updates while ensuring continued consistency. Addressing these challenges requires a collaborative approach that brings together experts from different fields to develop effective solutions. Enterprises and industrial parties must also stay vigilant and informed about key technologies in their respective fields and beyond to identify opportunities for integrating different technologies and systems.

The Integrated Systems Design and Technology (ISDT) conference aims to facilitate collaboration among diverse disciplines in various sectors. The event features renowned scientists from around the globe who specialize in a wide range of fields, with the goal of fostering a deeper understanding of other disciplines and the approaches required for multidomain research and projects. The theme of 2023 conference revolved around the pivotal role of data in engineering and knowledge technology. As technology continues to advance, data driven approaches have become increasingly critical in shaping the future of engineering design and production. This year's conference aimed to delve into the various ways in which data and knowledge can be utilized to enhance and optimize engineering processes, improve precision, and facilitate innovation.

The ISDT community is committed to establishing a network of professionals worldwide to foster international collaboration. This network will facilitate the exchange and sharing of ideas, best practices, and innovations among experts from different disciplines, resulting in the development of effective solutions for complex challenges, including those encountered in Industry 4.0 and Society 5.0. By bringing together scientists from different fields, ISDT aims to promote interdisciplinary collaboration and innovation, ultimately yielding improved outcomes and more efficient problem-solving.

Berkeley, CA, USA — Mohammad-Reza Alam
Siegen, Germany — Madjid Fathi

Contents

Intelligent Robotics and Embedded Controls

Knowledge Technology in Material Engineering

Knowledge-Based Learning

Automated Photonic Waveguide Loss Measurement Using Out-scattering Light Method

Xiangjian Zeng, Jay W. Reddy, and Maysamreza Chamanzar

Abstract Characterization of optical waveguide propagation loss is fundamental to the practical realization of integrated photonics. Evaluating waveguide performance throughout the fabrication process enables rigorous quality control and rapid fabrication process development and optimization. Conventional measurement techniques such as the cutback and modified cutback methods directly measure the out-coupling light intensity from optical fibers and photonic waveguides of different lengths, and subsequently estimate the waveguide propagation loss. These methods are generally laborious, requiring many measurements for each device. Additionally, they are destructive to the device under test, as in the case of the cutback method, or require specially-designed arrays of various lengths, as in the case of the modified cutback method. In this work, we describe a method utilizing the out-scattering light intensity profile to efficiently assess multiple waveguides non-destructively and in a fully-automated manner to determine the individual waveguide propagation loss and identify faulty waveguide regions. The accuracy of the out-scattering method is validated using traditional modified cutback methods. Moreover, a comparative analysis of the two methods is presented, along with strategies to optimize their performance. This study is performed using Parylene photonic waveguides, which we have recently demonstrated for light delivery into the brain for neural stimulation.

X. Zeng · J. W. Reddy · M. Chamanzar (✉)
Department of Electrical and Computer Engineering, Carnegie Mellon University, 5000 Forbes Ave, Pittsburgh, PA 15213, USA
e-mail: mchamanz@andrew.cmu.edu

X. Zeng
e-mail: xiangjiz@andrew.cmu.edu

J. W. Reddy
e-mail: jreddy1@andrew.cmu.edu

M.-R. Alam and M. Fathi (eds.), *Integrated Systems: Data Driven Engineering*,
https://doi.org/10.1007/978-3-031-53652-6_1

1 Introduction

Optical waveguides are widely utilized in domains such as optical communications, integrated photonics, and biophotonics [1, 2]. A fundamental property of an optical waveguide is how quickly the optical power decays as light propagates through the waveguide (i.e., the propagation loss). The propagation loss ultimately determines the suitability of a given waveguide architecture for its intended application, depending on the requirements for the input/output power and the length of the waveguide. Therefore, characterizing the waveguide propagation losses is an essential part of quality control and performance optimization.

Numerous waveguide propagation loss characterization methods have been demonstrated in the past. The direct cut-back method, in which a manufactured waveguide is cut shorter sequentially for each output power measurement, provides a straightforward probe into propagation loss by measuring the optical power attenuation between each measurement [3]. However, such an approach is destructive to the waveguide sample and hence undesirable under many circumstances. For instance, in a research setting, fabricating many samples for repeated measurements can be inefficient. In addition, in a production line, destructive testing can only be used for spot-checking quality control, but can't be used to validate the performance of every single device before use. The latter is especially problematic for critical systems, i.e., in biophotonics, where a defective device can put an implanted animal model or human being at risk.

To preserve the samples under test, the modified cut-back method compares the output light intensity of multiple waveguides with different lengths [4]. However, these measurements require repeated alignment to many waveguides, which can introduce experimental error from trial-to-trial variance. Additionally, it requires specially-designed cutback structures of various lengths to be fabricated. Furthermore, it only measures the average propagation loss across a group of waveguides, meaning that the measurement accuracy will suffer from device-to-device variations, as well as variations in the coupling efficiency. Although cutback methods are the most prevalent, various other measurement techniques have been introduced in recent years to address their limitations, including using ring resonators [5], Fabry-Pérot resonators [6–8], and electro-optic Mach-Zehnder interferometers [9] to estimate the propagation loss. Although these techniques are nondestructive and accurate, they nonetheless require the construction of a specialized structure in order to measure the propagation loss in an equivalent waveguide. Consequently, they cannot assess the in-situ performance of a functional device. More importantly, the performance characteristics of functional devices must be inferred from the specialized device under test, so these methods are poorly suited for characterizing the intra-batch variance during the fabrication, or for isolating localized defects within a single sample waveguide. To fill the gap of these existing methods, there is an outstanding need for a method to allow the accurate, high-throughput, in-situ characterization of the waveguide propagation losses.

Due to the above constraints, the out-scattering method has been explored as a way to improve waveguide propagation loss characterization capabilities [10]. By observing the intensity profile of out-scattering light, this method estimates the optical power decay within the waveguide with high spatial resolution while being nondestructive. The propagation loss of individual regions within the waveguide can be obtained by a localized measurements of out-scattering light. Additionally, this method directly observes the waveguide itself, without the need for specialized structures. Because of these benefits, the out-scattering method has seen increasing used in recent years [11–14].

However, key limitations to the out-scattering method have prevented its widespread adoption. The reliability of the out-scattering method hasn't been validated using a rigorous comparison to the gold-standard techniques such as the cutback method. Prior published work has shown that the out-scattering light method and modified cutback method have good agreements for some cases [15]. However, to the best of the authors' knowledge, there has not been extensive characterization of the regime under which out-scattering light methods can be considered reliable. In addition, the out-scattering light technique doesn't inherently eliminate problems of repetitive re-coupling for each waveguide. These two issues need to be addressed to render the out-scattering method suitable for wider adoption.

Here, we demonstrate the out-scattering method to characterize Parylene photonic waveguides. Parylene photonics is a flexible biocompatible material platform for biophotonic devices such as implantable optical brain-computer interfaces. Previous work has shown that the fabrication of low-loss (i.e., < 5 dB/cm) Parylene photonic waveguides requires optimized etching conditions and an additional smoothing step to reduce scattering losses [13]. Since the out-scattering method is non-destructive, it may be used for intermediate process characterization between individual fabrication process steps, i.e., before the smoothing step to characterize the waveguide performance in the middle of the fabrication process. This technique was previously used to characterize Parylene photonic waveguide performance [14] demonstrating losses of (3.2 dB/cm at λ = 680 nm, 4.1 dB/cm at λ = 633 nm, 4.9 dB/cm at λ = 532 nm, 6.1 dB/cm at λ = 450 nm) after smoothing. Here, we provide a detailed description of the out-scattering method and demonstrate its utility for wafer-scale characterization during intermediate steps of the fabrication process.

In this work, we validate the out-scattering light method by performing a rigorous characterization of integrated photonic waveguides using both the out-scattering light and modified cutback methods. Furthermore, we provide guidelines and constraints on when to utilize the out-scattering light method in order to obtain accurate results. This out-scattering characterization method, combined with a custom-designed hardware and software system enables high-throughput and fully-automated measurement of photonic waveguides on the wafer scale.

2 Method

To perform a rigorous comparison of the out-scattering method and the modified cutback method for assessing propagation loss, we constructed a custom-built robotic characterization system to perform high-throughput and low-variance measurements of micro-fabricated integrated photonic waveguides on the wafer scale (Fig. 1). A custom software pipeline was built to perform the automated measurements as well as the analysis. In this section, both the waveguide wafer design and acquisition setup are explained in detail. Subsequently, we discuss the software system and techniques for automated fiber alignment and imaging to obtain waveguide path images for propagation loss fitting. Finally, the experimental procedures for two propagation loss measurement methods—the modified cutback method and the out-scattering method—are presented.

2.1 Waveguide Fabrication

In this work, we utilized the micro-fabricated Parylene photonic waveguides as testing samples to validate the performance of the proposed method. Here, a specific wafer layout is designed to facilitate the measurement process.

Parylene photonic waveguides were fabricated using the microfabrication process described in [14]. Briefly, 45-degree micromirror molds are etched in a silicon wafer at the input and output ports of the waveguides via anisotropic KOH etching to enable out-of-plane input and output coupling. Low refractive index polydimethylsiloxane (PDMS) is spin-coated as a lower cladding of the waveguides to a thickness of 1 μm, and high-index Parylene C is deposited via chemical vapor deposition to a thickness of 3.5 μm across the wafer surface, then etched in oxygen plasma to form the waveguide core using a lithographically-defined Chromium hard mask. Notably, to demonstrate the feasibility of nondestructive wafer-scale characterization of the waveguides during an intermediate process step, measurements are performed midway through the process–before depositing the subsequent loss-reducing Parylene C smoothing layer described in previous work [13, 14] is applied.

For the purposes of the analyses performed in this paper, we utilize a wafer layout that can be measured via either the modified cutback method or the out-scattering method. This design allows results for both methods to be collected from the same samples to verify the accuracy of the out-scattering method. The wafer layout contains 200 waveguides in total. The waveguides are organized in groups of 20 waveguides of the same length. Each group contains 4 sets of 5 waveguides with widths of 5 μm, 10 μm, 15 μm and 30 μm each. Each group of 20 waveguides are referred to as a waveguide batch and are linearly arranged on the wafer so that all of the input and output ports are aligned. Each subsequent batch of waveguides (5 mm longer) is placed adjacent to the previous batch, protruding 2.5 mm longer in both directions. This pattern is iterated until the 10th batch of waveguides reaches 50 mm

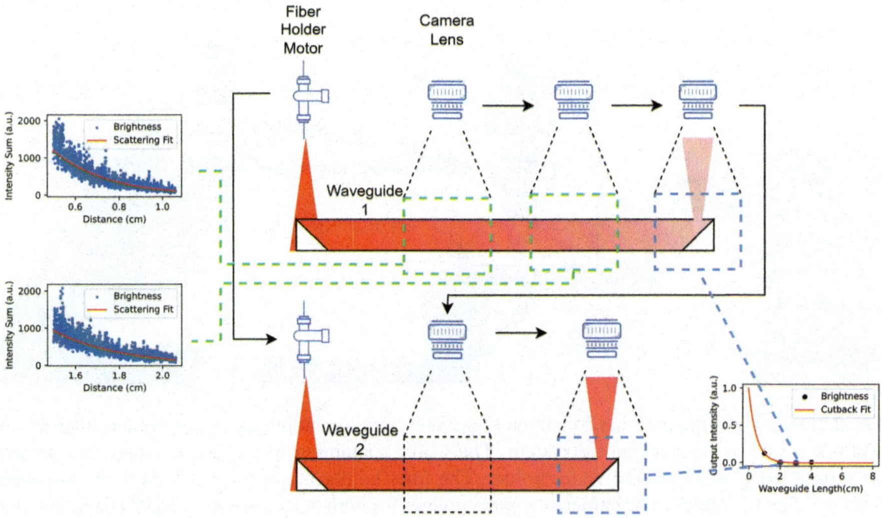

Fig. 1 Diagram showing the full waveguide characterization pipeline. Initially, the input optical fiber is aligned to a single waveguide input, and the camera traverses the entire length of the waveguide to capture multiple images for out-scattering light loss measurements, which are shown in inset plots to the left. Once the out-scattered light of the entire waveguide has been measured, the output intensity is captured. The fiber is then moved to align to the next waveguide. The process is repeated until all waveguides are imaged. Multiple output measurements from waveguides of different lengths can be used to calculate the loss via the modified cutback method, which is shown in the inset figure on the right

in length, providing a range of waveguide lengths to measure. Each waveguide on the wafer features monolithically-integrated micromirrors at the waveguide input and output ports to enable out-of-plane input and output coupling. Out-of-plane output coupling permits both the output port and out-scattering light path to be observed by a vertically-oriented camera, whereas out-of-plane input coupling allows coupling to a vertically-oriented input optical fiber. As will be discussed in later sections, a sufficiently long waveguide is essential for obtaining accurate results with the out-scattering method. In this work, we have tailored our experimental protocol to our particular wafer layout design. However, in principle, this method is generalizable to any wafer that features out-of-plane coupling and waveguides of various lengths.

2.2 Acquisition Setup

To achieve robust, high throughput measurements, the waveguide characterization system is capable of automatically observing and adjusting the alignment of this input optical fiber onto the wafer via the integrated camera and motorized stages.

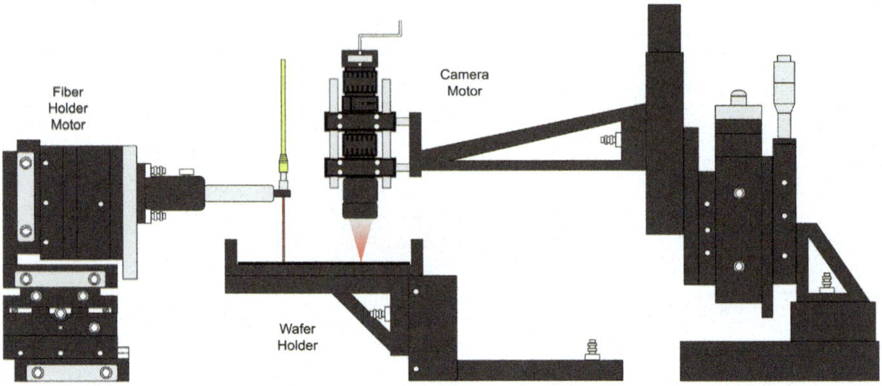

Fig. 2 Diagram of the automatic characterization system, including the wafer holder, fiber holder motor system, and camera motor system. The wafer containing waveguides being characterized is secured using a 3D printed wafer holder. The fiber holder motor positions the fiber to couple light into the waveguide. The camera captures images for both automated alignment and automated imaging steps in the acquisition pipeline

The waveguide characterization system consists of three modules: (1) wafer holder, (2) fiber holder stage and motor system, and (3) camera stage and motor system (Fig. 2). The fiber holder motor and camera holder motor each can move in the XYZ directions, and the camera holder can additionally rotate along the center axis. The input optical fiber is fixed in a vertical orientation above the wafer surface by the fiber holder. The fiber motor allows the fiber to not only travel in X and Y direction to align with individual waveguides, but also to adjust the distance to the wafer surface to efficiently couple to the waveguide input port. Light is coupled into the waveguides from a 633 nm 50 mW pigtailed laser diode (LP633-SF50, ThorLabs). Likewise, the camera holder moves the camera to locate and capture the output ports of the waveguides, while the rotation allows the camera to pivot to prevent collision with the input optical fiber while imaging near the input port. The camera system used for imaging consists of a CCD Camera (EO-5012M, Edmund Optics) and VZM^{TM} 600i Zoom Imaging Lens with a 60 mm working distance. This CCD camera was chosen since it provides low noise imaging in low-light conditions. The chosen lens provides an adjustable field of view (1.0–6.4 mm), suitable to image small segments of the waveguides (5–50 mm long).

For each round of measurements, the waveguide wafer under test is secured using a 3D printed wafer holder. The initial position of the fiber holder is manually set to ensure all waveguides fall within the motor travel range.

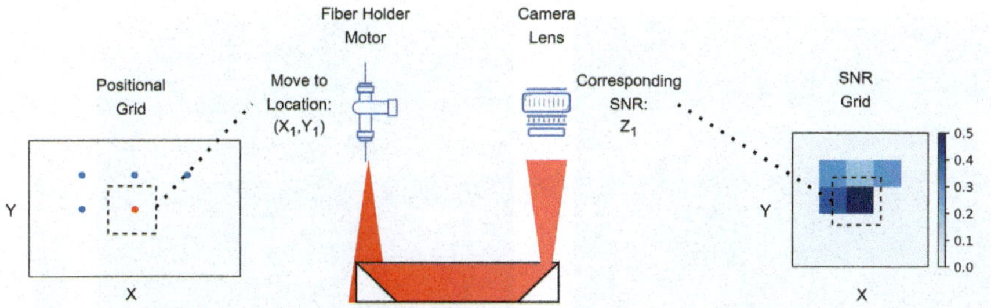

Fig. 3 Diagram of the automated alignment step. The fiber holder scans through a grid to find the optimum coupling position. For each position, an image containing waveguide path is taken by the camera to evaluate the signal-to-noise ratio (SNR). The system chooses the position corresponding to the highest SNR value for subsequent automated imaging

2.3 *Acquisition Pipeline*

After securing the waveguide wafer and setting the initial fiber location, the software system begins the image acquisition pipeline for each waveguide, which consists of two phases: automated alignment and automated imaging.

During automated alignment, the system couples the fiber to the micro-mirror input port of the waveguide to maximize output intensity (Fig. 3). The fiber holder scans a positional grid around the starting location while monitoring the waveguide out-scattering brightness to discover the optimal waveguide coupling position, as described in our previous publication [16]. The system performs three rounds of grid search around the optimal coupling position with decreasing grid spacing each time to gradually approach the optimum coupling position. For our waveguide with a width of up to 30 μm, we chose 6 μm, 1 μm, and 0.2 μm grid spacing for each iteration and 6 by 4 grid positions in X and Y directions. For each position the holder moves to, the camera captures one image containing out-scattering light 5 mm from the input port.

In our previous work [16], we used the intensity sum along the waveguide path as a measure of the coupling quality. Further investigation showed that the previous strategy would favor images with strong stray light reflections from the input port (Fig. 4 right) over high coupling efficiency (Fig. 4 left).

To overcome this limitation, here the quality of the coupling in each image is evaluated via the Signal-to-Noise Ratio (SNR). Here SNR is defined as the average intensity of pixels in the waveguide path region over the average intensity of pixels in the non-path region (defined as more than 35 pixels away from the waveguide center):

$$SNR = \frac{\frac{\sum_{n \subset WG} I_n}{|n \subset WG|}}{\frac{\sum_{n \notin WG} I_n}{|n \notin WG|}} \tag{1}$$

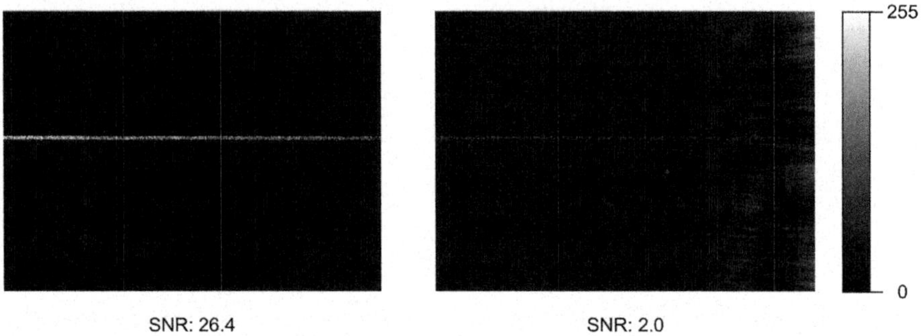

Fig. 4 Comparison between waveguide path image with good coupling (left) and high stray light intensity (right). As shown here, SNR can effectively distinguish the good coupling scenario

The SNR metric helps with the identification of fiber positions with high stray light intensity since the background light intensity is much higher than that of fiber positions with high coupling efficiency. After the positions are ranked, the fiber is moved to the optimal position.

Once the input coupling is optimized, the system proceeds to the automated imaging step. Propagation loss estimation via the out-scattering method requires imaging of regions of interest along the entire waveguide. The system automatically scans through the length of the waveguide, taking piecewise images. Each image covers 5.6 mm of the waveguide length. These piecewise images are offset by increments of 1 mm along the length of the waveguide, yielding a partially overlapping sequence.

Optical power decays exponentially within the waveguide, so the corresponding out-scattering light and intensity at the output port will vary by several orders of magnitude along the length of long waveguides. Thus, it's necessary to automatically adjust the camera exposure time to ensure the image intensity falls within the dynamic range of the camera sensor (0–255 pixel intensity in 8-bit encoding). To perform accurate high dynamic range imaging, we characterized the linearity of the sensor. We projected a 12 μW power laser light directly into the camera lens and varied the exposure time while measuring the peak brightness. As shown in Fig. 5, the peak brightness in the image exhibits nonlinear saturation behavior well before it reaches the peak 255 pixel intensity value. Thus, values in the image close to 255 may underestimate the true brightness level of the waveguide. Intensity ranging between 50 to 200 appears to follow a linear relationship with the exposure. Thus, to ensure linearity across a high dynamic range, the exposure time of the camera is automatically adjusted for each image to maintain a peak pixel intensity between 50–200.

During automated imaging, for each new camera location, a trial image is taken for exposure adjustment. Exposure time is adjusted according to the maximum image intensity via a binary search. If the intensity is greater than 200, then the exposure time is halved; if the intensity is less than 50, the the exposure time is doubled for the next iteration. This process is repeated until an optimum brightness is reached or the

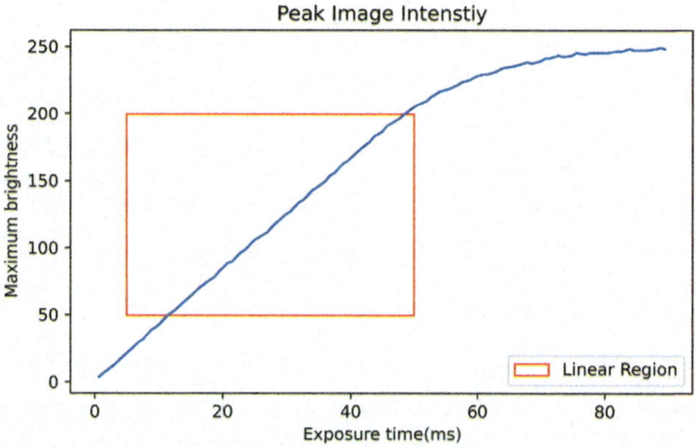

Fig. 5 Pixel intensity does not scale linearly with the exposure time for the CCD camera used in this study. Saturation occurs for intensity values greater than 200, before the dynamic range maximum (255) of the image is reached. The linear region of the sensor is indicated in red

camera exposure time reaches the hardware limit (0.6 μs–2.7 s). The final exposure time is stored alongside each image. In the following analyses, the system utilizes adjusted brightness $I_{adjusted}$ to represent the true brightness of the waveguide, which is a product of the image brightness I_{image} and the camera exposure time $T_{exposure}$:

$$I_{adjusted} = I_{image} \cdot T_{exposure} \tag{2}$$

After the characterization system automatically aligns to and images each waveguide, the fiber and camera motor will move to the position of the next waveguide based on the wafer layout to start another iteration of the acquisition pipeline. Due to the small size of the waveguide port and pitch (less than 40 μm), any in-plane angle mismatch between the fiber motor and wafer axes could significantly impact the movement accuracy and lead to misalignment for the subsequent input ports. Considering uni-directional target movement over distance L, the resultant distance mismatch in two directions given angle mismatch θ should be smaller than half of the maximum search grid range L_G:

$$L \sin\theta < \frac{L_G}{2} \tag{3}$$

$$L(1 - \cos\theta) < \frac{L_G}{2} \tag{4}$$

For our setup with travel distance between waveguide batches of 2.5 mm in X direction, 0.7 mm in Y direction, and a maximum search grid range of 36 μm in X and 24 μm in Y, the tolerance of angle mismatch is 0.27° (Constrained by Y mismatch: $2.5mm \sin\theta < \frac{24\mu m}{2}$), which is infeasible to reliably achieve via manual alignment. Thus, the pipeline employs automatic angle mismatch compensation. As shown in

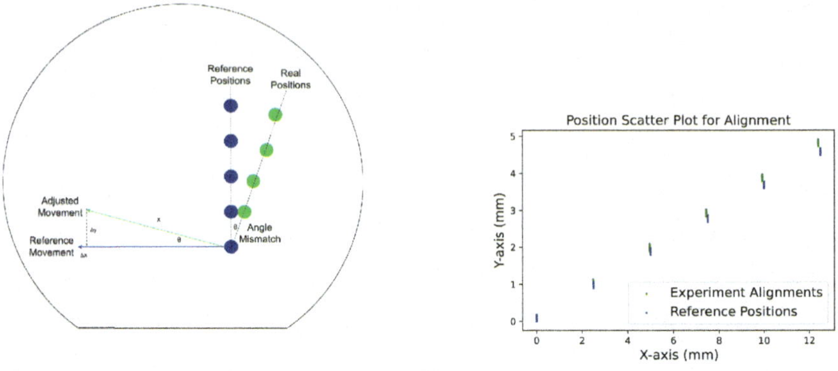

(a) (b)

Fig. 6 **a** After acquisition of waveguides close to the initial fiber position, the actual alignment positions of waveguides on the wafer are captured. These positions are then compared with the reference position from the wafer layout. A mismatch angle θ is computed accordingly, which will be continuously updated for all future motor movements. **b** An example position scatter plot showing the effect of angle adjustment. An alignment trial across 30 waveguides in 6 batches is performed to generate the above plot. The positional difference due to angle mismatch becomes more and more pronounced after each iteration. Thus, angle adjustment is necessary to ensure valid alignment for long-travel-range trials

Fig. 6, after imaging initial sets of adjacent waveguides, the final alignment positions of waveguide input ports are recorded, which are compared to the known wafer layout to calculate the angle mismatch to correct subsequent motor movements. To learn the initial wafer orientation, the design features a batch of waveguides of the same length, which are close together (0 µm, 60 µm pitch for X and Y direction respectively). The fiber motor traverses these 5 initial ports that are 240 µm apart without adjustment for an initial orientation estimation. This relaxes the manual alignment constraint to 4.30° (Constrained by X mismatch: $240\mu m \sin\theta < \frac{36\mu m}{2}$). Following this initial batch, the system continually refines its estimate of the rotation mismatch as it aligns to additional waveguides.

Subsequently, the acquisition procedure is iterated until all waveguides of interest on the wafer have been imaged. As in Fig. 6, with this method, the system is able to account for the angle mismatch to perform reliable alignment to many waveguides with input ports distributed across the wafer surface.

2.4 *Analysis Pipeline*

In order to validate the reliability of the out-scattering method, we performed measurement of the test sample using both the modified cutback method and the out-scattering method and compared their results. In addition, we investigate the performance of both measurement techniques under various imaging conditions and identify criteria (i.e., SNR and waveguide length) to ensure accurate measurement of the device under test.

2.4.1 Modified Cutback Method Fitting

We used the modified cutback method fitting as a reference method to measure the average waveguide propagation loss. The method measures the output intensity of waveguides of different lengths on the same wafer to estimate the propagation loss. Within the waveguide, the light intensity follows an exponential decay:

$$I = I_0 e^{-\alpha L}, \tag{5}$$

where I and I_0 are output and input intensity, L is the light travel distance, and α is the propagation loss of the waveguide. With a constant input intensity, the output intensity will exponentially decay with waveguide length. Here, the output intensity of the waveguide is measured by imaging the output port and calculating the total intensity sum, subject to the linear camera sensor regime discussed previously.

To obtain the output intensity necessary for calculating the propagation loss, the system first identifies the output port in acquired images containing the end of the waveguide. The system uses a predefined output port template to perform template matching to locate the center of the output port [17].

In an initial calibration step that only needs to be performed once per wafer layout, the input fiber is first manually coupled into one waveguide. An image of the output port with out-coupling light is captured, manually cropped to the region of interest, and defined as an output port template (Fig. 7). The template is cross-correlated with the image to obtain the corresponding correlation value map. The maximum value in that map represents the coordinates where the image region best resembles the template and is deemed the center of the output port. Subsequently, a circular mask of radius 22 pixels (empirically chosen to produce repeatable intensity readings from a single waveguide) is defined around the detected center. The final output intensity value is the sum of all pixel brightness values over the mask. Output intensities from different waveguides (of different lengths) are then fitted to an exponential curve to obtain the average propagation loss value of all waveguide batches on the wafer. In this method, we are assuming that different waveguides with the same widths but different lengths have the same propagation loss, which is a good assumption when the variation between the waveguides due to fabrication imperfections is relatively small.

2.4.2 Out-Scattering Method Fitting

The out-scattering method provides an estimation of the propagation loss from individual waveguide images instead of an aggregate estimate from an entire batch of

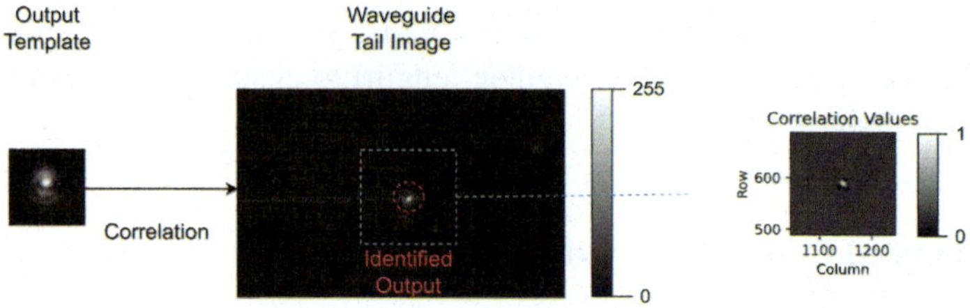

Fig. 7 Template matching to locate the waveguide output in the image. During template matching, predefined output template is cross-correlated pixel-wise with the raw image captured at the end of the waveguide. The maximum correlation value indicates the detected output port coordinates

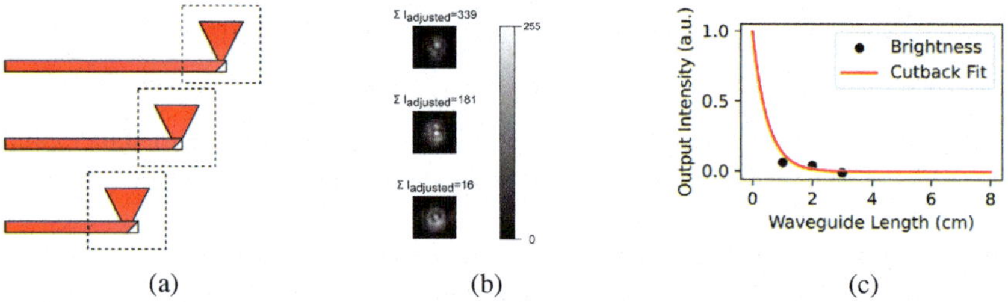

Fig. 8 Pipeline performing the modified cutback method. **a** Example image containing the output region of waveguides under characterization. Output light reflected by the micromirror appears as a bright spot in the recorded image. **b** Output port detection using template matching. Its intensity is summed to measure the optical power. **c** For waveguides with various lengths, their output pixel intensity sums are fitted to an exponential decay to calculate the propagation loss

waveguides (Fig. 8). Here, the out-scattering light from the waveguide is assumed to follow an exponential decay according to the propagation loss. In this analysis, the software utilizes a series of images taken along the waveguide path. For each image, the 1D Hough Transform is applied to identify the waveguide region [18]. In this step (Fig. 9), the image is first thresholded (at 70% of max intensity) to isolate bright pixels of interest. Then, the number of bright pixels is counted row-wise to vote for the waveguide region center row in the image. Its surrounding region containing 70 rows of pixels in total (empirically chosen based on the waveguide width and magnification level) can be identified as the waveguide area in the image. Thus the intensities of the waveguide pixels are summed column-wise to fit an exponential decay curve.

However, captured waveguide images may contain issues that skew the propagation loss measurement. Consequently, more processing is required to obtain an accurate estimate of the propagation loss of waveguides.

First, images captured within 15 mm distance of the input port are observed to have an unusually high propagation loss (Fig. 12). Since such high loss fails to agree with loss measured with the modified cutback method fitting, we hypothesize that it may be related to the out-scattering light associated with the rapid decay of spurious or higher-order modes of the waveguide which possess a higher propagation loss.

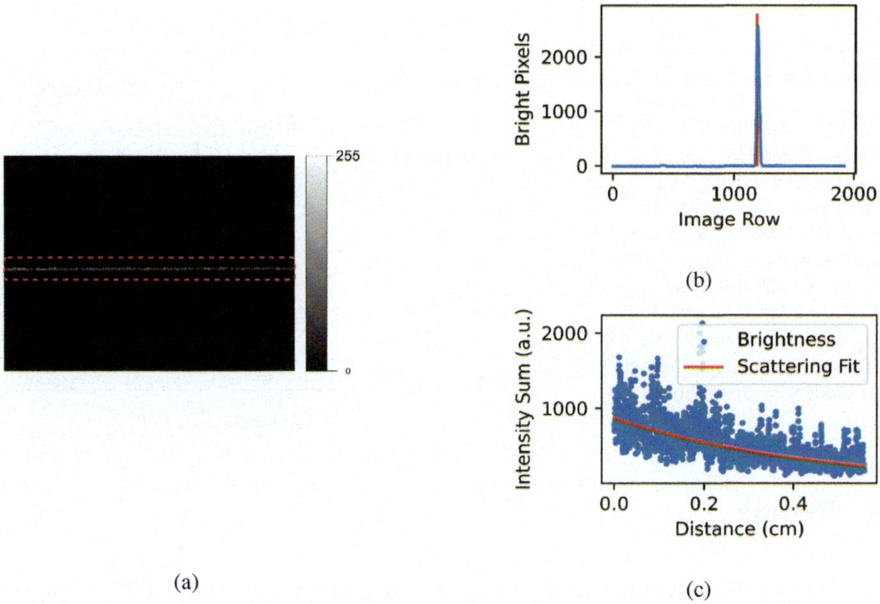

Fig. 9 Out-scattering method fitting on a single waveguide image. **a** Raw image containing the detected waveguide path. **b** 1D Hough transform of the raw image where bright pixels are counted row-wise. The peak row indicates the waveguide center. **c** Out-scattering method fitting result of the shown image. Pixel intensities in the region of interest are summed column-wise and fitted to an exponential decay curve. The fitted out-scattering intensity sum loss corresponds to the overall propagation loss of the waveguide

Some optical power may be coupled into these modes at the input port, but due to their rapid decay, their affect is localized to near the input coupling region. Optical power lost due to coupling into higher-order modes should not be considered when measuring the propagation loss experienced by light that is effectively coupled to efficiently guided lower order modes. Else, the out-scattering fitted propagation loss will be heavily overestimated due to the rapid decay of higher-order mode power in images. Hence, we choose to ignore these near-input images for out-scattering fitting.

Furthermore, we also need to remove artifact-contaminated and faulty waveguide images from fitting since these outliers can skew the overall loss estimates. This step is done by detecting and removing exponential fitting results with R-squared values smaller than 0.4. Lastly, we heuristically remove noisy images with SNR lower than 20 since they tend to underestimate the fitting loss by averaging background noise. A SNR threshold of 20 ensures minimal level of background noise in images, which is shown by numerical simulation to give a high confidence ($p < 0.05$) of accurate estimation within 10% despite noise. Further examples and discussion of these conditions are included in Sect. 3.

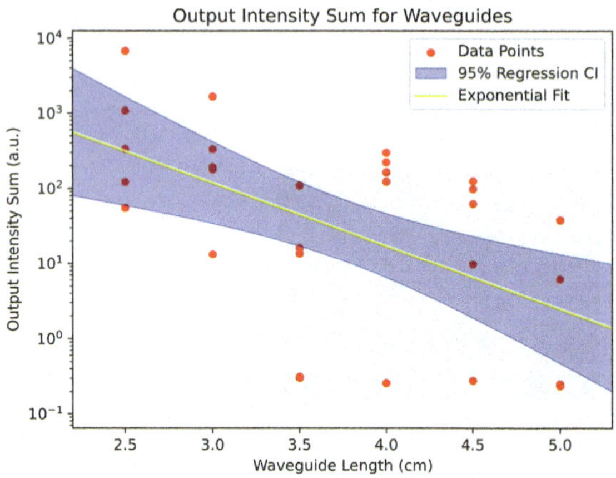

Fig. 10 Modified cutback method fitting result is shown above. 30 total 30 μm wide waveguides were measured (5 each of lengths of 25, 30, 35, 40, 45, and 50 mm). Their output intensity are plotted in red in logarithmic scale. The fitted propagation loss plotted in yellow for the batch was 8.36dB/cm with a 95% confidence interval (CI) of [3.60 dB/cm, 13.19 dB/cm] (CI calculated based on Student's t distribution)

3 Results

To validate the performance of out-scattering method, modified cutback method and out-scattering method were separately used to measure the propagation loss of the same wafer. Individual propagation losses obtained from out-scattering method were in good agreement with that of the modified cutback method. Other factors potentially compounding out-scattering measurement accuracy, such as alignment variance, SNR, and fitting R-squared value, are characterized and addressed in the following section.

3.1 Modified Cutback Method Result

We performed the characterization on the aforementioned wafer containing Parylene photonic waveguides [13]. We imaged thirty 30-μm-wide waveguides ranging from 25 to 50 mm in length. We first obtained results using the modified cutback method. This widely-used method serves as a baseline to evaluate the accuracy of the out-scattering method fitting. The modified cutback method fitting result is shown in Fig. 10. Thirty total waveguides were measured (5 each of lengths 25, 30, 35, 40, 45, and 50 mm). The fitting shows an average loss of 8.36 dB/cm with a 95% confidence interval of [3.60 dB/cm, 13.19 dB/cm] (Fig. 10). Confidence interval is computed as $\overline{x} \pm t \cdot SE \cdot \overline{x}$, where $\overline{x}$, t, SE are mean fitting coefficient, t-score and standard error, respectively.

The wide confidence interval of the modified cutback method fitting is likely due to the large variance in sample measurements. To characterize the automated characterization system measurement accuracy, a repeated alignment trial was performed. The system repeatedly coupled to four waveguides for 20 times each and measured the output port intensity profile. Between each run, a random perturbation up to 2

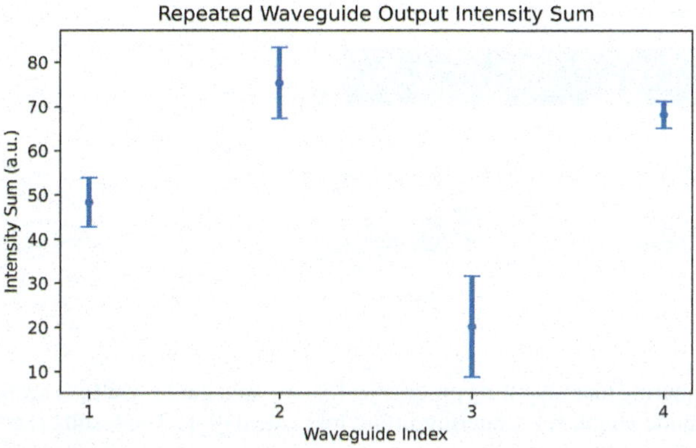

Fig. 11 Repeated alignment over the same waveguides shows the stability of the system. Four waveguides were each aligned to the input port for 20 times each and the corresponding output intensity sum was computed as a measurement of the relative coupling efficiency. Vertical bars indicate the intensity standard deviation for each waveguide. Random perturbation is introduced between each alignment. The average measurement variance of the waveguides is 21%

μm in X and Y direction was imposed on the fiber holder and the input coupling is restored via automated alignment. Figure 11 shows the intensity sum of output profile results. Intensity sum variance was computed for each waveguide based on 20 different measurements. The average variance measured intensity sum for each waveguide is 21%. Thus, there is approximately 1 dB of variance in insertion loss due to automated alignment in the system. Notably, the waveguide-to-waveguide variation is also significant (40%, 2.24 dB), due to device-to-device variation of the fabrication process. Other factors contributing to large variance might also include wafer degradation over time and potential damage to individual waveguides due to particulates outside of a cleanroom environment. The modified cutback method computes the average loss across many waveguides regardless of coupling variation, which makes the cutback measurement vulnerable to errors. While this issue can be mitigated with measuring a large number of samples, the intra-batch waveguide performance variation further motivates the use of the out-scattering method, which has the unique advantage of assessing the individual waveguide performance.

3.2 *Out-Scattering Method Result*

The out-scattering method results for one example waveguide were shown in Fig. 12, collected using the automated procedure described in Fig. 9. Sliding window sections with 1 mm intervals are taken along the waveguide path to produce localized measurements of the out-scattering method fitting propagation loss. This way, the localized propagation loss can be plotted along the distance from input port.

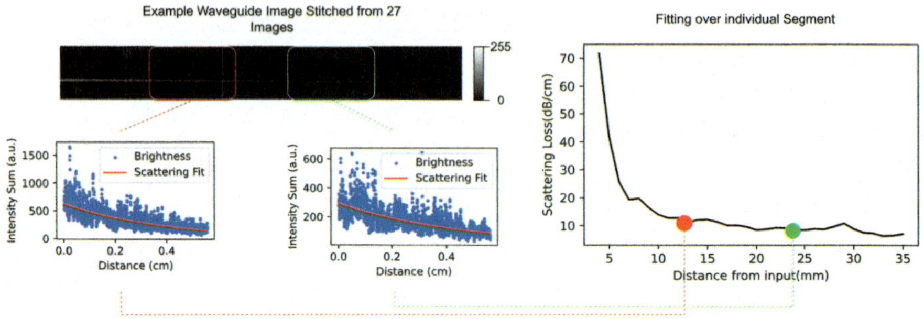

Fig. 12 Out-scattering method fitting processing for one waveguide. Multiple sections were imaged along the waveguide length for scattering fitting loss calculation. The results were aggregated to a single line in plot

For all Parylene photonic waveguides, a very high out-scattering fitting propagation loss around 80–100 dB/cm dominates the images captured near the input port, and declines very quickly until reaching 15 mm from the input port (Fig. 12). Prior work in the literature has also observed the same polymer waveguide propagation loss behavior [19]. The SNR of these images is in line with images further away from input port, which suggests that such a high loss is not introduced by stray light from the input port. We hypothesize that this is due to the higher-order modes which dominate near the input region of the waveguide as discussed in Sect. 2. Considering this loss is confined to the input region and does not scale with the waveguide length, we classify it as the insertion loss rather than propagation loss. Thus, the propagation losses obtained near the input region via the out-scattering method were not considered in the propagation loss evaluation.

In addition, the out-scattering method fitting is capable of identifying faulty waveguide regions by using the fitting R-Squared metric. Such faulty waveguide regions can result from a defect in the waveguide, fabrication imperfections or external debris on the surface of the waveguide, causing a localized outscattering of light. Faulty waveguide regions do not follow a standard out-scattering exponential decay, exhibiting a low R-Squared fitting. One example image having low R-Squared value is shown in Fig. 13a. In the corresponding out-scattering method fitting (Fig. 13b), two peaks are visible in the intensity profile which contribute to the low R-Squared value. Subsequent manual inspection of wafer surface revealed that the two intensity peaks are because of the surface debris on the waveguide which caused out-coupling of light from the waveguide. Thus, thresholding R-Squared value of out-scattering measurements can identify particles or debris affecting the waveguide. This method provides a promising direction for future quality control to detect variations in material quality and identify fabrication defects.

Furthermore, high SNR is also essential to ensure accurate out-scattering method fitting. Due to the exponential intensity decay, the brightness of out-scattering light decreases significantly along the waveguide propagation direction. Since the baseline noise level in the CCD camera is constant, the SNR decreases in proportion to the

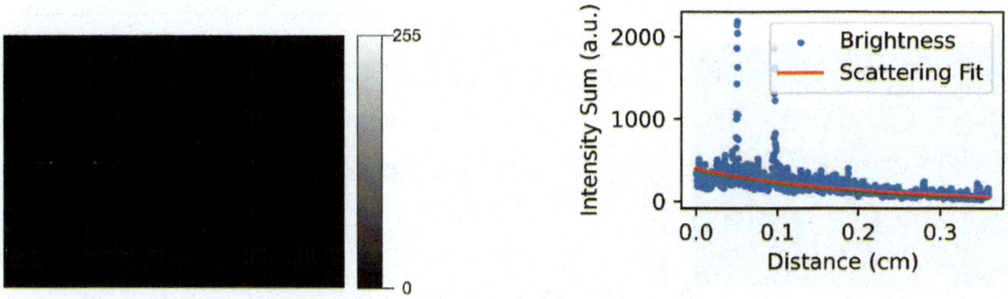

(a) (b)

Fig. 13 Example of R-squared threshold for defect detection. **a** Waveguide path image containing a faulty region. Two bright points on the waveguide path were identified to be surface debris causing out-coupling and optical power loss. **b** Out-scattering fitting of the left image. The fitting R-Squared value is 0.30, indicating a large deviation from an exponentially-decaying curve

decreasing out-scattering light intensity level. When SNR is low, the out-scattering method fitting curve will be affected by a uniform baseline noise, which tends to skew the fit and underestimate the actual propagation loss. In the following section, we provide an analytical relationship to illustrate the effect of low SNR. We assume a waveguide with true propagation loss α, an initial out-scattering intensity of I_0 at given length $L(z)$. Its out-scattering intensity at point z is given by

$$I(z) = I_0 e^{-\alpha L(z)}. \tag{6}$$

Then to model the impact of background noise, we assume noise as a positive half normal distribution with mean μ and variance σ^2 on intensity. The observed intensity is the sum of the two:

$$I(z) = I_0 e^{-\alpha L(z)} + |\mathcal{N}(\mu, \sigma^2)|. \tag{7}$$

Since the background noise is non-negative, we want to ensure a lower bound propagation loss $\eta\alpha$ where $0 < \eta < 1$ such that our observed intensity at that point will not be larger than a waveguide with propagation loss $\eta\alpha$. Its corresponding intensity is defined as the bounded intensity.

$$I_0 e^{-\alpha L(z)} + |\mathcal{N}(\mu, \sigma^2)| \leq I_0 e^{-\eta\alpha L(z))}. \tag{8}$$

Then at the given point L_0, the probability of the observed intensity being smaller than the bounded intensity is:

$$P(|\mathcal{N}(\mu, \sigma^2)| \leq I_0(e^{-\eta\alpha L(z)} - e^{-\alpha L(z)})). \tag{9}$$

This equation applies to every point L_0 over the imaged waveguide length L. The relationships between L, α, μ, and σ demonstrate the trends of fitting perfromance

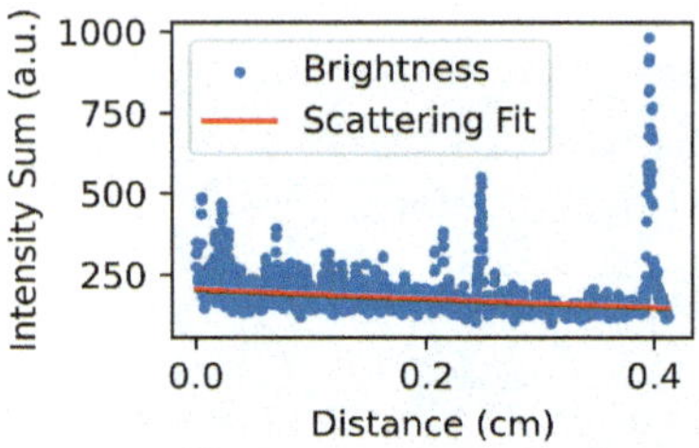

Fig. 14 Example of low-SNR image biasing out-scattering method fitting. **a** Waveguide path image containing low-SNR region. The background noisy spots have intensity comparable with waveguide path due to low SNR. **b** Out-scattering fitting of the left image. The fitting is dominated by background noise. The underestimated out-scattering fitting loss is 5.8 dB/cm

with background noise. Note that smaller propagation loss α and shorter imaging distance L are more easily skewed by background noise, requiring higher SNR.

As a numerical example to demonstrate the effect of background noise on loss fitting, we chose $L = 5.6\,\text{mm}$ and $\alpha = 10\,\text{dB/cm}$. For a distance x that is small, the value of exponential term $e^{-\alpha x}$ is close (1) Thus we can model an SNR of 20 by using a positive Gaussian noise model with zero mean and standard deviation $\frac{\sqrt{\pi}}{SNR\cdot\sqrt{2}}$ (which has an expected value of $\frac{1}{SNR}$):

$$I(x) = e^{-\alpha x} + |N(0, \left(\frac{\sqrt{\pi}}{SNR \cdot \sqrt{2}}\right)^2|$$

We simulated 10000 waveguides over L with 457 data points/mm to match our imaging system. We chose an η of 0.9 as the lower tolerance bound. In this setup, with a SNR of 20, 99.97% of the simulation trials produced an out-scattering propagation loss fitting result within the 10% error tolerance. Thus, with an SNR of 20, measurement results using the scattering method are highly unlikely (less than 0.01% chance) to deviate by more than 10% from the true value under the simulated conditions.

Consequently, to accurately perform out-scattering method measurements, the following steps should be followed:

(1) Imaging waveguides longer than 35 mm to maximize regions unaffected by near-input high loss, (2) excluding waveguides with R-Squared values lower than 0.4, and (3) excluding waveguide regions with SNR smaller than 20.

Figure 14 shows an example image with low SNR. The image is taken 35 mm from input port, where out-scattering light has decayed to a level difficult to detect even for the maximum camera exposure time. Here, the background noise skews the exponential decay, which results in a fitted propagation loss lower than estimation (5.8 db/cm versus 8.36 db/cm from the cutback method).

Seven 30 μm wide waveguides with lengths ranging from 35 mm to 50 mm passed the aforementioned criteria along their entire length. Their out-scattering

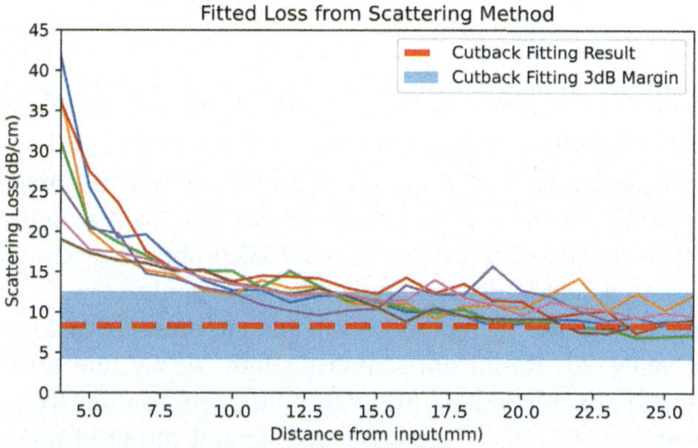

Fig. 15 Aggregated out-scattering method fitting result is shown above. 7 waveguides passing all aforementioned criteria are demonstrated in the plot. The majority of waveguide propagation losses converge within the 3 dB of modified cutback measurement after travelling 15 mm from input port

Table 1 Measurement time estimation for each waveguide

Step	Alignment 1–3	Imaging
Time (min)	2/ea	3–10

method fitting results are plotted for up to 27 mm from the input port (Fig. 15). The out-scattering method fitting for all qualifying waveguides have a majority of propagation losses stabilize to a value within the 3 dB of the cutback fitting result after travelling 15 mm from input port. This result demonstrates agreement between the the out-scattering fitting method compared to the gold-standard modified cutback method results.

The processing speed of the automated waveguide characterization system was also assessed. Since the light intensity captured by the camera during alignment is fixed, no exposure time adjustment is necessary in this phase. The time for each step is shown in Table 1, when the exposure time for each image is 0.5 s, the input coupling step takes 2 min each for three grid iterations and the automated waveguide imaging step takes 3 to 10 min for each waveguide depending on its length. On average, characterizing each waveguide takes around 15 min and the system can finish characterizing the propagation losses of 30 waveguides on a wafer in 12 hours. This is all done in an automated fashion using the custom-designed robotic system and after initializing the process, no human supervision is required.

4 Discussion

We have demonstrated an automated waveguide characterization system capable of measuring propagation loss of many (greater than fifty) waveguides without human intervention using either the out-scattering method or modified cutback method. The out-scattering fitting provides inherent advantages due to it's ability to characterize individual waveguides and provide a localized probe into individual waveguide propagation loss to detect faulty regions on the waveguide.

However, the out-scattering method is sensitive to multiple factors that have been identified and analyzed. Rapid out-scattering light decay due to the presence of higher-order modes overestimates the waveguide propagation loss near the input port. Particles and other defects skew the exponential fitting of intensity decay but can be automatically detected via a low R-squared value. Lastly, low SNR in the captured image can cause underestimation of measured propagation loss values, necessitating a suitably high SNR. Therefore, we define a set of quality metrics to ensure accurate measurements: (1) Imaging at least 15 mm from the input port, (2) R-squared value less than 0.4, (3) SNR greater than 20. A majority of waveguides propagation losses that satisfied our quality metrics converge within a 3 dB of the modified cutback measurement result. Therefore, under proper conditions, the out-scattering method provides a reliable assessment of propagation losses for waveguide samples compared to the gold-standard modified cutback method.

Two remaining limitations to the characterization system demonstrated here are the requirement for initial manual alignment of the fiber to the wafer and speed of imaging. Currently, the fiber must be manually lowered to the wafer surface and aligned to the first waveguide input port. Additional capabilities to perform coarse fiber alignment via computer vision detection of fiducal marks on the wafer surface, and proximity detection for the fiber to avoid collision would allow for wafers to be automatically loaded and unloaded in a production environment. Furthermore, the imaging speed may also be a bottleneck for the performance of the system in a high-throughput setting. The current 15-minute-per-waveguide trials may not scale well with foundry-level production speed. To increase the processing speed of automated alignment step, exposure time can be further reduced, at the cost of higher noise provided the waveguide path images are bright enough to maintain sufficient SNR. In addition, adopting a low-noise camera would reduce the need for a long integration time and expedite the acquisition process in automated characterization pipeline. Furthermore, there is a trade-off between the number of images taken on each waveguide and the acquisition time. Increasing the field of view for each image and reducing the overlap overlap between adjacent images during the automated imaging step is also a viable way to improve the throughput.

5 Conclusion

To summarize, the out-scattering method provides a unique capability to directly investigate the propagation loss of individual waveguides under characterization, thereby enabling assurance of device integrity in critical systems development, particularly in biophotonics. It is worth noting that the specificity in characterization offered by the out-scattering method has not been possible in the modified cutback or resonators methods, as their implementation relies on additional structures, nor in the direct cutback method, which is destructive to the sample under test. Hence, the out-scattering characterization method is desirable for high-throughput quality assurance in a production environment.

The present work showcases a multi-stage system that can automatically characterize propagation losses of waveguide samples with high throughput. Additionally, we have validated the measurements by comparing the results of the out-scattering method to the gold-standard modified cutback method. Importantly, the out-scattering method accuracy is susceptible to image quality and waveguide defects. Therefore, we have developed quality metrics which can be used to determine the reliability of a measurement using the out-scattering method. With the methods demonstrated here, the out-scattering method can benefit the field of integrated photonics by providing an expedited characterization technique to analyze yield and performance of a microfabrication process. Furthermore, the out-scattering method possesses novel capabilities such as the ability to locally probe the propagation loss in a segment of a waveguide, which can be further developed and explored to aid in quality control and fabrication process development. The individualized and specific characterization capability offered by the out-scattering measurement method is uniquely helpful in biophotonics applications such as developing optical neural probes for brain stimulation, where the reliability of each individual device is critical.

Appendix

The data and code used in this work are available from the corresponding author, X.Z., upon reasonable request.

References

1. Meng, Y., Chen, Y., Lu, L., Ding, Y., Cusano, A., Fan, J.A., Hu, Q., Wang, K., Xie, Z., Liu, Z., Yang, Y., Liu, Q., Gong, M., Xiao, Q., Sun, S., Zhang, M., Yuan, X., & Ni, X. (2021). *Light: Science and Applications*, *10*, 1. https://doi.org/10.1038/s41377-021-00655-x
2. Pan, T., Lu, D., Xin, H., & Li, B. (2021). *Light: Science and Applications*, **10**, 1. https://doi.org/10.1038/s41377-021-00561-2
3. Tynes, A. R., & Keck, D. B. (1972). *Applied optics, 11*(7), 1502–1506. https://doi.org/10.1364/AO.11.001502

4. da Silva, D. S., Wetter, N. U., de Rossi, W., Kassab, L. R. P., & Samad, R. E. (2018). *Optical Materials, 75*, 267. https://doi.org/10.1016/J.OPTMAT.2017.10.033
5. Moresco, M., Romagnoli, M., Boscolo, S., Midrio, M., Cherchi, M., Hosseini, E.S., Coolbaugh, D., Watts, M.R., & Dutt, B. (2013). *Optics Express, 21*. https://doi.org/10.1364/oe.21.005391
6. Dumon, P., Member, S., Bogaerts, W., Wiaux, V., Wouters, J., Beckx, S., Campenhout, J.V., Taillaert, D., Luyssaert, B., Bienstman, P., Thourhout, D.V., Baets, R., & Member, S. (2004). *Ieee Photonics Technology Letters, 16*. https://doi.org/10.1109/LPT.2004.826025
7. Toyoda, N., Liang, H., Alshehri, B., Siew, S. Y., Cheung, E. J. H., Bettiol, A., Dogheche, E., & Danner, A. J. (2018). *Optics Express, 26*(4), 4421–4430. https://doi.org/10.1364/OE.26.004421
8. Hu, Y.W., Zhang, Y., Gatkine, P., Bland-Hawthorn, J., Veilleux, S., & Dagenais, M. (2018). *IEEE Journal of Selected Topics in Quantum Electronics, 24*. https://doi.org/10.1109/JSTQE.2018.2827663
9. Lin, J., Zhou, J., Wu, R., Wang, M., Fang, Z., Chu, W., Zhang, J., Qiao, L., & Cheng, Y. (2019). *Micromachines, 10*, 612. https://doi.org/10.3390/MI10090612
10. Okamura, Y., Yoshinaka, S., & Yamamoto, S. (1983). *Applied Optics, 22*(23), 3892–3894. https://doi.org/10.1364/AO.22.003892
11. Skorka, O., Meyler, B., & Salzman, J. (2004). *Applied Physics Letters, 84*, 3801. https://doi.org/10.1063/1.1741025
12. Krause, M., Müller, J., Harke, A., Krause, M., & Mueller, J. (2014). researchgate.net. https://doi.org/10.1049/el:20052387
13. Reddy, J. W., & Chamanzar, M. (2018). *Optics Letters, 43*(17), 4112–4115. https://doi.org/10.1364/OL.43.004112
14. Reddy, J.W., Lassiter, M., & Chamanzar, M. (2020). *Microsystems & Nanoengineering 6*. https://doi.org/10.1038/S41378-020-00186-2
15. Zorzos, A. N., Fonstad, C. G., & Boyden, E. S. (2010). *Optics Letters, 35*(24), 4133–4135. https://doi.org/10.1364/OL.35.004133
16. Venkateswaran, R., Reddy, J.W., & Chamanzar, M. (2021). In *2021 43rd Annual International Conference of the IEEE Engineering in Medicine & Biology Society (EMBC)* (pp. 7502–7505). https://doi.org/10.1109/EMBC46164.2021.9630691
17. Brunelli, R. (2009). *Template matching techniques in computer vision: Theory and practice*. Wiley Publishing
18. Stockman, G., & Shapiro, L. G. (2001). *Computer Vision* (1st ed.). USA: Prentice Hall PTR.
19. Helke, C., Reinhardt, M., Arnold, M., Schwenzer, F., Haase, M., Wachs, M., Goßler, C., Götz, J., Keppeler, D., Wolf, B., Schaeper, J., Salditt, T., Moser, T., Schwarz, U.T., & Reuter, D. (2023). *Materials, 16*(1). https://doi.org/10.3390/ma16010106

Interpretable Prototype Discovery in Deep Learning-Based Time Series Classification

Gaurav R. Ghosal and Reza Abbasi-Asl

Abstract Multivariable time series classification problems are increasingly prevalent and complex across various domains, such as biology and finance. While deep learning methods are an effective tool for these problems, they often lack interpretability. In this work, we propose a modular prototype learning framework for multivariable time series classification. In the first stage of our framework, we use contrastive learning to train a set of long short-term memory (LSTM) encoders for each variable. We then use prototype matching layers to learn representative prototypes for the variables that contribute to the classification task. The visualization of these prototypes offers an interpretation of the representative patterns within each variable. We validated our framework on a simulated dataset with embedded patterns. On this simulated dataset, we found that our model produces interpretations consistent with the embedded patterns. Overall, our findings pave the road for constructing interpretable deep learning-based models for time series classification.

1 Introduction

Multivariate time series classification aims to leverage the measurement of multiple variables over a period of time to assign data points to classes. As such, these problems arise naturally across various domains, including biology [1], finance [2], and activity recognition [1]. Along with their increased prevalence, multivariate time series classification datasets have increased in size and complexity, making their analysis challenging [3]. Internet-of-Things technologies, for example, allow for the connection of multiple sensing devices [4]. These device networks generate data sets and classification tasks involving a large number of sensors, many with markedly different characteristics [5]. Similarly, developments in bio-sensing technologies have given rise to the *multi-modal biosensing* problem, where time series readings from a diverse set of bio-sensors are used to characterize the mental and physical state of a subject. Biosensing has been applied in sleep staging [6], emotion recognition

G. R. Ghosal · R. Abbasi-Asl (✉)
University of California, San Francisco, CA, USA
e-mail: Reza.AbbasiAsl@ucsf.edu

M.-R. Alam and M. Fathi (eds.), *Integrated Systems: Data Driven Engineering*,
https://doi.org/10.1007/978-3-031-53652-6_2

[7, 8], stress detection [9], attention studies [10, 11], brain-computer interfaces [12], and wellness monitoring [13, 14]. These problems predominantly arise in research settings, with the goal of deriving scientific knowledge from the data. As such, model interpretability is of central importance.

Multivariate time series classification, due to its broad applicability, remains an active area of research with a plethora of existing approaches. These approaches can be divided into deep-learning-based methods and feature-based methods. Many feature-based time series classification methods utilize hand-crafted feature extraction techniques. For example, in [1], discrete features learned from the multivariate time series are used to construct a bag-of-patterns model for classification. Deep learning-based methods do not rely on hand-crafted features, instead learning an end-to-end model. In [15] for example, a combination of recurrent and convolutional neural networks is proposed for the problem of EEG classification in a brain-computer interface setting. Deep learning techniques are an increasingly powerful tool for dealing with the rising complexity of multivariate time series classification problems. Although deep learning methods are able to learn complex relationships, it is difficult to reveal these relationships in a human-understandable way. This is primarily due to the high number of parameters of deep learning models, and the large space of functions that they could represent [16, 17].

This study introduces a modular, prototype learning framework for interpretable classification on multivariable time series data. Our framework provides model-based interpretability, as it is designed to produce classifications transparently [18]. In this framework, encoders are trained by contrastive learning to extract meaningful features from each variable. Then, prototype matching layers determine representative prototypes for each relevant variable.

The contributions of this work are threefold. First, we present an interpretable method for multivariable time series classification which explicitly models both individual variable patterns as well as cross variable patterns. Second, we provide a step-wise training procedure for effectively training the multi-level prototype model. Finally, we design a synthetic data set with heterogeneous variables and cross-variable patterns to verify the validity of the interpretations.

2 Related Literature

Multivariable time series classification has been studied extensively and various classification techniques have been proposed in different domains. In this section, we introduce some of these related works with a specific emphasis on interpretability. We divide these prior works into feature-based approaches and neural network-based approaches.

Feature-based Approaches Feature-based methods primarily rely on manual feature extraction techniques and discretization to derive features from time series streams. One prominent category of such methods is the Bag-of-Patterns (BOP) approach.

BOP generates symbolic features from subsequences of time series and represents signals using a symbol histogram, akin to a bag-of-words representation. For instance, in [1], Symbolic Fourier Approximation (SFA) is employed for each dimension of a multivariable time series classification problem, and the univariate symbolic representations are combined to form a multivariable Bag-of-Patterns representation. In another study [19], a random forest learner is utilized to generate symbols for a multivariable Bag-of-Patterns method.

Another notable class of feature-based approaches are Shapelet methods. These methods involve extracting time series subsequences, known as shapelets, that are highly informative for classification. In [20], an efficient method is proposed for extracting shapelets from large and multivariable datasets by pruning similar shapelet candidates. In [21], multivariate shapelets are employed for early classification of signals. Shapelet-based methods often possess favorable interpretability properties, as the shapelets themselves directly represent characteristic patterns. However, both shapelet and discretization approaches may be computationally intensive and generate high-dimensional feature spaces.

Neural Network Approaches Neural network-based approaches harness the capability of neural networks to learn complex functions, serving as feature extraction techniques. However, some of these methods combine neural networks with discretization and feature extraction approaches. For instance, in [22], SFA features are fed through a neural network to project them into an encoded space, followed by prototype learning for final classification. In [23], prototype learning is used to interpret the latent space of deep autoencoders. In another work, [24] employs a random dimension permutation with an attentional prototype network for multivariable time series classification, however, interpretability is not explicitly addressed in this study. Finally, [25] proposes learning interpretable multivariate shapelets in an embedded space encompassing all dimensions of the multivariable classification problem.

Here, we build upon existing studies and present a multivariable time series classification network based on recurrent neural networks that are trained using contrastive learning. We then introduce a prototype matching layer to the architecture of the network and train this layer using cross-entropy loss objective to produce representative prototypes for each variable.

3 Model

We propose a two-level encoder-prototype learning method for multivariate time series classification.

Problem Formulation Consider a multivariable time series classification task containing d dimensions and n time samples per dimension and a dataset $D = \{(\mathbf{X^{(i)}}, y_i)\}$, where $y_i \in \mathbb{Z}$ denotes the label for training example i and $\mathbf{X^{(i)}} \in \mathbb{R}^{n\times d}$ is a matrix for time series sample point i. Each row of $\mathbf{X^{(i)}}$ is a time point and each column represents a variable. In particular, we can write $\mathbf{X^{(i)}} = \left[\mathbf{x_1^{(i)}} \dots \mathbf{x_d^{(i)}}\right]$, where $\mathbf{x_k^{(i)}} \in \mathbb{R}^n$

-represents the time series corresponding to the k-th variable. The multi-variable time series classification problem aims to learn a function $f : \mathbb{R}^{n \times d} \rightarrow \mathbb{Z}$, which maps instances of the multi-variable time series to their correct class.

Encoder Space First, our method performs feature extraction independently for each variable. Feature extraction is accomplished by a set of encoder networks each of which maps a particular variable's time series signal to an encoded vector representation. In this work, we utilize a recurrent neural network and specifically a long short-term memory (LSTM) encoder for time series feature extraction, however, the choice of encoding network is entirely flexible and can be tuned using domain knowledge. Concretely, our model learns a set of encoder functions $\{E_1, ..., E_d\}$, $E_k : \mathbb{R}^n \rightarrow \mathbb{R}^{f_k}$ mapping raw time series data for variable k into a feature space of dimension f_k. We denote the encoding for variable k for some sample point $\mathbf{X} = \left[\mathbf{x}_1, ..., \mathbf{x}_d\right]$ as $\mathbf{e_k} \in \mathbb{R}^{f_k}$ and have that $\mathbf{e_k} = E_k(\mathbf{x_k})$.

Prototype Matching Layer Following the prototype learning method that was previously proposed for the deep autoencoders [23], our method learns prototype patterns but for our recurrent neural network in the encoder layer. This is achieved by a prototype matching layer. We define a prototype matching layer for the k-th variable as a function $P_k : \mathbb{R}^{f_k} \rightarrow \mathbb{R}^{n_k}$ parametrized by a set of n_k prototype vectors $\{\mathbf{p}_1^{(k)}, ..., \mathbf{p}_{n_k}^{(k)}\}$ in the feature space for variable k. P_k maps vectors from variable k's feature space to a vector of similarity scores with each of the prototype vectors. Explicitly, for the encoded vector $\mathbf{e_k}$, we have that $P_k(\mathbf{e_k}) = \left[\text{sim}(\mathbf{e}_k, \mathbf{p}_1^{(k)}), ..., \text{sim}(\mathbf{e}_k, \mathbf{p}_{n_k}^{(k)})\right]$, for a similarity function sim. In this work, we consider $\text{sim}(\mathbf{v_1}, \mathbf{v_2}) = \frac{1}{||\mathbf{v_1} - \mathbf{v_2}||_2}$.

3.1 Training Procedure

Encoder Pretraining In order for meaningful prototypes to be selected for each variable, the encoder should be able to extract information from the samples that are meaningful in the classification task. To this end, we propose a pre-training step for the encoders based on contrastive learning [26]. During the contrastive learning stage, pairs of training examples are input into the encoder and their encodings are computed. These pairs of training examples are used to compute the contrastive loss function shown in Eq. 1 and the encoder is trained to minimize this loss function. This contrastive loss function encourages training examples of the same class to be mapped close to each other in the encoded space, and training examples from different classes to be mapped farther away.

$$L_{contrastive} = \frac{1}{2}(1 - y)||f(x_1) - f(x_2)||_2^2 + \frac{1}{2}(y) \max\left(0, m - ||f(x_1) - f(x_2)||_2\right)^2 \tag{1}$$

Prototype matching layer training In the next part of the training process, the prototype matching layers are trained, resulting in a set of representative prototypes being learned for each variable. In order to train the prototype matching layers, we construct a new network where the outputs of all the prototype matching layers are concatenated and fed through a fully connected layer to determine a final classification output. This new network is then trained with the cross-entropy loss objective. This new network allows the prototype matching layers to be trained taking into account the interactions between variables, while avoiding the challenges posed by end-to-end training.

4 Evaluation

We evaluate our framework on the following data sets to verify its ability to both classify with high performance and reveal meaningful interpretations. For these experiments, we utilize the PyTorch deep learning library [27]. All experiments were conducted on an Apple 2015 MacBook Pro with an Intel i7 processor.

Generation of the Dataset Our simulated dataset contains four variables. Three of these variables are designed to contain meaningful information for the classification task, whereas one of the variables is designed to be irrelevant noise. In order to test the flexibility of our framework in processing time series signals, each of the three relevant variables contains a different type of time series pattern. The first variable contains patterns that are localized to a subsection of the total time series, but are shift-invariant–the location of the pattern sub-sequence relative to the entire time series should not affect classification. The second variable contains patterns that are also restricted to a sub-sequence of the time series but are shift-variant–the location of the pattern sub-sequence with respect to the entire time series is meaningful. Finally, the last variable contains a frequency domain pattern. Each of the three meaningful variables has four pattern states that it can exhibit, as shown in Fig. 1. We refer to these states as variable-level patterns.

A particular class in the simulated data set is determined by the combination of the patterns found in the three variables. We create a class corresponding to each combination of variable-level patterns in the relevant variables, resulting in $4^3 = 64$ classes. As a result of this design, the contributions of all three variables are necessary to correctly classify a point. During sample generation, 100 points are generated per class and appropriate types of noise are added to each variable.

Encoded Spaces and Prototypes On the simulated dataset, our framework is able to consistently achieve 1.00 hold-out test set classification accuracy, as is expected from the prescribed structure present in the dataset. In order to investigate interpretability we begin by visualizing the encoded space corresponding to each variable using UMAP [28] in Fig. 2a. The encoded spaces for the three relevant variables each contain 4 clusters and each cluster corresponds to one of the variable-level patterns

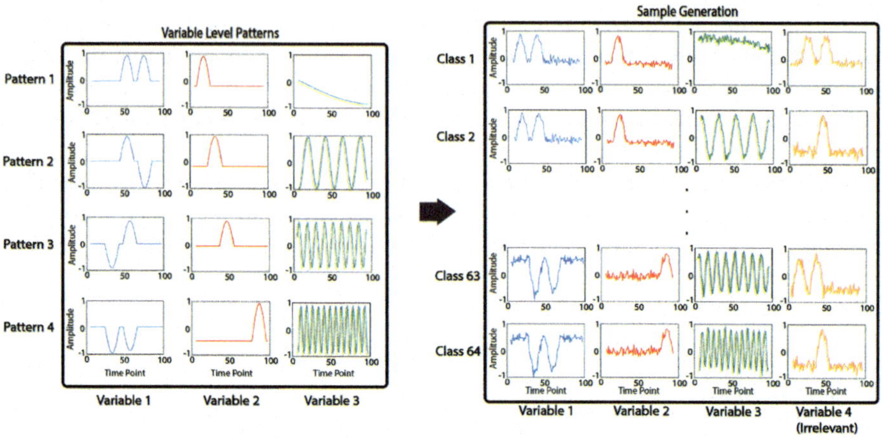

Fig. 1 Schematic showing the construction of the simulated dataset. Each of the relevant variables has 4 possible patterns, as displayed in the variable level pattern box (left). Each class is a unique combination of one variable level pattern taken from each variable, as shown in the sample generation box (right). The sample generation box also shows the addition of sample noise to the dataset, as well as the randomly sampled irrelevant variable

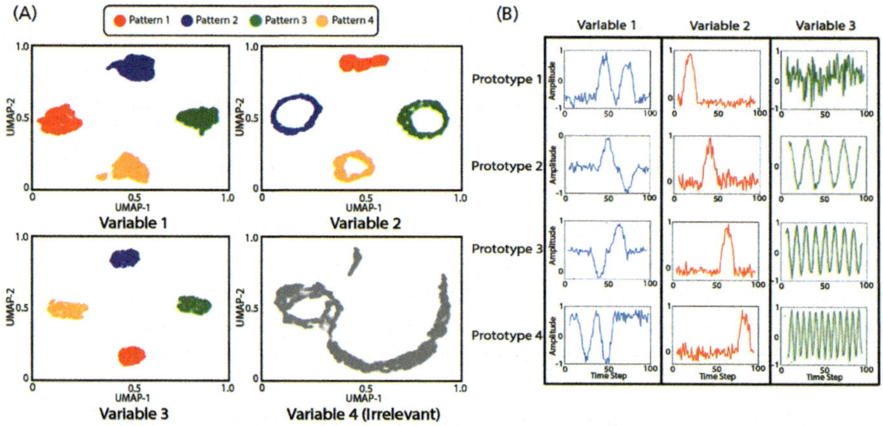

Fig. 2 **a** 2-dimensional UMAP visualization of the training set encodings corresponding to each variable. Variables 1,2, and 3 are relevant for classification and thus sample points are colored by the variable-level pattern exhibited in the variable. **b** Representation of the prototypes by projection onto the dataset. Each prototype is represented by the training point that lies closest to it in the encoded space

implanted in the dataset. The irrelevant variable, on the other hand, does not exhibit the same structure because it does not contain any relevant information for classification. Therefore, Variable 4 is not included in the subsequent analysis for prototype learning.

In the simulated data set, the time series signal in any given variable alone is insufficient to classify the data point. As such, it is notable that the contrastive loss training step, which only takes into account the classes of the data points and treats each variable independently, is able to learn about the underlying patterns present in each variable. This demonstrates the effectiveness of contrastive loss training even in the case where classes can only be determined by integrating information across several variables.

For the simulated dataset, we set the number of prototypes to four. The output of the prototype learning module is depicted in Fig. 2b. Upon comparing Fig. 2b with Fig. 1, it becomes evident that each relevant variable contains one prototype corresponding to each of the implanted variable-level patterns. The prototypes for Variable 4 are not visualized because the UMAP analysis of the feature space demonstrated the irrelevancy of this variable in the classification. These results indicate that our model effectively identifies prototypes within each variable while achieving a high classification accuracy.

5 Discussion

Interpretable multivariable time series classification is an essential task for a variety of scientific and industrial applications. In this study, we have presented a prototype learning-based model that allows for the discovery of important patterns at the variable level, as well as easily interpretable rules encoding the relationships between variables that are important for classification. Our method is modular in both its structure and its training procedure, allowing for a large degree of flexibility in handling diverse variable types as well as the opportunity to deploy domain-specific knowledge. On our simulated dataset, our framework is able to fully detail the patterns characterizing each class, offering a set of interpretable prototypes for each relevant variable.

Our future work includes expanding this framework to identify multivariable prototypes to reveal the interactions amongst the variables with additional validation in real-world time-series datasets such as biosensor data [29]. One limitation of our current work is the need to determine the number of prototypes manually. Future work will examine methods to automate or make this choice more systematic. Another potential area for future work is enhancing the complexity of cross-variable patterns that can be represented by our framework. This could potentially be achieved by introducing metric learning in the multivariable representation space, instead of using our current similarity measure. Finally, for more complex or noisy variables, the single variable prototypes alone could be insufficient to identify the meaningful features of the time series. Thus, interpretability could be enhanced by introducing the attention mechanism to the encoders, potentially providing fine-grained identification of meaningful time points and subsequences within the prototypes.

Acknowledgements We gratefully acknowledge the feedback and comments provided by the members of the Neuroscape Center at UCSF.

References

1. Schäfer, P., Leser, U. (2017). Multivariate time series classification with weasel+ muse. *arXiv preprint* arXiv:1711.11343
2. Kwon, D.-H., Kim, J.-B., Heo, J.-S., Kim, C.-M., & H, Y-H. (2019). Time series classification of cryptocurrency price trend based on a recurrent lstm neural network. *Journal of Information Processing Systems, 15*(3), 694–706.
3. Xu, H., Duan, Z., Bai, Y., Huang, Y., Ren, A., Yu, Q., Zhang, Q., Wang, Y., Wang, X., Sun, Y., & Wang, W. (2020). Multivariate time series classification with hierarchical variational graph pooling
4. Azar, J., Makhoul, A., & Couturier, R. (2020). Using densenet for iot multivariate time series classification. In *2020 IEEE Symposium on Computers and Communications (ISCC)*, (pp. 1–6). IEEE
5. Gupta, A., Gupta, H.P., Biswas, B., & Dutta, T. (2020). A divide-and-conquer-based early classification approach for multivariate time series with different sampling rate components in iot. *ACM Trans. Internet Things*, *1*(2).
6. Zhai, B., Perez-Pozuelo, I., Clifton, E.A.D., Palotti, J., Guan, Y. (2020). Making sense of sleep: Multimodal sleep stage classification in a large, diverse population using movement and cardiac sensing. *Proc. ACM Interact. Mob. Wearable Ubiquitous Technol.*, *4*(2).
7. Ayata, D., Yaslan, Y., Kamasak, M.E. (2020). Emotion recognition from multimodal physiological signals for emotion aware healthcare systems. *Journal of Medical and Biological Engineering*, 1–9.
8. Yixiang, D., Xue, W., Pengbo, Z., & Weihang, Z. (2017). Wearable biosensor network enabled multimodal daily-life emotion recognition employing reputation-driven imbalanced fuzzy classification. *Measurement, 109*, 408–424.
9. Lee, J. H., Gamper, H., Tashev, I., Dong, S., Ma, S., Remaley, J., Holbery, J. D., & Yoon, S. H. (2020). Stress monitoring using multimodal bio-sensing headset. *Extended Abstracts of the 2020 CHI Conference on Human Factors in Computing Systems, CHI EA '20* (pp. 1–7). Association for Computing Machinery: New York, NY, USA.
10. Ziegler, D. A., Simon, A. J., Gallen, C. L., Skinner, S., Janowich, J. R., Volponi, J. J., Rolle, C. E., Mishra, J., Kornfield, J., Anguera, J. A., et al. (2019). Closed-loop digital meditation improves sustained attention in young adults. *Nature Human Behaviour, 3*(7), 746–757.
11. Mishra, J., Lowenstein, M., Campusano, R., Hu, Y., Diaz-Delgado, J., Ayyoub, J., Jain, R., & Gazzaley, A. (2021). Closed-loop neurofeedback of alpha synchrony during goal-directed attention. *Journal of Neuroscience*.
12. Wen, D., Liang, B., Zhou, Y., Chen, H., Jung, T.P. (2020). The current research of combining multi-modal brain-computer interfaces with virtual reality. *IEEE Journal of Biomedical and Health Informatics*, 1–1.
13. Jung, Y., Ik Yoon, Y. (2016). Monitoring senior wellness status using multimodal biosensors. In *2016 International Conference on Big Data and Smart Computing (BigComp)* (pp. 435–438). IEEE.
14. Jung, Y., & Ik Yoon, Y. (2017). Multi-level assessment model for wellness service based on human mental stress level. *Multimedia Tools and Applications, 76*(9), 11305–11317.
15. Zhang, D., Yao, L., Zhang, X., Wang, S., Chen, W., Boots, R., & Benatallah, B. (2018). Cascade and parallel convolutional recurrent neural networks on eeg-based intention recognition for brain computer interface. In *Proceedings of the AAAI Conference on Artificial Intelligence*, 32.

16. Abbasi-Asl, R., & Yu, B. (2017). Structural compression of convolutional neural networks. *arXiv preprint* arXiv:1705.07356.
17. Abbasi-Asl, R., Chen, Y., Bloniarz, A., Oliver, M., Willmore, B.D.B, Gallant, J.L., & Yu, B. (2018). The deeptune framework for modeling and characterizing neurons in visual cortex area v4. *bioRxiv*, pp. 465534.
18. James Murdoch, W., Singh, C., Kumbier, K., Abbasi-Asl, R., & Yu, B. (2019). Definitions, methods, and applications in interpretable machine learning. *Proceedings of the National Academy of Sciences, 116*(44), 22071–22080.
19. Gokce Baydogan, M., & Runger, G. (2015). Learning a symbolic representation for multivariate time series classification. *Data Mining and Knowledge Discovery, 29*(2), 400–422.
20. Grabocka, J., Wistuba, M., & Schmidt-Thieme, L. (2016). Fast classification of univariate and multivariate time series through shapelet discovery. *Knowledge and Information Systems, 49*(2), 429–454.
21. Ghalwash, M. F., & Obradovic, Z. (2012). Early classification of multivariate temporal observations by extraction of interpretable shapelets. *BMC Bioinformatics, 13*(1), 1–12.
22. Tang, W., Liu, L., & Long, G. (2020) Interpretable time-series classification on few-shot samples. In *2020 International Joint Conference on Neural Networks (IJCNN)* (pp. 1–8). IEEE.
23. Gee, A.H., Garcia-Olano, D., Ghosh, J., & Paydarfar, D. (2019) Explaining deep classification of time-series data with learned prototypes. *arXiv preprint* arXiv:1904.08935.
24. Zhang, X., Gao, Y., Lin, J., & Lu, C-T. (2020). Tapnet: Multivariate time series classification with attentional prototypical network. In *The Thirty-Fourth AAAI Conference on Artificial Intelligence, AAAI 2020, The Thirty-Second Innovative Applications of Artificial Intelligence Conference, IAAI 2020, The Tenth AAAI Symposium on Educational Advances in Artificial Intelligence, EAAI 2020, New York, NY, USA, February 7–12, 2020* (pp. 6845–6852). AAAI Press.
25. Li, G., Choi, B., & Jianliang, X. (2021). *Sourav S Bhowmick, Kwok-Pan Chun, and Grace LH Wong*. Shapenet: A shapelet-neural network approach for multivariate time series classification. AAAI.
26. Hadsell, R., Chopra, S., & LeCun, Y. (2006) Dimensionality reduction by learning an invariant mapping. In *2006 IEEE Computer Society Conference on Computer Vision and Pattern Recognition (CVPR'06)* (vol. 2, pp. 1735–1742). IEEE.
27. Paszke, A., Gross, S., Massa, F., Lerer, A., Bradbury, J., Chanan, G., Killeen, T., Lin, Z., Gimelshein, N., Antiga, L., Desmaison, A., Kopf, A., Yang, E., DeVito, Z., Raison, M., Tejani, Sasank Chilamkurthy, & A., Steiner, B., Fang, L., Bai, J., & Chintala, S. (2019). Pytorch: An imperative style, high-performance deep learning library. In H. Wallach, H. Larochelle, A. Beygelzimer, F. d'Alché-Buc, E. Fox, & R. Garnett (Eds.), *Advances in Neural Information Processing Systems 32* (pp. 8024–8035). Curran Associates Inc.
28. McInnes, L., Healy, J., & Melville, J. (2018). Umap: Uniform manifold approximation and projection for dimension reduction. *arXiv preprint* arXiv:1802.03426.
29. Navani, D., Block, V., Cree, B., & Abbasi-Asl, R. (2023). Diurnal step count patterns in progressive multiple sclerosis (p3-3.011).

One-Shot Defect Fingerprint Comparability Using Siamese Networks for Wafer Map Similarity in Semiconductor Manufacturing

Christian Weber, Sathvik Dembale Krishnappagowda, Ralf Montino, Peter Czerner, and Madjid Fathi

Abstract In the field of semiconductor manufacturing, the complex nature of the fabrication processes poses a significant challenge in maintaining a steady and high-quality yield. Microchips are fabricated on a silicon circular tray, known as a wafer, which can show specific defect patterns, based on the patterns of failures of groups of chips. Traditional classification algorithms are effective to classify distinct visual wafer patterns. However, in manufacturing, patterns are rarely fully distinct but represent composition of patterns which are significant for composing complex scenarios. Decomposing such patterns into feature vectors, can be utilized to identify and distinguish defect root causes. To address this challenge, the focus of this research is the development of a similarity-based classification, utilizing Siamese Networks and a comparison methodology, aiming to decompose patterns into composing core patterns as a defect fingerprint, their degree of intensity and, finally, enabling to compare current to past patterns for which defect root causes are already known.

Keywords Siamese networks · Cosine similarity · Pattern recognition · Semiconductor manufacturing · Wafer map

---

C. Weber (✉) · M. Fathi
Institute of Knowledge Based Systems and Knowledge Management, University of Siegen, Siegen, Germany
e-mail: christian.weber@uni-siegen.de

M. Fathi
e-mail: fathi@informatik.uni-siegen.de

S. Dembale Krishnappagowda
University of Siegen, Siegen, Germany

R. Montino · P. Czerner
Elmos Semiconductor SE, Dortmund, Germany
e-mail: ralf.montino@elmos.com

P. Czerner
e-mail: peter.czerner@elmos.com

M.-R. Alam and M. Fathi (eds.), *Integrated Systems: Data Driven Engineering*,
https://doi.org/10.1007/978-3-031-53652-6_3

1 Introduction

Defect pattern recognition is a technique used in semiconductor manufacturing, that aims to automate the process of identification and categorizing flaws or defects [2]. The use can be threefold. Firstly, the identification of patterns at early stages, enabling an early exclusion of products in cases in which patterns are too scarce to be recognized by human engineers. Secondly, to detect patterns even if the exact pattern isn't known, but similar patterns were seen in the past. This, thirdly, enables to consider root causes of past cases and prevent such defects through maintenance or repair of causing machines. However, manufacturing may not always lead to distinct patterns. Especially in semiconductor manufacturing, where process chains can be extensive and complex, resulting patterns will not represent singular, distinct patterns. They rather present overlays or combinations of significant patterns, which are frequent enough to be identified as parts of compositions. Significant compositions in turn are not frequent enough to represent classes with ample cases to train a classifier. The pressure attached will steadily increase with the rising impact of sub-nanometer designs, which in turn leads to increasingly powerful fault detection and root cause algorithms [11].

This work will look into defect pattern recognition of complex semiconductor wafer images. In focus is solving how complex patterns can be distinguished even if historical cases may be scarce. To do so, patterns are taken as composition of distinct core patterns, which are frequent and distinct enough for a reliable recognition. The composition is seen as a vector of degree or strength to which core patterns are occurring within a single pattern. The intuition of the vector is to represent a unique fingerprint, which supports not only the description of complex patterns, but also the comparison of such.

As a source for core patterns, the work of Wu, Jang and Chen is used [21], who compiled the widely used WM-811k wafer map reference data set. Wafer map in this case refers to images which show patterns of information on a circular representation of a real wafer. Information is differentiated using colors and can represent data with several intentions. In this work each microchip is represented as a collection of pixels, where each collection represents the information of existence of a failure with the color black and the non-existence with the color white. Wafer maps are generated based on a chip level end quality control. The benefit of the use of generated wafer maps is the flexibility of the representation of information, which furthermore enable a reuse of algorithms for a range of problems, as long as the underlying information is available for integration. The data set used here is a curated data set, generated from existing chip manufacturing data and labeled manually reusing core pattern classes of the WM-811IK data set: Center, Donut, Edge-Loc, Edge-Ring, Loc, Random, Scratch, Near-full and None.

In a well-established product line, well distinguished defects are rare. Far more regular, only light patterns of defects are visible, which in turn may represent mixes or compositions [8]. As such certain classes are considerably more or less frequent than other, leading to a highly imbalanced dataset. For example, is the class None with no or very light patterns the most frequent pattern. Furthermore, continuous changes in machine conditions will lead to the continuous occurrence of new types of defect pattern over time. These conditions present significant limitations for the selection of algorithms for creating defect fingerprint vectors.

To overcome this issue, this research will employ one-shot learning approaches with Siamese Networks. We present a comparative study of Siamese Network approaches, alongside fitting strategies to translate outputs into fingerprints. While each wafer is labeled with the most dominant core class of its pattern, no reference pattern vectors exist to quantitatively evaluate created vectors against a pre-known ground truth. This is mitigated with a reflection and estimation of the quality of the created vector. This work therefore is considered as a starting point to establish the concept and main argument of fingerprint vectors, along the implications of its composing methods and foundational evaluations. Future studies will have to utilize strategies to compose and reliably evaluate fingerprint vectors. These will be utilizing expert driven labeling procedures of vectors, as well as larger quantitative studies evaluating parameter optimizations on that very base. With this a powerful and flexible tool for analysis is primed, which can be integrated into sustainable strategies for defect handling [20].

2 Background

In the following, metrics, as well as similarity or distance oriented approaches for neuronal networks are explored. Metrics harness the distance between distributions not only for classification of the most likely class of a given individual but also can be stored as part of distance vectors. These distances in vectors can be seen properties of a specific pattern and represent classes as and class composing fingerprints.

2.1 Similarity Learning and Metric Learning

Similarity learning is a supervised machine learning technique that deals with the comparison of how similar or different two data objects are. Unlike traditional machine learning tasks such as regression and classification, this approach strives to learn the similarity function between two objects [12].

Metric learning is a subset of similarity learning, focusing on the learning distance function to measure similarity between the pair of objects. The basic principle here is to reduce the distance between similar and increase the distance between dissimilar objects [10].

There are several distance metrics used to measure the degree of similarity or dissimilarity. Well established metrics are here the Manhattan Distance, Euclidian Distance, and the Cosine Similarity, as a non-distance based similarity.

Using distance measures, metric learning can describe a new representation of the data samples. However, these metrics can be limited in their capturing of distance or similarities in multidimensional spaces in which the strength of variation of distances will deteriorate with an increase of dimensionality. To mitigate this problem, deep neural networks are employed in metric learning for automatic feature extraction and for better interpreting data. The concept of employing deep neural networks to learn distance functions for measuring similarity between similar and dissimilar data points is called Deep Metric Learning [13]. The use of deep neural networks provides solutions based on the problem by choosing the most prominent features from the data.

2.2 Siamese Neural Networks

A Siamese Neural Network (SNN) is a type of neural network architecture, used to measure the similarity between two different instances of data. Siamese networks were first introduced by Bromley et al. to solve a signature verification problem, comparing instances of signatures with different degrees of shape deviations against a limited sample of similar distances [4].

SNNs are also called "twin neural networks" since they contain two or more identical subnetworks, which together enable the comparison. These subnetworks have similar architecture, weights, and biases to ensure similar inputs are mapped to similar spaces. When each of the subnetworks receives inputs, they work in parallel and each of them is responsible for generating a meaningful representation of the input data. Each of the subnetworks are designed to reduce the size of the input and to have a compact representation with the most prominent features, which are then used to quantify the differences of inputs based on a measure of similarity. Recently, the Siamese architectures have proven to be efficient in applications such as face verification [6, 16], object tracking [3, 5], and many more application domains in which the comparison to base individuals is important.

2.3 N-Shot Learning

Conventional machine learning models may require a larger amount of data and many iterations of training to learn meaningful representations from data. However, if there is availability of large amounts of unstructured data, the task of data labeling is challenging. Reversely, this large data learning contradicts how humans learn from a few samples, as humans are able to understand the abstract essence of data with few examples and apply this knowledge to generalize on the new data. N-shot Learning can be defined as a learning approach that involves the transfer of learned knowledge to new data using a few (N) labeled samples per class [9]. In this approach, we will have N number of labeled examples for K classes. The goal is to classify the Query set Q into K classes.

In the basis of number of samples training samples used per class, the N-shot learning can be used in multiple domains and is classified into [15]:

1. **Zero-shot Learning**: In this approach, the model makes accurate predictions for a new class, which was not used during training [17].
2. **One-shot Learning**: In this type, the model recognizes the class of a given test image by comparing it with a single training image of each class [7].
3. **Few–Shot Learning**: Similar to One-shot learning, but here the test image is compared with more than one training image of each class [18].

3 Creating Fingerprints for Wafer Defect Pattern Comparisons

This research work is carried out utilizing a labeled semiconductor wafer map image dataset that was provided by the Elmos Semiconductor SE, where each label denotes the singular most dominant defect pattern for each given wafer map. The dataset consists of a total of 10,268 semiconductor wafer map images, sorted into the nine classes of the WM-811IK data set [21], which are shown with examples in Fig. 1.

The dataset contains 64 × 64 size wafer map images with 3 channels and these channels combine to produce the shades of gray. The tiny black squares represent the defective microchips, while the white color indicates microchips without any defects. In the given dataset, the necessary data pre-processing steps like subtraction of the outer mesh, grid removal, and removal of outer circles have been performed, as shown in Fig. 2. The dataset is imbalanced regarding the number of images in each class. Both, the data set and the pre-processing steps are the same as used in [19].

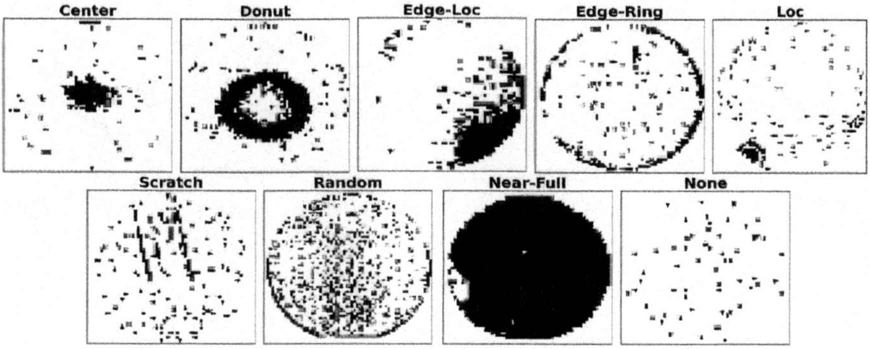

Fig. 1 The different labelled defect patterns in the given dataset

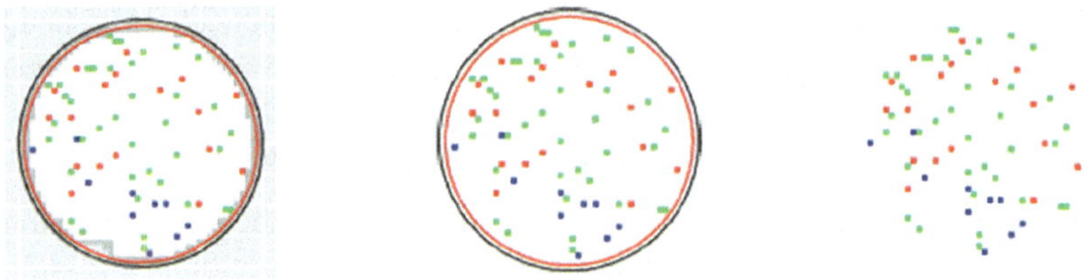

a) Raw wafer map b) Grid removed c) Outer rings removed

Fig. 2 Pre-implemented data pre-processing techniques [19]

3.1 Visualization Techniques for the Internal Distribution of Raw Images in the Dataset

The images in the dataset are of size 64 × 64, resulting in 4096 features. The direct visualization of such high-dimensional images can lead to sparse visuals, with only a small portion containing meaningful information. This phenomenon is termed the "curse of dimensionality" where distances between the data points become less meaningful in high dimensional space [1]. To address this issue this work uses t-distributed Stochastic Neighbor Embedding (t-SNE) as a dimensionality reduction technique In Figs. 3 and 4 is the plot of all the 10,268 raw wafer map images using t-SNE [14].

The plots show that the images are scattered and there are overlap between the images of different class. This is due to the raw image data containing noise, which is not relevant to distinguish between the different classes of defect patterns. To counter it, this work employs Siamese Neural Networks to learn meaningful embedding representations, with an emphasis on retaining the necessary features relevant to the similarity between the images.

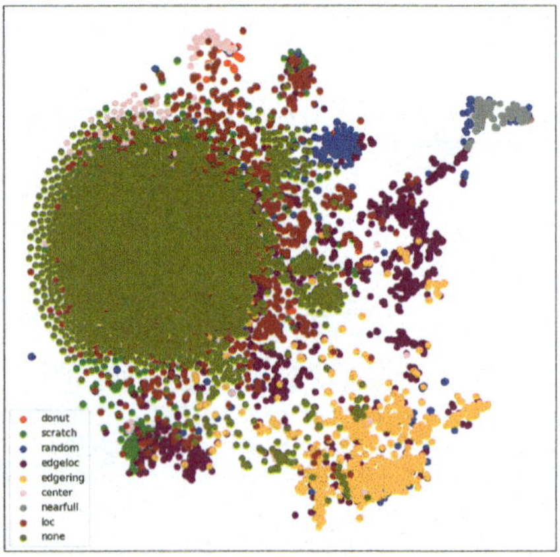

Fig. 3 A 2D t-SNE plot of raw images

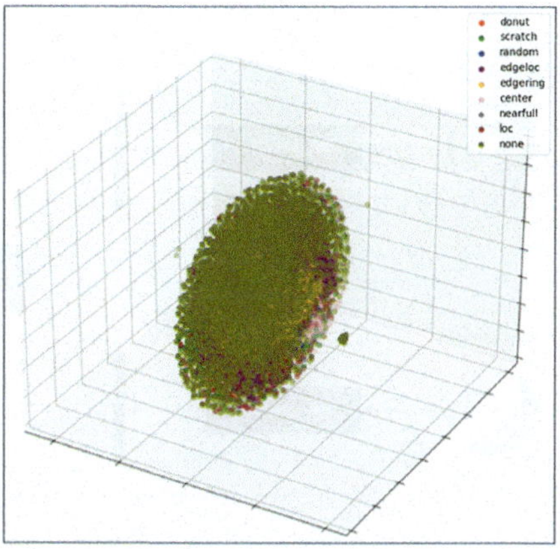

Fig. 4 A 3D t-SNE plot of raw images

4 A Comparative Study for Utilizing Siamese Networks

In the following, this section sheds light on the Siamese Network's architectural versatility, as well as their applicability to the given use case, combined with different loss functions for defect pattern recognition. The choice of loss the function is crucial in Siamese Neural Networks and has a direct influence on similarity learning. In this study various loss functions for defect pattern recognition are investigated and are evaluated along tailored concepts for comparable fingerprint vectors.

4.1 Experiment 1: Siamese Architecture with Binary Cross Entropy Loss

4.1.1 Creating Image Pairs

The architecture in Fig. 5 accepts input as pairs with the label indicating whether the images are similar or different. The images are randomly picked within the same class and are matched together to form positive pairs and are labeled as '1', indicating similarity. On the other hand, the images randomly picked from two different classes constitute negative pairs and are labeled as '0', indicating dissimilarity. However, though images are randomly picked for pairing, proper care is taken to ensure that each class contributes to the formation of pairs.

4.1.2 Defining the Architecture

The architecture is based on the general structure of the Siamese neural network, consisting of two parallel subnetworks with shared weights. In the architecture

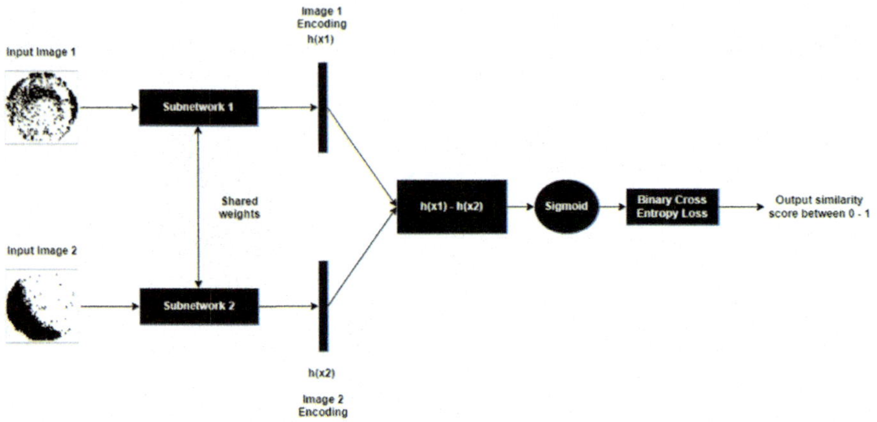

Fig. 5 Schematic diagram of siamese network architecture with BCE loss

utilized for this project, each of the subnetworks was built with three convolutional layers, as seen in Fig. 6, followed by three fully connected layers. The ReLU activation function is added at each layer to break the linearity.

Regularization techniques are employed such as batch normalization at each layer, dropout with random removal of 10% of neurons, and the addition of a weight decay penalty of 0.00001 to the loss function. The encodings generated from subnetworks are combined to compute the similarity between the inputs. Multiple methods are suitable for combining embeddings. In this study, experiments are conducted

Fig. 6 Subnetwork architecture: Siamese network + BCE loss

Layer (type)	Output Shape	Param #
ReflectionPad2d-1	[-1, 1, 66, 66]	0
Conv2d-2	[-1, 10, 57, 57]	1,010
ReLU-3	[-1, 10, 57, 57]	0
BatchNorm2d-4	[-1, 10, 57, 57]	20
Dropout2d-5	[-1, 10, 57, 57]	0
ReflectionPad2d-6	[-1, 10, 59, 59]	0
Conv2d-7	[-1, 10, 55, 55]	2,510
ReLU-8	[-1, 10, 55, 55]	0
BatchNorm2d-9	[-1, 10, 55, 55]	20
Dropout2d-10	[-1, 10, 55, 55]	0
ReflectionPad2d-11	[-1, 10, 57, 57]	0
Conv2d-12	[-1, 10, 55, 55]	910
ReLU-13	[-1, 10, 55, 55]	0
BatchNorm2d-14	[-1, 10, 55, 55]	20
Dropout2d-15	[-1, 10, 55, 55]	0
Linear-16	[-1, 800]	24,200,800
ReLU-17	[-1, 800]	0
Dropout-18	[-1, 800]	0
Linear-19	[-1, 400]	320,400
ReLU-20	[-1, 400]	0
Dropout-21	[-1, 400]	0
Linear-22	[-1, 100]	40,100

Total params: 24,565,790
Trainable params: 24,565,790
Non-trainable params: 0

using concatenating. For concatenating, the embeddings from each subnetwork are concatenated to form a long vector. Here all the information from both inputs is retained for similarity computation. Based on experimental results, the concatenation and differences between the embeddings were found to be effective in handling the dataset. The resulting embedding is passed through the sigmoid activation function to squeeze the output to a range of 0–1. The use of sigmoid activation ensures that the output similarity score can be interpreted as probabilities.

4.1.3 Computing BCE Loss and Backpropagation

The output similarity score from the network after the influence of sigmoid activation function confines to the range of 0–1. This similarity scores are then subjected into Binary Cross-Entropy (BCE) Loss (Sect. 2.4.2.1), and the loss is calculated my contrasting the similarity scores with similarity labels (1 for similar pairs and 0 for dissimilar pairs). The loss calculated is backpropagated to adjust the weights and biases of the network, so that the loss is minimized. Experiments are performed with multiple optimizers such as Stochastic Gradient Descent (SGD), RMSprop, Adam (Adaptive Moment Estimation) to perform backpropagation, and it is observed that the Adam optimizer was most effective in reducing the loss, thereby enhancing the performance of the network.

4.1.4 Training Configuration

The network was configured to get trained for 200 epochs, but with the used early stopping technique, the training was stopped at the 189th epoch when the validation loss surpassed the training loss. The batch size of 100 images per single pass was able to manage the memory constraints and the learning rate of 0.00001 was found to be optimal for learning speed and loss minimization. Data augmentation techniques such as random horizontal and vertical flipping were used to diversify the training data. However, the addition of rotational augmentations did not improve the learning process, instead, it had an adverse effect on similarity learning.

4.1.5 Loss Graph Analysis

In the initial epochs, the graph showed a rapid decrease in the loss since no initialization technique was used for weights and biases. With the training, the loss is gradually reduced, providing proof that the network is learning a similarity metric to differentiate between similar and dissimilar images. In addition, the validation loss is consistently lowered till the end of the training process, suggesting that the model is continuously learning without overfitting, as shown in Fig. 7.

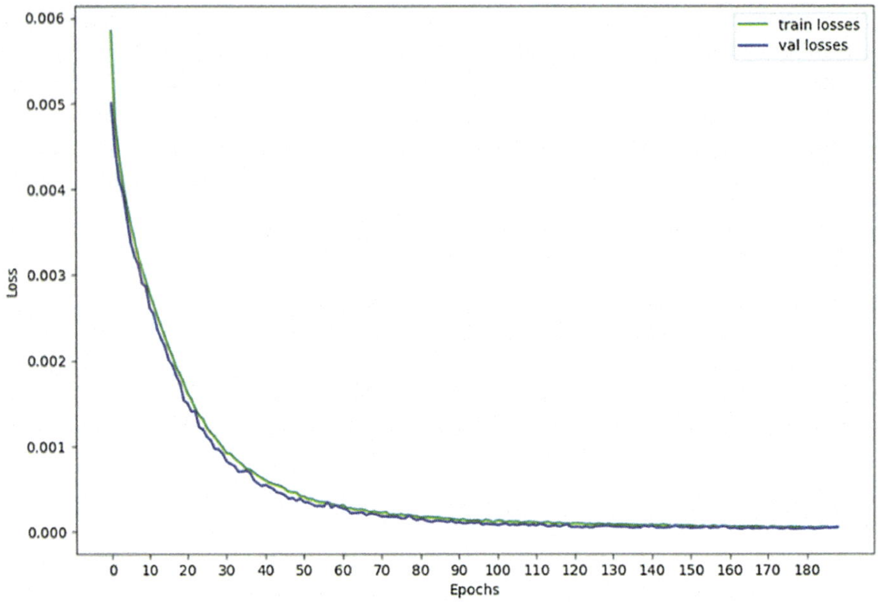

Fig. 7 Loss graph of Siamese network with binary cross-entropy

4.1.6 Visualization of the Feature Embeddings

The embeddings produced by each subnetwork have a length of 100, each element representing a feature that differentiates between positive and negative pairs. These high-dimensional features are visualized in two or three-dimensional space using a t-SNE plot, as shown in Figs. 8 and 9.

Fig. 8 2D t-SNE visualization of encodings

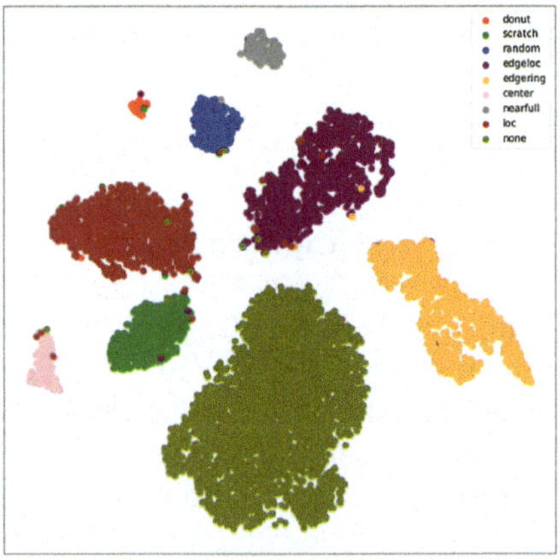

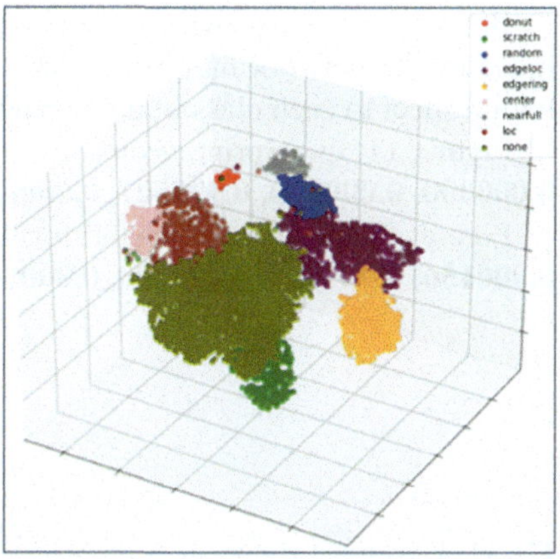

Fig. 9 3D t-SNE visualization of encodings

In comparison with the configuration of the raw data, the embeddings generated by SNN are distinct, and the data points belonging to similar classes are grouped together, enabling classification capability. Using the BCE loss, the network has been able to learn representations that provide class separability.

4.1.7 Generating Fingerprint Vectors

After the Siamese neural network is trained and validated, it is used to generate a similarity score between any two image inputs. For a given test image, the similarity scores are generated by comparing randomly picked images from each class. This process is repeated multiple times for the same test image to have an average similarity vector, which is also referred as a **similarity-based fingerprint vector**, where an example is visualized in Fig. 11.

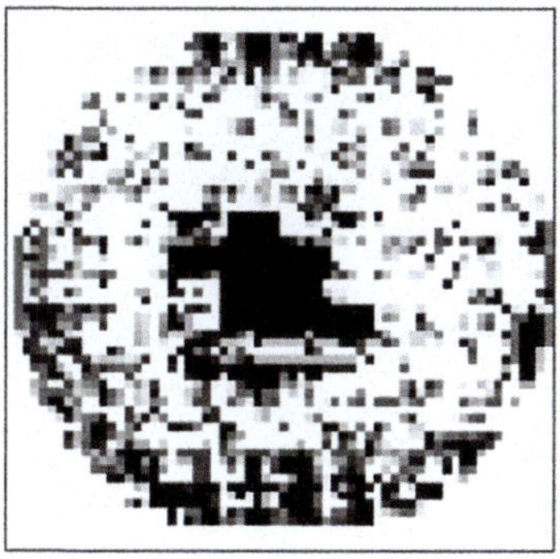

Fig. 10 Test image with center defect

Below, bar charts provide a visualization of example fingerprint vector predictions made by the network on test images, depicting the value of the fingerprint vectors with respect to each class of defect patterns.

Figure 11 fingerprint vector for Fig. 10 is [**0.000001, 0.006172, 0.000000, 0.000000, 0.000000, 0.999507, 0.000000, 0.000005, 0.000032**].

Figure 13 fingerprint vector for Fig. 12 is [**0.000000, 0.000010, 0.000025, 0.000156, 0.994008, 0.000000, 0.000000, 0.000000, 0.000004**].

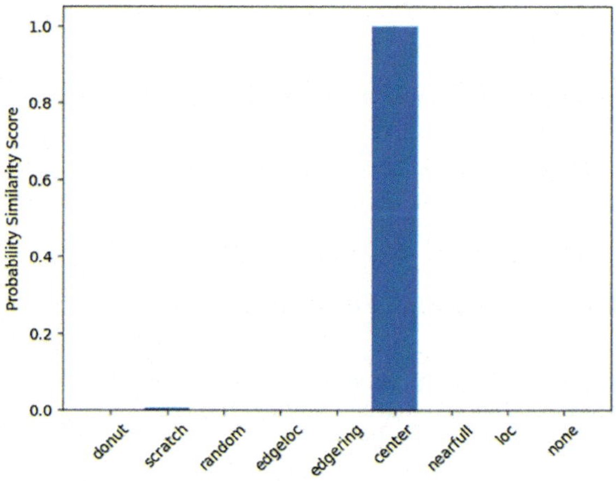

Fig. 11 Fingerprint vector magnitudes by class

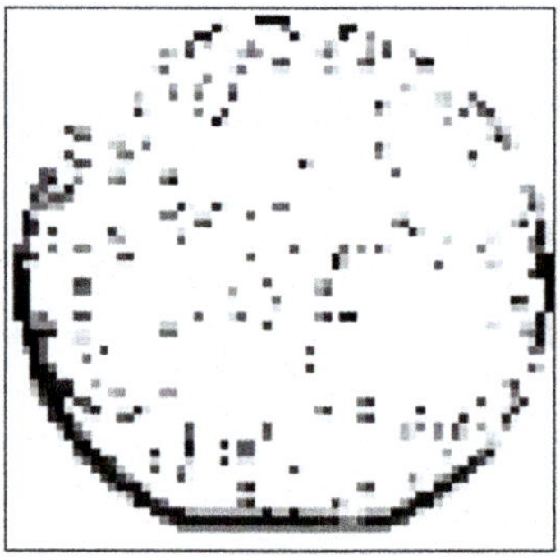

Fig. 12 Test image with edge-ring defect

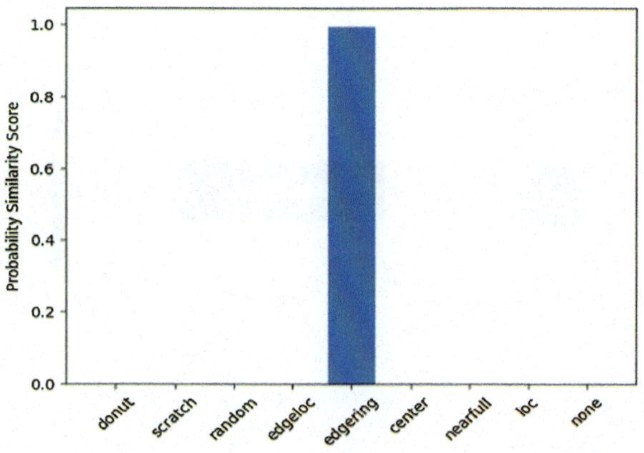

Fig. 13 Fingerprint vector magnitudes by class

Fig. 14 Classification report for Siamese network with BCE loss

	precision	recall	f1-score	support
donut	0.67	0.67	0.67	9
scratch	0.72	0.55	0.62	77
random	0.82	0.93	0.87	40
edgeloc	0.84	0.88	0.86	159
edgering	0.93	0.97	0.95	145
center	0.71	0.93	0.81	27
nearfull	0.96	0.96	0.96	23
loc	0.85	0.81	0.83	142
none	0.97	0.96	0.96	410
accuracy			0.89	1032
macro avg	0.83	0.85	0.84	1032
weighted avg	0.89	0.89	0.89	1032

4.1.8 Evaluation of Results with Performance Metrics

Figure 14 shows that the performance of the classification model is moderate in identifying certain classes. The overall classification accuracy of 89% is satisfactory.

4.1.9 Strengths and Limitations of the Approach

The shown approach can be advantageous for classification under data scarcity. Using Binary Cross-Entropy Loss, helps in detecting less prominent patterns in the dataset with high confidence, as it emphasizes these patterns by applying severe penalties to incorrect predictions with the use of the logarithm function. However, across predictions, the network assigns similarity scores close to 1 for similar images, and scores close to 0 for dissimilar images, due to the use of the sigmoid activation function, which pushes values towards either 1 or 0 for strengthening the distinctiveness of the decision threshold, which here is in conflict with the need for a continuous and homogenous distributed similarity score (Fig. 15).

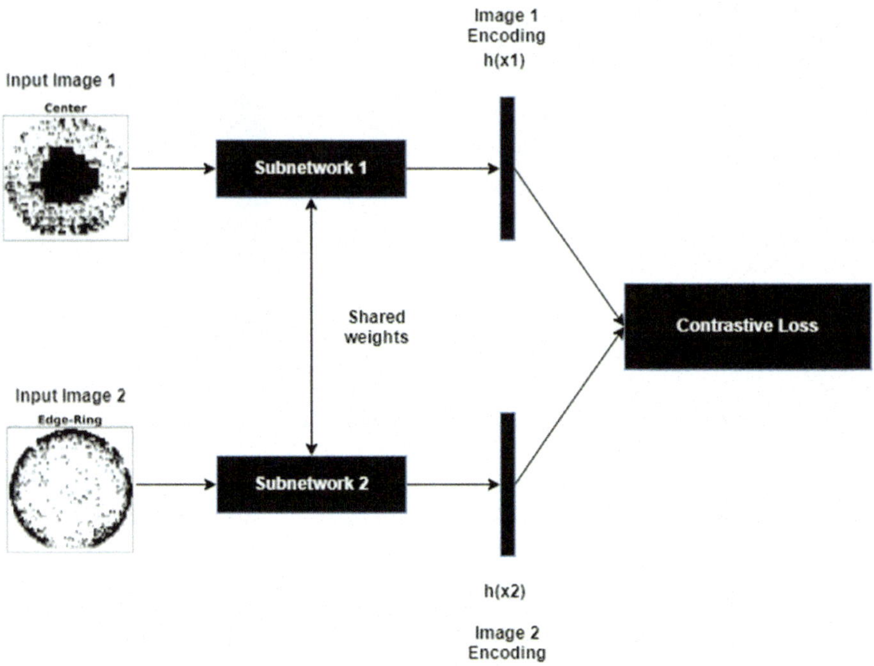

Fig. 15 Schematic diagram of Siamese network architecture with contrastive loss

4.2 Experiment 2: Siamese Architecture with Contrastive Loss

For this approach the contrastive loss function is used, which uses distance metric to differentiate between the two input pairs.

4.2.1 Creating Image Pairs

The Siamese architecture with contrastive loss requires an image pairing method similar to the first model, where image pairs are randomly selected for the dataset with the images from the similar class forming positive pairs (label: '1') and images from two different class form negative pairs (label:'0').

4.2.2 Defining the Architecture

The architecture contains two identical and parallelly functioning subnetworks with weight sharing. It processes the pair of inputs to generate the two encoded versions

Fig. 16 Subnetwork architecture: Siamese network and BCE loss

```
----------------------------------------------------------------
        Layer (type)               Output Shape         Param #
================================================================
   ReflectionPad2d-1            [-1, 1, 66, 66]               0
            Conv2d-2           [-1, 30, 57, 57]           3,030
              Tanh-3           [-1, 30, 57, 57]               0
       BatchNorm2d-4           [-1, 30, 57, 57]              60
         Dropout2d-5           [-1, 30, 57, 57]               0
   ReflectionPad2d-6           [-1, 30, 59, 59]               0
            Conv2d-7           [-1, 20, 55, 55]          15,020
              Tanh-8           [-1, 20, 55, 55]               0
       BatchNorm2d-9           [-1, 20, 55, 55]              40
        Dropout2d-10           [-1, 20, 55, 55]               0
  ReflectionPad2d-11           [-1, 20, 57, 57]               0
           Conv2d-12           [-1, 10, 55, 55]           1,810
             Tanh-13           [-1, 10, 55, 55]               0
      BatchNorm2d-14           [-1, 10, 55, 55]              20
        Dropout2d-15           [-1, 10, 55, 55]               0
           Linear-16                  [-1, 800]      24,200,800
             Tanh-17                  [-1, 800]               0
          Dropout-18                  [-1, 800]               0
           Linear-19                  [-1, 500]         400,500
             Tanh-20                  [-1, 500]               0
          Dropout-21                  [-1, 500]               0
           Linear-22                  [-1, 100]          50,100
================================================================
Total params: 24,671,380
Trainable params: 24,671,380
Non-trainable params: 0
```

of the data. Each subnetwork architecture, as seen in Fig. 16, contains three convolutional layers and three fully connected layers, delivering the best possible results for the given dataset. The Tanh (hyperbolic tangent) activation function was used between each layer to introduce non-linearity with batch normalization, dropout (10%), and weight decay with the loss penalty of 0.00001 as the regularization technique.

4.2.3 Computing Contrastive Loss and Backpropagation

The contrastive loss is computed by calculating the distance metric between the embeddings obtained from the subnetworks. Distance metrics such as Euclidean distance, and Manhattan distance can be used to calculate the distance between embeddings. The contrastive loss tries to minimize the distance between embeddings of positive input pairs while increasing the distance between negative pairs by at least the margin. The loss remains zero for embeddings that have a distance greater than the margin between them. The obtained loss is backpropagated using an optimizer (Adam optimizer).

4.2.4 Training Configuration

The network was trained for 100 epochs with a batch size of 100 images per single pass. The learning rate of 0.0001 was used to reach the global minimum. Experiments on the margin has shown that a small margin did not encourage the network to learn

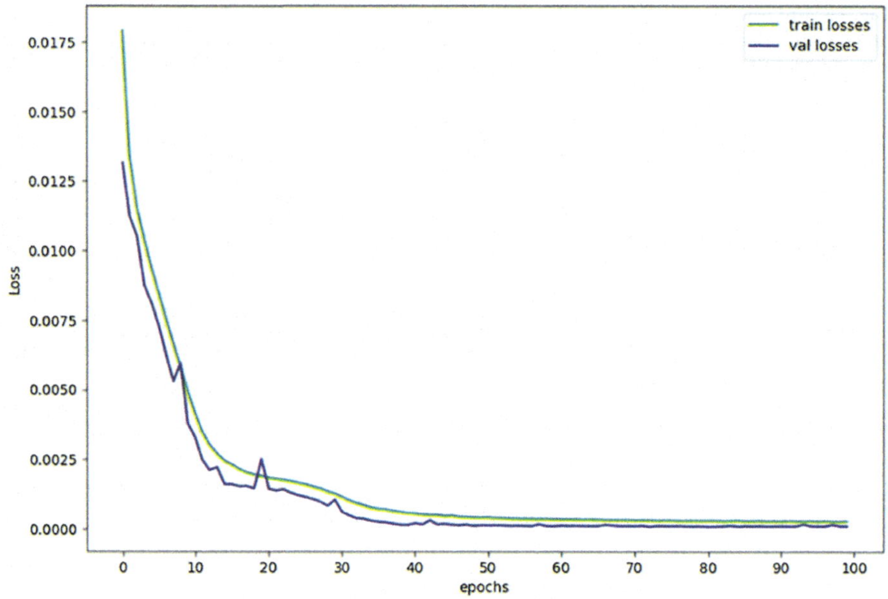

Fig. 17 Loss graph of siamese network with contrastive loss

the differentiation between the inputs, leading to overfitting. A margin distance of 3.0 has shown to differentiate well between embedding pairs.

4.2.5 Loss Graph Analysis

The loss graph characteristics in Fig. 17 show that the network is well-trained. However, despite training the model for additional number of epochs, there was not much reduction in loss beyond the 100th epoch.

4.2.6 Visualization of Feature Embeddings

In the plot Figs. 18 and 19, embeddings obtained from the Siamese neural networks show a better separation between classes. The margin value in the contrastive loss works effectively to differentiate the embeddings. Even though certain classes had a limited number of images, the network was able to differentiate well.

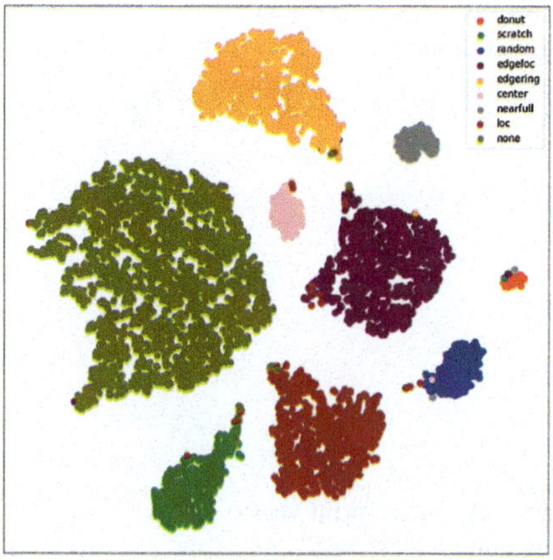

Fig. 18 2D t-SNE visualization of encodings

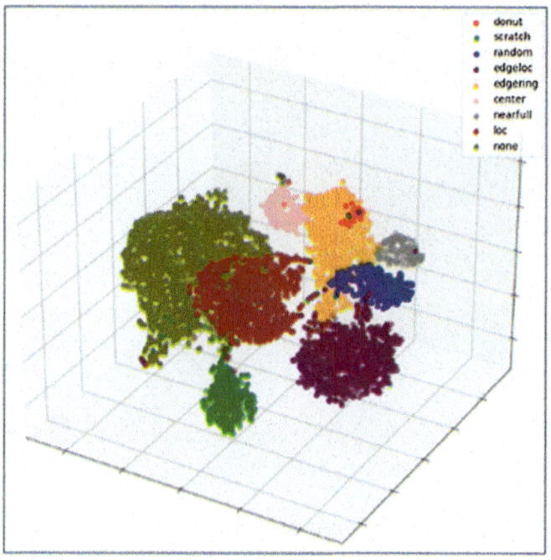

Fig. 19 3D t-SNE visualization of encodings

4.2.7 Generating Fingerprint Vectors

The output similarity scores obtained from the above combination of Siamese architecture will be the distances between the embeddings. To generate the fingerprint vector, the subnetwork of the trained Siamese network is used to generate lower dimension embeddings for a given test image. The generated embedding is compared with the mean of embeddings as a reference for the specific class. The distance similarity scores are generated with respect to each class and are then combined to form a **distance-based fingerprint vector**, with **lesser distance indicating high similarity and vice versa**. Figures 20 and 23 show examples of generated fingerprint vectors over test images:

Figure 21 distance fingerprint vector is [**0.0624, 5.1595, 2.8367, 3.4341, 4.4789, 2.9148, 2.2696, 2.9740, 4.5858**] (Fig. 22).

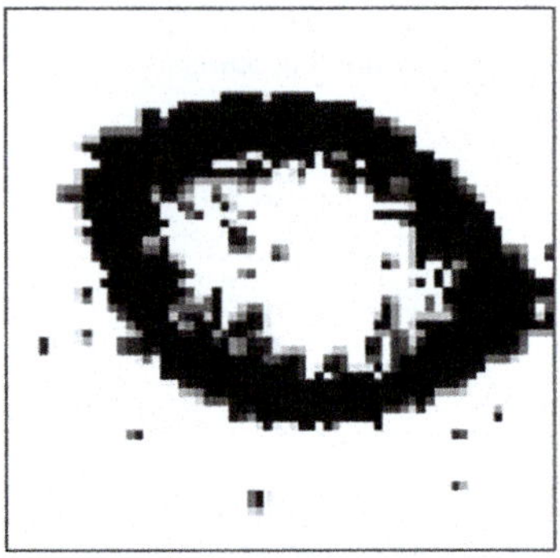

Fig. 20 Test image with donut defect

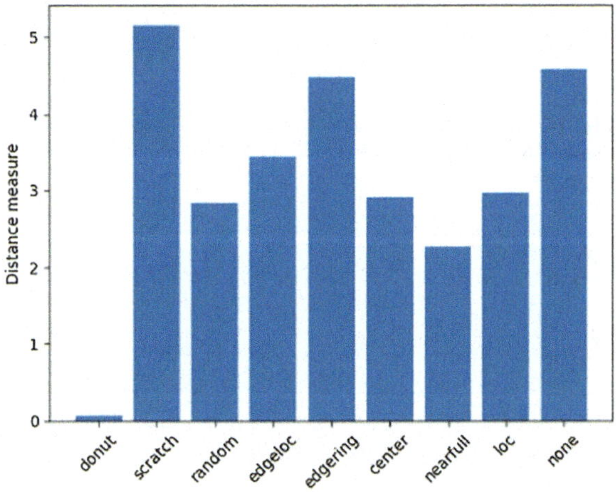

Fig. 21 Fingerprint vector magnitudes by class

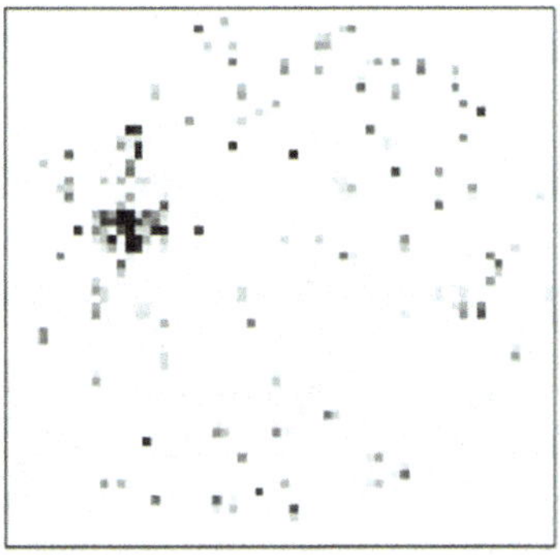

Fig. 22 Test image with Loc defect

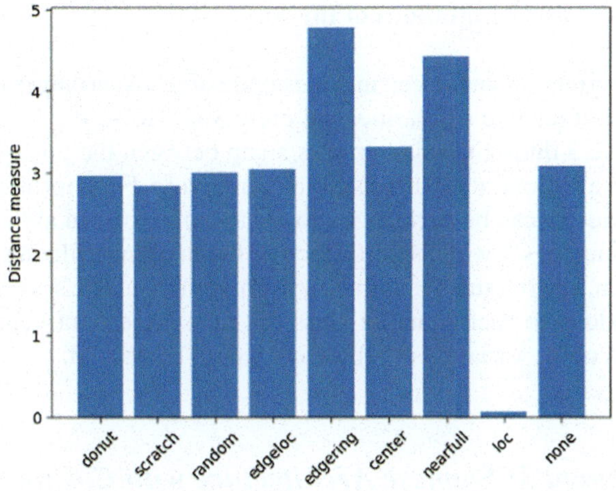

Fig. 23 Fingerprint vector magnitudes by class

Figure 23 distance fingerprint vector is [**2.9585, 2.8356, 2.9917, 3.0441, 4.7844, 3.30457, 4.4190, 0.0552, 3.08573**].

4.2.8 Evaluation of Results with Performance Metrics

The classification report shows that the model has performed well with overall accuracy of 90%. Overall, the Siamese architecture with contrastive loss has performed well in differentiating defect patterns (Fig. 24).

	precision	recall	f1-score	support
donut	0.78	0.78	0.78	9
scratch	0.80	0.52	0.63	77
random	0.79	0.82	0.80	40
edgeloc	0.89	0.91	0.90	159
edgering	0.93	0.99	0.96	145
center	0.77	0.85	0.81	27
nearfull	0.96	0.96	0.96	23
loc	0.81	0.81	0.81	142
none	0.95	0.98	0.97	410
accuracy			0.90	1032
macro avg	0.85	0.85	0.85	1032
weighted avg	0.90	0.90	0.90	1032

Fig. 24 Classification report for Siamese network with contrastive loss

4.2.9 Strengths and Limitations of the Approach

From the fingerprints, it can be seen that the contrastive loss arranges the embeddings in such a way that relative similarities are preserved. However, the use of a margin value of distance 3 during training creates a gap between the positive and negative pairs, leading to a discontinuity in the similarity measure. For improved visualization, the fingerprint vector can be normalized to a range of 0 to 1. In above plots, the raw values are retained to show the significance of the margin. Utilizing the knowledge about the margin, a correcting factor below the margin could be scaled to the selected margin to be added to each winning class for each fingerprint vector. However, a careful analysis of the variation of all factors would be needed.

4.3 Experiment 3: Siamese Architecture with Cosine Embedding Loss

This architecture finds the similarity between input pairs based on the cosine similarity. The advantage of this architecture is that the output similarity is confined to a range between -1 to 1. This is due to output similarity scores being based on the cosine similarity, which varies between these values (Fig. 25).

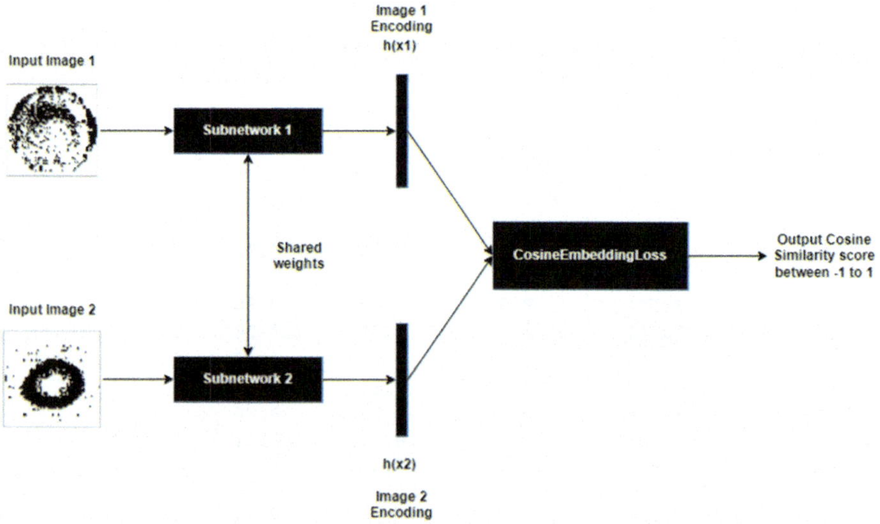

Fig. 25 Schematic diagram of Siamese network architecture with cosine embedding loss

Fig. 26 Subnetwork architecture: Siamese network and contrastive loss

```
----------------------------------------------------------------
        Layer (type)               Output Shape         Param #
================================================================
            Conv2d-1           [-1, 30, 55, 55]           3,030
              SiLU-2           [-1, 30, 55, 55]               0
       BatchNorm2d-3           [-1, 30, 55, 55]              60
         Dropout2d-4           [-1, 30, 55, 55]               0
            Conv2d-5           [-1, 20, 51, 51]          15,020
              SiLU-6           [-1, 20, 51, 51]               0
       BatchNorm2d-7           [-1, 20, 51, 51]              40
         Dropout2d-8           [-1, 20, 51, 51]               0
            Conv2d-9           [-1, 10, 49, 49]           1,810
             SiLU-10           [-1, 10, 49, 49]               0
      BatchNorm2d-11           [-1, 10, 49, 49]              20
        Dropout2d-12           [-1, 10, 49, 49]               0
           Linear-13                  [-1, 600]      14,406,600
             SiLU-14                  [-1, 600]               0
          Dropout-15                  [-1, 600]               0
           Linear-16                  [-1, 300]         180,300
             SiLU-17                  [-1, 300]               0
          Dropout-18                  [-1, 300]               0
           Linear-19                  [-1, 100]          30,100
================================================================
Total params: 14,636,980
Trainable params: 14,636,980
Non-trainable params: 0
```

4.3.1 Creating Image Pairs

The architecture receives inputs in pairs. In this case, the labels are assigned according to cosine similarity, where the similar pairs are labeled with '1' indicating the highest cosine similarity measure, while dissimilar pairs are labeled with '-1' indicating the lowest. Like previous approaches, the images are randomly picked for pairing with care taken to include images from all the classes, such that images from all the classes are equally included in the image pairs.

4.3.2 Defining the Architecture

The architecture consists of two subnetworks with shared weights. The experimented architecture comprises three convolution layers, as seen in Fig. 26, followed by three fully connected layers. The batch normalization and dropout with 20% removal of neurons are implemented in each layer. The SiLU (Sigmoid Linear Unit) activation performed well for this architecture.

4.3.3 Computing Cosine Embedding Loss and Backpropagation

The cosine embedding loss is calculated by computing the cosine similarity between the two output embeddings. The main objective of the cosine embedding loss is to maximize the cosine similarity between the similar pairs to 1 and to separate dissimilar pairs with a margin cosine similarity. The loss calculated is backpropagated with Adam optimizer to perform backpropagation.

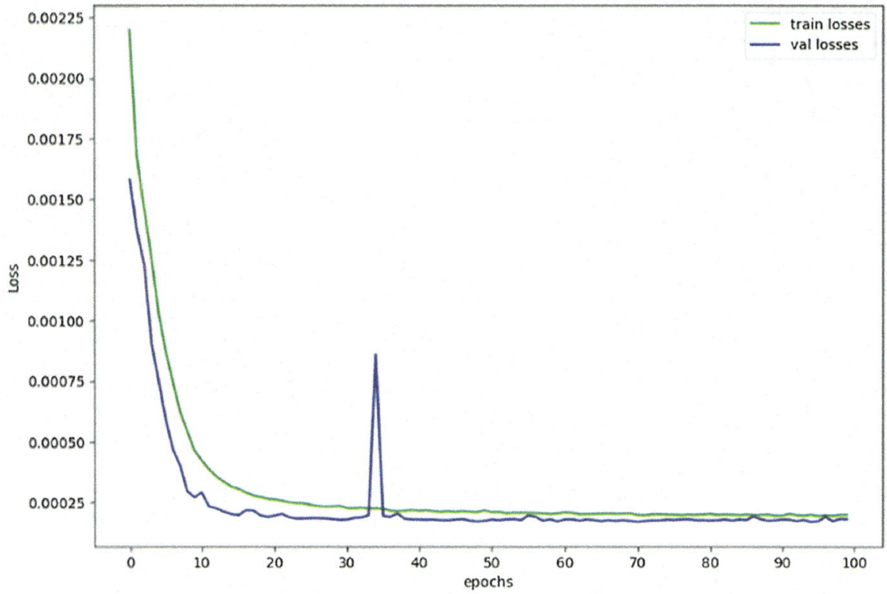

Fig. 27 Loss graph of siamese network with cosine embedding loss

4.3.4 Training Configuration

The network was trained for 100 epochs with a batch size of 100 images. The learning rate of 0.0001 with a weight penalty of 0.00001 was able to reach the global minimum. Data augmentation techniques like random horizontal and vertical flipping are used during training. However, the addition of slight rotation augmentations to the training images had a worsening effect on the learning process.

4.3.5 Loss Graph Analysis

The loss graph shows that the network was able to learn during each epoch. Initially, the network was trained with the Tanh activation function which resulted in the fluctuation in the training loss, but with the use of the SiLU activation function, the network had a good control over these fluctuations (Fig. 27).

4.3.6 Visualization of Feature Embeddings

The embeddings appear to be uniquely organized due to the cosine similarity, considering the cosine of the angle between the embeddings instead of actual magnitudes. The 2D and 3D plots show that similar data points cluster together and are arranged at

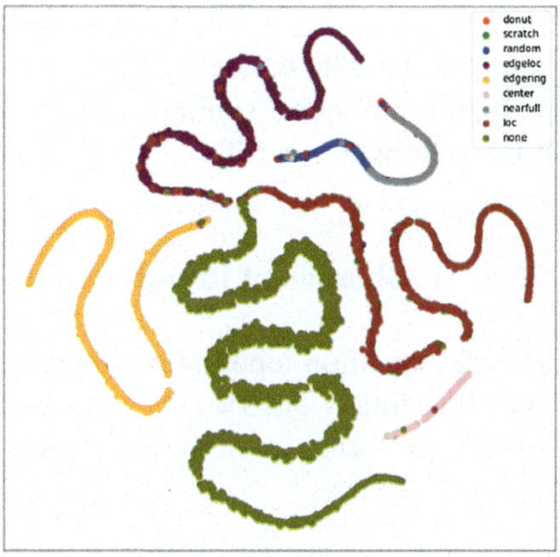

Fig. 28. 2D t-SNE visualization of embeddings

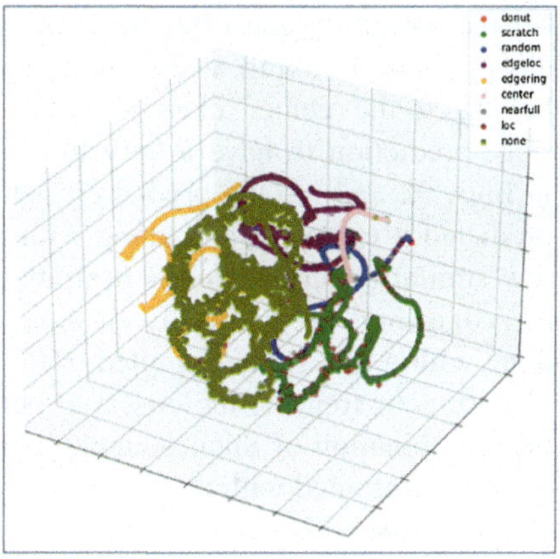

Fig. 29 3D t-SNE visualization of embeddings

a certain angle. In addition, a significant separation is observed between the embeddings, which proves that the architecture was able to successfully distinguish between the embeddings of different classes (Figs. 28 and 29).

4.3.7 Generating Cosine Similarity-Based Fingerprint Vectors

The trained network can now differentiate the inputs, and the subnetworks can produce embeddings such that they have high cosine similarity for similar inputs and lower cosine similarity for dissimilar pairs. For a given test image, one of the subnetworks is used to generate embeddings, which are then compared with the mean of embeddings or centroid of each defect type. The cosine similarity score is calculated in each instance with respect to each class and the combined scores form a **cosine similarity-based fingerprint vector**.

Figure 31 cosine similarity fingerprint vector is [**0.0975, 0.1345, 0.0929, −0.0369, −0.5995, 0.9998, 0.0909, 0.1293, 0.1313**].

Figure 33 cosine similarity fingerprint vector is [**−0.1162, −0.7634, −0.1092, 0.1087, 0.9999, −0.5904, −0.1030, −0.7601, −0.0656**].

4.3.8 Evaluation of Results with Performance Metrics

The classification report shows that the network performs moderately well with an overall accuracy of 81%. Overall, the Siamese architecture with cosine embedding loss has performed well in differentiating defect patterns.

4.3.9 Strengths and Limitations of the Approach

Cosine similarity can offer more information by considering the direction of the embeddings. Looking at the similarity ratings in Figs. 30 and 31, the test image belongs to the 'center' class, and the network predicted it correctly with the 'edge-ring' prediction to have a very low similarity score. To further substantiate this, in Figs. 32 and 33, the prediction belonged to the 'edge-ring' class. The network predicted it correctly, and along with it, the 'center' class was predicted to have a low similarity score. This is desirable as a 'center' prediction is contradictory to an 'edge-ring' prediction. Even though no large scale vector evaluation could be done within this study, sampled trials across the bank show a strong indication of a continuous trend of the observed behavior. From these results, it is believed that this combination of Siamese architecture with cosine embedding loss can be very effective if trained with the large balanced dataset (Fig. 34).

Due to the cosine-based similarity rating more prominent changes in the rating of dimensions with proportionally less prominent changes in visual patterns is expected. However, to the observed extent this is desirable, as based on manufacturing experience a majority of composite patterns for similar root causes behaviors have a

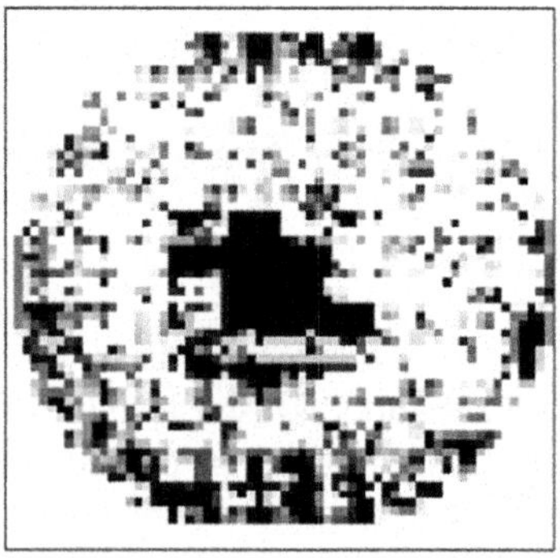

Fig. 30 Test image with center defect

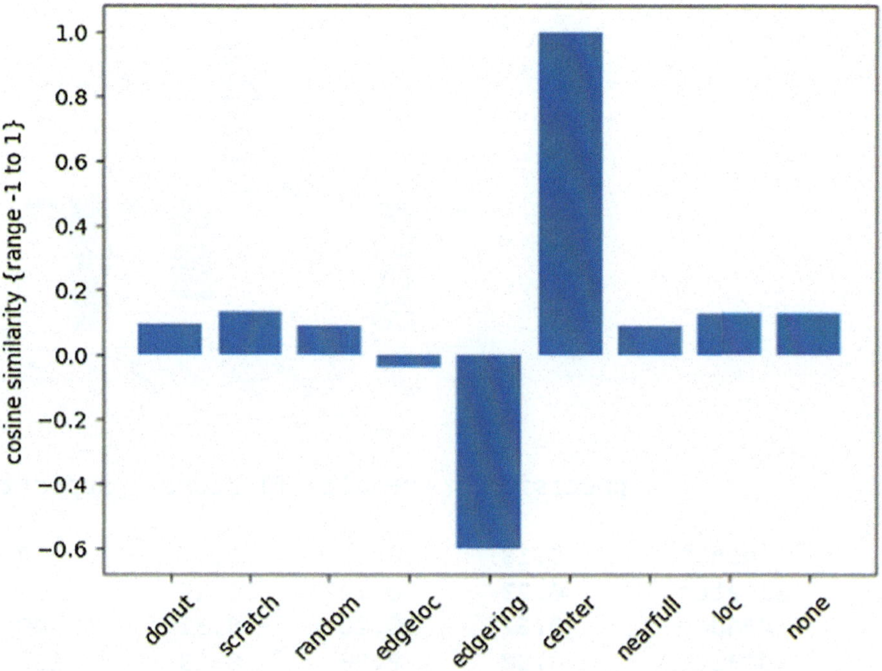

Fig. 31 Fingerprint vector magnitudes by class

Fig. 32 Test image with edge-ring defect

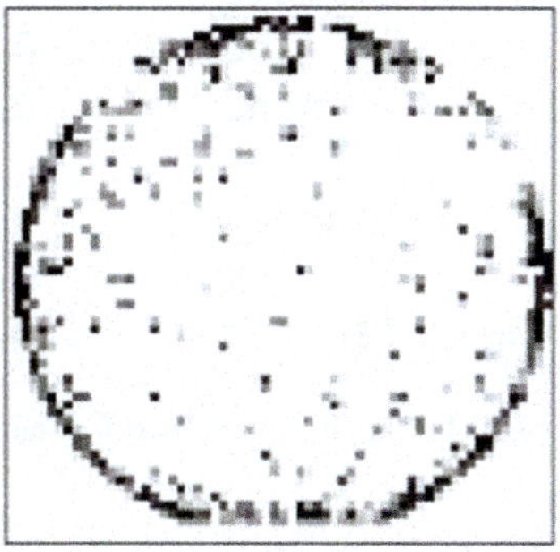

tendency to also show signs of only one of the 9 distinct pattern prominently. Differences between the remainder factors within the fingerprint vector will then have to explain the variation of observations. Values close to 0 will exhibit more often a "flipping" behavior between the positive and negative range but simultaneously explain only a fraction of the overall variation as they also indicate inconclusive patterns, while factors with larger (positive/negative) ratings will contribute stronger to the explanation, while not having a strong risk of flipping.

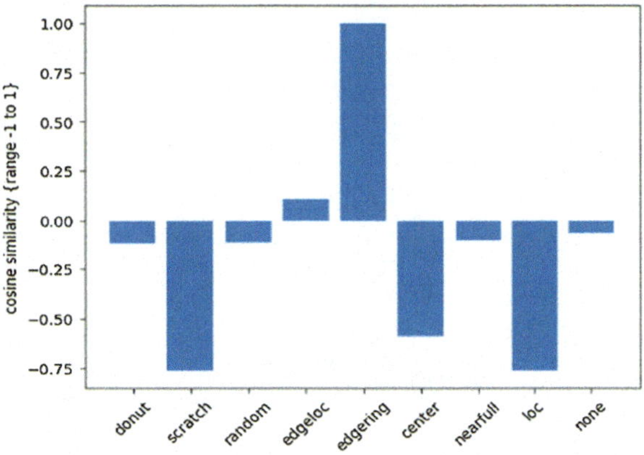

Fig. 33 Fingerprint vector magnitudes by class

	precision	recall	f1-score	support
donut	0.18	0.22	0.20	9
scratch	0.23	0.12	0.15	77
random	0.67	0.20	0.31	40
edgeloc	0.80	0.90	0.85	159
edgering	0.92	0.99	0.95	145
center	0.75	0.89	0.81	27
nearfull	0.47	0.87	0.61	23
loc	0.61	0.62	0.62	142
none	0.96	0.97	0.96	410
accuracy			0.81	1032
macro avg	0.62	0.64	0.61	1032
weighted avg	0.79	0.81	0.79	1032

Fig. 34 Classification report for siamese network with cosine embedding loss

5 Conclusion

We could show across three experiments that Siamese Network implementations are highly suitable for classifying wafer map based defect patterns even if selected classes are underrepresented in a data set. The inherent ability of Siamese Networks makes them candidates for expressing similarities or distances between wafer maps showing complex patterns, if classes are implementing prominent, distinct patterns as composing core patterns. However, not all configurations of Siamese Networks were able to provide similarity or distance based ratings in such a way that changes in ratings exhibit a close to linear behavior, which does not ensure comparability between all possible combinations of pattern compositions.

Within our experiments, Siamese Architectures with Cosine Embedding Loss were able to exhibit a desired comparability of patterns, in which classes with indecisive patterns will contribute less to the overall explanation of variation, in comparison to classes with a strong presence or strong absence within the pattern will contribute strongly. Therefore, the Cosine Embedding Loss is a convincing base for creating fingerprint vectors for wafer defect comparability and consequently create a base for identifying past root cause findings which are most likely similar to new defect root causes, described by limited samples of wafer map patterns as a novel one-shot fingerprint vector detection.

Acknowledgements This work was conducted as part of the European iRel40 project. iRel40 is a European co-funded innovation project that has been granted by the ECSEL Joint Undertaking (JU) under grant agreement No. 876659. The funding of the project comes from the Horizon 2020 research program and participating countries. National funding is provided by Germany, including the Free States of Saxony and Thuringia, Austria, Belgium, Finland, France, Italy, the Netherlands, Slovakia, Spain, Sweden, and Turkey.

References

1. Aggarwal, C. C., Hinneburg, A., & Keim, D. A. (2001) On the surprising behavior of distance metrics in high dimensional space. In J. Van den Bussche, & V. Vianu (Eds.), *Database theory—ICDT 2001*. Springer (Lecture Notes in Computer Science) (pp. 420–434). https://doi.org/10.1007/3-540-44503-X_27.
2. Batool, U., et al. (2021). A systematic review of deep learning for silicon wafer defect recognition. *IEEE Access, 9*, 116572–116593. https://doi.org/10.1109/ACCESS.2021.3106171.
3. Bertinetto, L., et al. (2016). Fully-convolutional siamese networks for object tracking. In G. Hua, & H. Jégou (Eds.), *Computer vision—ECCV 2016 workshops*. Springer International Publishing (Lecture Notes in Computer Science) (pp. 850–865). https://doi.org/10.1007/978-3-319-48881-3_56.
4. Bromley, J., et al. (1993). Signature verification using a "siamese" time delay neural network. *International Journal of Pattern Recognition and Artificial Intelligence, 7*(04), 669–688. https://doi.org/10.1142/S0218001493000339.
5. Cui, Z., & Lu, N. (2023). Feature-comparison network for visual tracking. *Applied Intelligence, 53*(15), 18263–18276. https://doi.org/10.1007/s10489-023-04466-y.
6. Diez-Tomillo, J., Alcaraz-Calero, J. M., & Wang, Q. (2023). Empirical comparison of face verification algorithms from UAVs. In *2023 International Conference on Software, Telecommunications and Computer Networks (SoftCOM). 2023 International Conference on Software, Telecommunications and Computer Networks (SoftCOM)* (pp. 1–6). https://doi.org/10.23919/SoftCOM58365.2023.10271666.
7. Fei-Fei, L., Fergus, R., & Perona, P. (2006). One-shot learning of object categories. *IEEE Transactions on Pattern Analysis and Machine Intelligence, 28*(4), 594–611. https://doi.org/10.1109/TPAMI.2006.79.
8. Geng, H., et al. (2023). Mixed-type wafer failure pattern recognition. In *Proceedings of the 28th Asia and South Pacific Design Automation Conference. New York, NY, USA: Association*

for Computing Machinery (ASPDAC '23) (pp. 727–732). https://doi.org/10.1145/3566097.3568363.
9. He, K., et al. (2023). Few-shot and meta-learning methods for image understanding: A survey. *International Journal of Multimedia Information Retrieval, 12*(2), 14. https://doi.org/10.1007/s13735-023-00279-4.
10. Kaya, M., & Bilge, H. Ş. (2019). Deep metric learning: A survey. *Symmetry, 11*(9), 1066. https://doi.org/10.3390/sym11091066.
11. Kim, T., & Behdinan, K. (2023). Advances in machine learning and deep learning applications towards wafer map defect recognition and classification: a review. *Journal of Intelligent Manufacturing, 34*(8), 3215–3247. https://doi.org/10.1007/s10845-022-01994-1.
12. Liu, K., Bellet, A., & Sha, F. (2015). Similarity learning for high-dimensional sparse data. In G. Lebanon & S. V. N. Vishwanathan (Eds.), *Proceedings of the Eighteenth International Conference on Artificial Intelligence and Statistics*. PMLR (Proceedings of Machine Learning Research) (pp. 653–662). https://proceedings.mlr.press/v38/liu15.html.
13. Lu, J., Hu, J., & Zhou, J. (2017). Deep metric learning for visual understanding: An overview of recent advances. *IEEE Signal Processing Magazine, 34*(6), 76–84. https://doi.org/10.1109/MSP.2017.2732900.
14. van der Maaten, L., & Hinton, G. (2008). Visualizing data using t-SNE. *Journal of Machine Learning Research, 9*(86), 2579–2605.
15. Song, Y., et al. (2023). A comprehensive survey of few-shot learning: Evolution, applications, challenges, and opportunities. *ACM Computing Surveys, 55*(13s), 271:1–271:40. https://doi.org/10.1145/3582688.
16. Taigman, Y., et al. (2014). DeepFace: Closing the gap to human-level performance in face verification. In *2014 IEEE Conference on Computer Vision and Pattern Recognition. 2014 IEEE Conference on Computer Vision and Pattern Recognition* (pp. 1701–1708). https://doi.org/10.1109/CVPR.2014.220.
17. Wang, W., et al. (2019). A survey of zero-shot learning: Settings, methods, and applications. *ACM Transactions on Intelligent Systems and Technology, 10*(2), 13:1–13:37. https://doi.org/10.1145/3293318.
18. Wang, Y., et al. (2020). Generalizing from a few examples: A survey on few-shot learning. *ACM Computing Surveys, 53*(3), 63:1–63:34. https://doi.org/10.1145/3386252.
19. Weber, C., et al. (2021). Clustering wafer defect patterns within the semiconductor industry based on wafer maps, using an agile unsupervised deep learning approach. In *2021 IEEE International Conference on Systems, Man, and Cybernetics (SMC). 2021 IEEE International Conference on Systems, Man, and Cybernetics (SMC)* (pp. 1913–1918). https://doi.org/10.1109/SMC52423.2021.9658907.
20. Weber, C., Czerner, P., & Fathi, M. (2023). Digit-DM: A sustainable data mining modell for continuous digitization in manufacturing. In *2023 IEEE International Conference on Electro Information Technology (eIT). 2023 IEEE International Conference on Electro Information Technology (eIT)* (pp. 197–202). https://doi.org/10.1109/eIT57321.2023.10187390.
21. Wu, M.-J., Jang, J.-S. R., & Chen, J.-L. (2015). Wafer map failure pattern recognition and similarity ranking for large-scale data sets. *IEEE Transactions on Semiconductor Manufacturing, 28*(1), 1–12. https://doi.org/10.1109/TSM.2014.2364237.

Opportunities of Data Driven Medicine: Collection, Transfer, and Processing of Vital Data in Outpatient Care

Alexander Keil, Rainer Brück, Kai Hahn, and Olaf Gaus

Abstract Data driven medicine can help to relieve physicians and medical personnel while maintaining or even improving the quality of medical care. An example in this context is self-measuring of medical vital signs by patients at home and the data transfer and processing for the physician to be a foundation of diagnoses. This procedure avoids unnecessary appointments in the practice and offers a better monitoring of the patients condition since the continuous measurement can early detect deviations so that therapies can be adapted in time. The article comprises a comprehensive model for such a process and highlights all the aspects from vital sign acquisition via transfer and data processing to doctor's access. Additionally this approach was successfully carried out in a study in Germany.

1 Introduction

In many developed countries the expenditures for the healthcare sector are steadily increasing and already consume more than 10% of the GDP. In Germany the share in 2021 was 13.2% coming from 10.2% 20 years earlier [1]. An ageing society in many industrialized countries will make the challenge even bigger [2, 3]. To guarantee a continued high quality medical care in this context without relying on

A. Keil · R. Brück · K. Hahn (✉)
Institute for Medical Informatics and Microsystem Engineering, University of Siegen, Am Eichenhang 50, 57068 Siegen, Germany
e-mail: kai.hahn@uni-siegen.de

A. Keil
e-mail: alexander.keil@uni-siegen.de

R. Brück
e-mail: rainer.brueck@uni-siegen.de

O. Gaus
Digitale Modellregion Gesundheit Dreiländereck, University of Siegen, Am Eichenhang 50, 57068 Siegen, Germany
e-mail: olaf.gaus@uni-siegen.de

M.-R. Alam and M. Fathi (eds.), *Integrated Systems: Data Driven Engineering*,
https://doi.org/10.1007/978-3-031-53652-6_4

more physicians can only be realized using innovative technologies for treatment and preventive medical care.

Data driven medicine can help to relieve physicians and medical personnel [4]. In this context the following article highlights an approach that uses the self-measuring of medical vital data by patients at home and their transfer and processing for the doctors to be assessed. This procedure avoids unnecessary appointments in the practice and offers a better monitoring of the patients condition since the continuous measurement can early detect deviations so that therapies can be adapted in time.

The article starts by modelling a health care process based on remote patient monitoring in Chap. 2. In the following chapters all necessary parts for the model are explained: Chap. 3 summarizes the challenges related to the vital data collection at patient's home. Chapter 4 describes the data transfer to designated cloud environments where data processing can take place. Since artificial intelligence (AI) will become more and more important in the medical field, an AI based disease detection by processing vital data is shown in Chap. 5. Since personalized medical data is highly sensitive the important legal boundary conditions for patients' data in Germany are summarized in Chap. 6. Chapter 7 highlight some practical challenges from studies that were carried out and a conclusion finishes the article.

2 Model for Health Care Process in Outpatient Care

Data driven medicine often requires changes in health care process. In the following sections, a novel digital care process in the context of outpatient care and the relation to GP practices will be presented, which was partially implemented and studied in detail. Three perspectives are in the focus: process view, health care perspective, and technical view [5, 6]. It is important to note that the process can only be carried out under the consultation and guidance of attending physicians to ensure proper support and acceptance.

2.1 Process View

Figure 1 illustrates the general process of the new medical care model. The first step involves the physician prescribing the measurement of vital signs based on the patient's clinical condition. Similar to a medication prescription, this specifies which vital signs need to be measured and at what frequency. In the second step, patients independently measure their vital signs at home, following the prescribed guidelines. The recorded vital data needs to be analyzed (e.g. AI-driven) as the third step and is therefore uploaded to a secure cloud environment. This analysis is crucial to manage the large volumes of data and ensure efficient processing. For further assessment as the fourth step, the analyzed data needs to be receipt and viewed by the physician. If no abnormalities are detected in the data, the physician can simply continue the

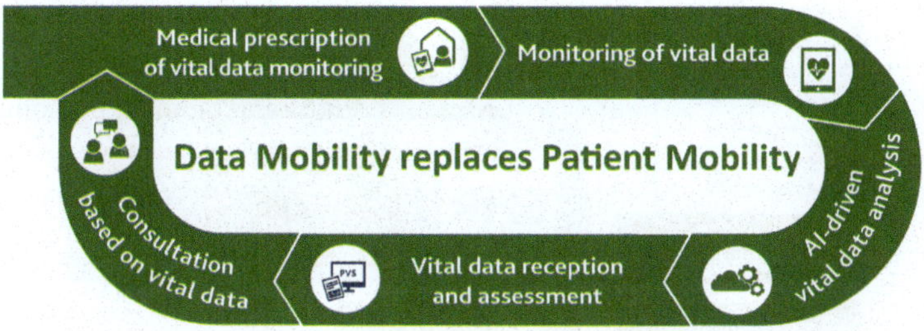

Fig. 1 Process view

measurements. However, if necessary the physician may arrange a consultation with the patient based on the vital data. Then adjustments to the prescription may be made, such as changing or continuing the same measurement requirements [6].

2.2 Component View

Figure 2 presents the care model from a component-oriented perspective. It starts with the physician prescribing vital sign monitoring. Next, the patient needs to be equipped with the necessary devices, and an initial setup must be performed. Once the pre-configured devices are received by the patient, recording of vital signs can start. These measurements are automatically transmitted to the cloud where the data can be integrated into the patient's electronic health record and evaluated using AI algorithms. Since the data is readily available in the electronic health record, it is automatically integrated into the medical practice management software used by the attending physician. Now, the physician or a trained assistant can consult the patient based on the vital data or the AI-summary to make any necessary adjustments to the treatment or vital sign prescription. Anther option is to end treatment and the recording of vital signs if it is not necessary anymore [6].

2.3 Health Care Perspective

Figure 3 provides an overview of how a physician-supervised monitoring process of vital signs could be structured. The first step always involves an initial examination at the doctor's office. In cases of acute or severe diseases, the physician may prescribe regular monitoring of specific vital signs that are relevant to the particular disease pattern. These vital signs are automatically transmitted to the physician, who continually monitors and reviews them. Additionally, trained doctor's office assistants may also have access to the data. If the vital signs only slightly deviate from the standard range or after a certain period of time, a follow-up examination with the patient can

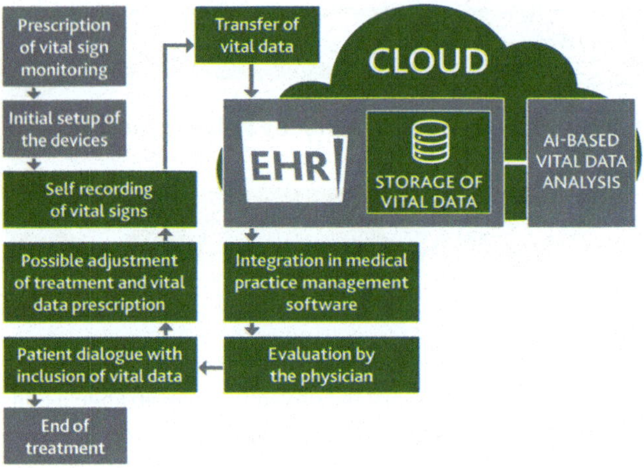

Fig. 2 Technical view

be conducted, either virtually or in person. This examination can be performed by the physician or delegated to assistants. In cases where there is a significant deviation in values or other important factors, the physician may call the patient to the doctor's office for a more thorough examination. After these examinations, adjustments to the vital sign prescription can be made if necessary. Once the disease is cured, treatment can be phased out after a final examination [5].

The following sections will take this model as a guideline starting with the measurement of vital signs at patient's home.

3 Collection of Vital Data by Telemonitoring

In this section the patient's frontend with the measurements of vital signs and their importance to monitor specific diseases will be in the focus.

3.1 Measurable Vital Signs

Nowadays patients have the capability to measure a broad range of vital signs from the comfort of their own homes. There is a diverse selection of affordable devices readily available in the market for this purpose. Depending on the area of interest different vital parameters are taken into account.

When it comes to the cardiovascular system, there are three significant vital signs to consider: heart rate, blood pressure, and electrocardiogram (ECG). Blood pressure monitors are already widely used in private households. Various devices, including

Fig. 3 Health care view

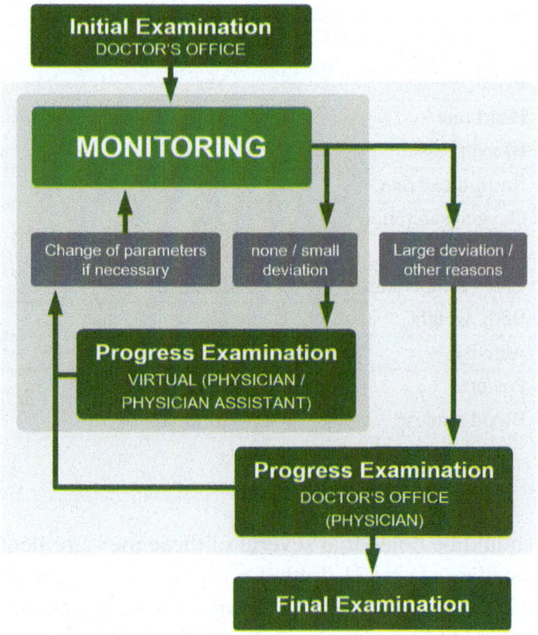

blood pressure monitors and pulse oximeters can measure heart rate, either in close proximity or potentially even at a distance in the future [7]. Commercially available devices for ECG measurement are also accessible at reasonable prices.

For the respiratory system, three vital signs are relevant: oxygen saturation, respiratory rate, and lung volume. Blood oxygen saturation can be easily measured using photoplethysmography, a function already integrated into a number of smartwatches. Respiratory rate can be measured using chest straps or, alternatively, the camera of a smartphone [8]. Spirometers, which have become smaller in recent years, can determine the volume of air inspired into the lungs.

Further vital signs include body temperature, weight, and blood glucose. Unfortunately, blood glucose measurement currently requires invasive methods involving blood extraction. However, in the future, spectroscopic and other non-invasive techniques may enable blood glucose measurements [9].

Furthermore, modern wearables and smartphones can measure physical activity, providing valuable insights into a patient's health status. Although most of the aforementioned vital signs are currently measured using separate devices and not continuously, many of them can already be measured continuously using wearables, such as smartwatches. For instance, the Apple Watch Series 8 can measure oxygen saturation, ECG, heart rate, and temperature [10], while the Samsung Galaxy Watch5 can measure ECG, heart rate, blood pressure, and body fat through bioelectrical impedance analysis [11]. Both devices can also measure activity. However, it

Table 1 Vital signs and their measurability with wearables [5]

Vital value	With wearable	Continuously measurable
ECG	Yes	No
Heart rate	Yes	Yes
Blood pressure	Limited	Yes
Respiratory rate	No	Yes
Oxygen saturation	Yes	Yes
Lung volume	No	No
Body Temperature	Limited	Yes
Body weight	No	No
Activity	Yes	Yes
Posture	Yes	Yes
Blood glucose	In future	In future

should be noted that several of these measurement methods do not yet meet medical standards [12, 13]. Table 1 presents an overview of the vital signs, their measurability with wearables, and the possibility of continuous measurement.

3.2 *Vital Signs and Diseases*

By utilizing the vital signs listed in Table 1, it is possible to monitor several common diagnoses and disease patterns for the purposes of prevention, treatment, post-treatment, and assessing potential consequences. Table 2 presents a compilation of common diseases and their prevalences in Germany. In addition, pneumonia is considered because it is the 6th leading cause of death in high-income countries, in low-income countries, it is even in second place [14].

In the following discussion, the disease patterns from Table 2 will be correlated with the vital signs from Table 1 in terms of prevention, treatment, aftertreatment, and potential consequences [5]. Table 3 presents an overview of these correlations.

Table 2 Prevalence of common diseases in Germany

Disease	Prevalence (%)	
Heart failure	4.7	[15]
Atrial fibrillation	0.4–2.0	[16]
Hypertension	31.8	[17]
Diabetes mellitus	7.2–9.1	[18]
COPD	5.8	[19]

Table 3 Vital signs and corresponding disease patterns. (P = Prevention, T = Treatment, A = Aftertreatment, C = (possible) Consequences) [5, 20]

	Heart failure				Arterial fibrillation				Hyper-tension				Type 2 diabetes				Pneu-monia				COPD			
	P	T	A	C	P	T	A	C	P	T	A	C	P	T	A	C	P	T	A	C	P	T	A	C
ECG	X			X		X	X					X												X
Heart rate		X				X	X					X								X				
Blood pressure	X	X			X	X			X	X						X				X				X
Respiratory rate				X														X			X	X		
Oxygen saturation				X														X				X		
Lung volume																		X			X	X		
Body temperature																	X	X				X		
Body weight	X				X				X							X						X		
Activity	X	X			X	X	X			X			X	X	X									
Blood glucose										X			X	X										X

When considering heart failure, all vital signs related to the heart are crucial for prevention and treatment, in conjunction with monitoring weight and activity. Acute deterioration of cardiac output can be a possible outcome, which can be monitored through ECG, respiratory rate, and oxygen saturation to assess pulmonary involvement [5, 20].

In the case of atrial fibrillation, cardiac values, weight, and activity play important roles in every stage of therapy. For hypertension, blood pressure in combination with weight and activity is necessary for both prevention and therapy. Possible consequences of hypertension may include cardiopathies, which can be detected through monitoring ECG and heart rate [5, 20].

For patients with type 2 diabetes, blood glucose levels and activity are significant for prevention, treatment, and post-treatment. Possible consequences may involve obesity and the development of heart disease, highlighting the importance of monitoring blood pressure and weight [5, 20].

Pneumonia often leads to an increase in body temperature, making monitoring a valuable indicator. During treatment, it is important to monitor lung functions such as respiratory rate, oxygen saturation, and lung volume, in addition to temperature measurement. Pneumonia can also result in cardiopathy, necessitating the monitoring of blood pressure and heart rate [5, 20].

For patients with chronic obstructive pulmonary disease (COPD), in prevention respiratory rate and lung volume is of important, to detect deterioration of the respiratory systems at an early stage. During the treatment, all values of the respiratory system are of importance in combination with temperature for the detection of Inflammations and weight as weight loss is common. COPD increases the risk of cardiovascular diseases and the metabolic syndrome, therefore ECG, blood pressure and blood glucose should be monitored [20].

In the case of multimorbid patients, the recorded vital signs complement each other. Table IV provides an example of a patient with type 2 diabetes, arterial hypertension, and obesity. This example demonstrates how six combined vital signs can be utilized to monitor three diseases, including potential consequences associated with them [5].

4 Transfer and Storage

After measuring, the vital data must be transferred to an accessible cloud storage system where on the one hand the data processing can take place and on the other hand physicians, medical personal and nursing services have access to the patients' data.

4.1 Medical Data Spaces

Medical data spaces (MDS) are emerging as a transformative approach in healthcare that aims to enable secure and interoperable data exchange while maintaining data sovereignty and privacy. This concept draws inspiration from two influential frameworks, namely the International Data Spaces (IDS) and the Gaia-X Cloud Reference Architecture [21].

The International Data Spaces initiative, developed by the International Data Spaces Association, emphasizes the creation of a decentralized and secure data ecosystem. It provides a framework for data sovereignty, data interoperability, and trusted data sharing among diverse stakeholders. The IDS architecture promotes secure data exchange through the use of standardized connectors, data usage policies, and secure data containers. By adopting IDS principles, medical data spaces can establish secure and controlled environments for exchanging sensitive healthcare data [22].

Gaia-X, on the other hand, is a European initiative that focuses on building a federated data infrastructure to enable secure and sovereign data sharing across industries. Gaia-X aims to create a decentralized network of cloud services based on common European values, such as data sovereignty, transparency, and privacy. By leveraging Gaia-X's cloud reference architecture, medical data spaces can benefit from standardized interfaces, secure data sharing mechanisms, and compliance with European data protection regulations [23].

4.2 AWS Cloud

Since the both concepts (MDS, GAIA-X) described above are hardly or not at all implemented so far it is necessary to concentrate on commercially available cloud solutions for proof of concept implementations.

Amazon Web Services (AWS) is one of the few cloud systems that has received certification for handling medical data and ensures that data remains within the jurisdiction of Germany. This certification, granted by the German Federal Office for Information Security (BSI), includes compliance with standards such as C5 (operational security), ISO/IEC 27001:2013 (security management controls), ISO/IEC 27018:2019 (personal data protection), and GxP (quality guidelines), allowing the storage of sensitive medical and personal information, including vital data [24]. For the implementation of the model as shown in Chap. 2 therefore AWS was chosen.

5 AI-Based Disease Detection

When a monitoring is implemented as previously described, an enormous amount of vital data will collected. Physicians will not be able to handle and assess all this data. Therefore an data preprocessing and evaluation based on artificial intelligence (AI) is needed.

5.1 General Approach

In a first attempt the interesting question of automatic detection of possible diseases (by their ICD-codes) by analyzing vital signs using AI algorithms was investigated in [25].

Within an examination of this problem in the MIMIC-III dataset was used and two strategies with several models were trained based on decision trees. The idea was to generate predictions regarding the diagnosis of patients using classical vital parameters. For data preparation and analysis, two AWS services, Amazon Athena and Amazon Sagemaker, were used to create datasets suitable for machine learning.

The previously created data was then further adjusted and prepared for model training. Models were created using XGBoost with a gradient boosting algorithm [26] and Scikit [27] in an python environment. Gradient boosting decision trees, an ensemble method that sequentially creates a large number of trees, were used to improve the errors of the previous iterations. The MIMIC-III dataset was used for training, which was divided into three versions based on the number of ICD codes. Additionally, the results were evaluated on the MIMIC-ED dataset.

It turned out that the multiclass strategy, did not work properly due to the uneven distribution of classes, the results are distorted, making the multilabel strategy better suited for the problem.

5.2 Multi-label Data Analysis

Especially in the multilabel strategy, tools such as feature importance and the classification report provide insights into the performance of individual classes, allowing for a more detailed analysis of diseases. The threshold at which a diagnosis is considered present can also be adjusted to make the models more generous and facilitate detection. While diseases like hypertension achieve a recall of up to 0.76 and a precision of up to 0.48, the average for all diseases is only 0.37 and 0.38. With the expansion of the data, these values increase to 0.42 as shown in Table 4.

Overall, the results show that it is difficult to generalize diagnosis statements. For diseases like hypertension or heart diseases such as atrial fibrillation, classical parameters like blood pressure or heart rate are highly informative. However, as

Table 4 Classification Report of 30 ICD code and extended version

Disease	Precision		Recall	
	30 codes	Extended	30 codes	Extended
Hypertension	0.48	0.47	0.76	0.81
Diabetes	0.28	0.30	0.25	0.20
COPD	0.22	0.30	0.10	0.08
Pneumonia	0.28	0.40	0.18	0.54
Average	0.38	0.42	0.37	0.42

diseases become more specific, it becomes more challenging for a model to draw accurate conclusions. Based on the models trained in this work, reliable diagnoses cannot be made. The evaluation on the MIMIC-ED dataset also shows that missing parameters can deteriorate performance, making the results always dependent on the quality of the data.

However, the expansion of the dataset with additional features has shown that despite optimizations, it could still be a useful tool to quickly gain initial insights into possible diagnoses.

To further improve the results, there are various options. One possibility would be to use a significantly larger number of features, incorporating information such as blood tests, pre-existing conditions, or medications used in the creation of a model. It would be sensible to gradually examine different diseases individually and consider what could be relevant for the diagnosis. This would eventually result in a feature database that can be continuously expanded with new diseases. Examples of additional inputs could include various laboratory tests, pre-existing conditions, medications used, or other notes from a medical history.

Automatic detection of diseases based on vital signs using machine learning remains an exciting topic that can improve many aspects of healthcare. The growing knowledge within medicine and advancements in artificial intelligence and available data offer the possibility of making diagnoses more reliable in the future, particularly with a growing number of patients and increasing workload, reducing the number of undetected diseases.

6 Regulations in Germany

In this section the legal boundary conditions in the German healthcare system related to data driven medicine are discussed. Vital data of patients is from its nature of course sensitive and vulnerable to misuse. Therefore data protection rules are of the utmost significance in today's digital society. Additionally, the patient's privacy is safeguarded by these regulations. Two regulations pertaining data protection will be presented in the following sections: the General Data Protection Regulation (GDPR) and the German Federal Data Protection Act. The public healthcare system in general uses additionally the telematics infrastructure (Telematikinfrastruktur—TI) as a closed VPN based network with strictly monitored access to all stakeholders.

6.1 Patients Data Protection

The **General Data Protection Regulation** (Datenschutz-Grundverordnung—DSGVO) [28] has been in place in Europe since May 25, 2018. It outlines the legal criteria for the handling of personal data by both private companies and government entities. This law sets obligations on business owners, among others, including reporting and accountability requirements, data security requirements, and the application of data subject rights. According to the law, the patient's consent is required before personal data under article 6(1d) DSGVO may be disclosed. In addition, according to article 7(3) DSGVO, the patient has the right to withdraw the consent at any time.

The **Federal Data Protection Act** (Bundesdatenschutzgesetz—BDSG) [29] applies to all companies and organizations that process personal data in Germany. This law regulates that personal data may only be collected, processed and used under certain conditions. In relation to telemedicine, it is of great importance, since health data has the highest confidentiality.

6.2 Health Care Regulations

Healthcare regulations in Germany encompass various aspects, including data protection, electronic health records (EHR), and interoperability. Germany has implemented the electronic health record system, aiming to centralize patient information for improved healthcare delivery. Interoperability plays a significant role, enabling seamless data exchange between different healthcare providers and systems, ensuring coordinated care and efficient communication. These regulations prioritize patient privacy, information security, and enhanced healthcare outcomes.

Telematics Infrastructure—TI

The core component of the German digital health system is the Telematics Infrastructure (Telematikinfrastruktur—TI), introduced in volume V of the German Code of Social Law (Fünftes Buch Sozialgesetzbuch—SGB V) [30] from § 334. The TI serves as a secure digital infrastructure that connects healthcare providers, health insurers, insured individuals, and other stakeholders in the German healthcare system.

Electronic Health Record

The electronic health record (Elektronische Patientenakte—ePA) is also defined in SGB V from §341 and health insurance companies must make it available to every insured person upon request since 2021. Individuals should be able to add their own health data to the ePA.

The permitted file types for uploading to the ePA, include documents (PDF), text files (TXT, RTF, DOCX, ODT), spreadsheets (XLSX, ODS), pictures (JPG, TIFF), and machine-readable formats (XML, HL7 CDA/R2 XML). While PDF files are suitable for vital record evaluations, XML or HL7 CDA/R2 XML formats are most suitable for storing structured vital data [31].

Digital Health Applications

A special feature of the German healthcare system are digital health care applications (Digitale Gesundheitsanwendungen—DiGA), which are either web applications or smartphone apps. They must have a proven enhanced care effect for a specific disease. The use of the DiGA can be prescribed by a physician in a similar way to a drug, and the cost of the DiGA is then paid by public health insurances. The legal basis for this is on the one hand SGB V again complemented by the digital health applications regulation (Digitale-Gesundheitsanwendungen-Verordnung—DiGAV) [32].

DiGAs, with their unique position in the German health system, are allowed to write data to the ePA. To facilitate interoperability, Medical Information Objects (Medizinisches Informationsobjekt—MIO) were developed. They are based on XML and were created in cooperation with HL7 (Health Level 7) on the basis of FHIR (Fast Healthcare Interoperability Resources) [33].

7 Implementation, Study, and Practical Challenges

Such a digital health solution based on remote patient monitoring and data evaluation can only be widely deployed if it is supported by all stakeholders and meets with widespread acceptance. For this reason, a study was conducted in 2021 and 2022 to assess the experiences of patients, physicians and caregivers using the system. At the same time, the technical infrastructure was permanently enhanced on the basis of feedback from the stakeholders and support structures were set up. The study had two phases, the first one in a nursing home and the second on at the patient's home.

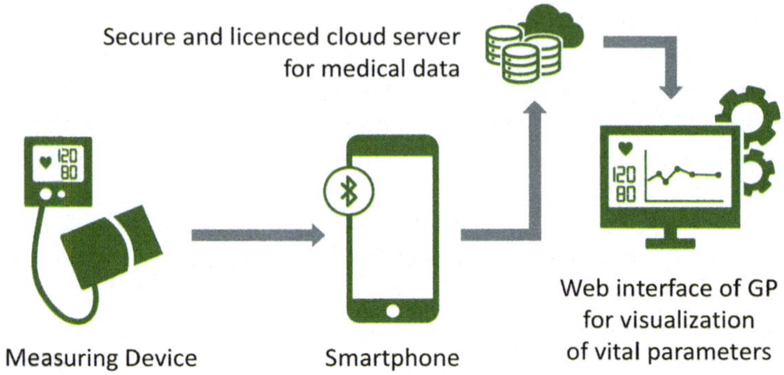

Fig. 4 Path of the data in the study

7.1 Hard- and Software Implementation

In the study, commercially available measurement devices for blood pressure, oxygen saturation and body weight with bluetooth connection were used in addition to Apple Watches for ECG. The patients were equipped with these devices depending on the individual prescription of measurements by their physician. The vital data is automatically transferred to a smartphone via bluetooth. Here the data is preprocessed and forwarded to an AWS cloud environment. A web interface was provided for the physicians to access the visualized data of their patients (see Fig. 4) [34, 35].

7.2 Acceptance by Patients

With regard to acceptance, a mixed picture has emerged among patients. The patients living at home were on average 69.4 years old and most of them were relatively fit, both physically and mentally. The patients in the nursing home however were on average about 8 years older. Furthermore, some of them were physically as well as mentally impaired.

Nursing Home

Some of the patients in the nursing home had a relatively high affinity for technology and were also in possession of computers, smartphones or voice assistants. These persons did not have any major problems in measuring their vital signs by themselves. Some of the residents were no longer mentally capable of familiarizing themselves with the devices due to their age. Among the other residents, there were some who took up the new technical challenge and found out how easy it was to use after an initial familiarization phase. However, there were also residents who rejected the

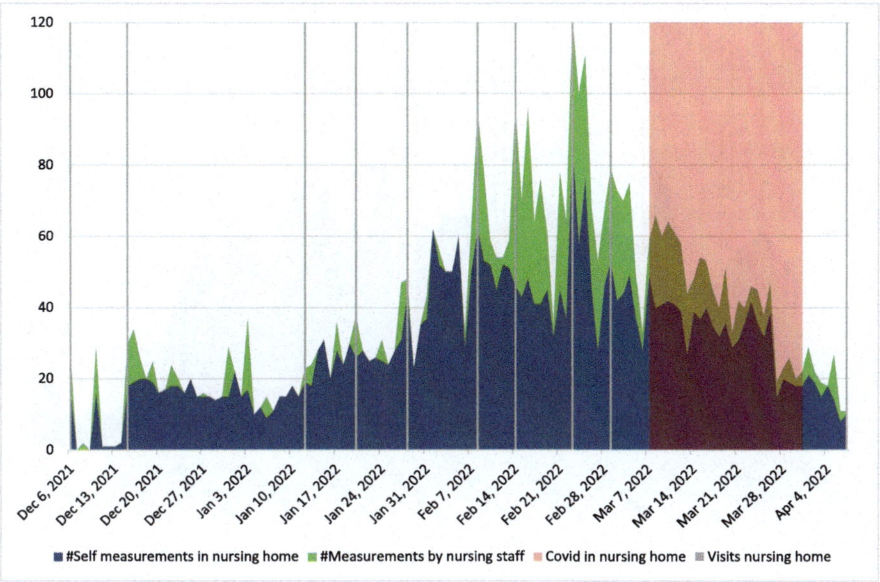

Fig. 5 Number of measurements in the nursing home [36]

new technology from the beginning and did not even want to try to operate it. In the end, 8 of the 19 residents of the nursing home measured their vital signs themselves, the rest had to require help from the nursing staff.

Figure 5 shows an overview of the measurements done in the nursing home. The blue area represents the number of incoming measurements of every day from the patients who measure their vital signs themselves. The green add-on area above shows measurements by the nursing staff. The grey vertical lines represent visits of the project team at the nursing facility. At the end of the project phase, there was an outbreak of COVID-19 which prohibited further visits marked in red [36].

Patient's Home

Due to the comparatively younger age of the patients at home, the acceptance was completely different. All project participants were used to smartphones, so that basic operations were not a problem. Since the participants were spread over a large rural area, it was no longer possible to visit them personally in the event of technical problems. Instead, a support helpline was set up so that project participants could contact technical project staff at any time.

Figure 6 shows an overview of the measurements at patient's home. The blue area is the number of measurements per day. The yellow lines mark visits at the patients home while the brown lines marks calls at the support helpline [36].

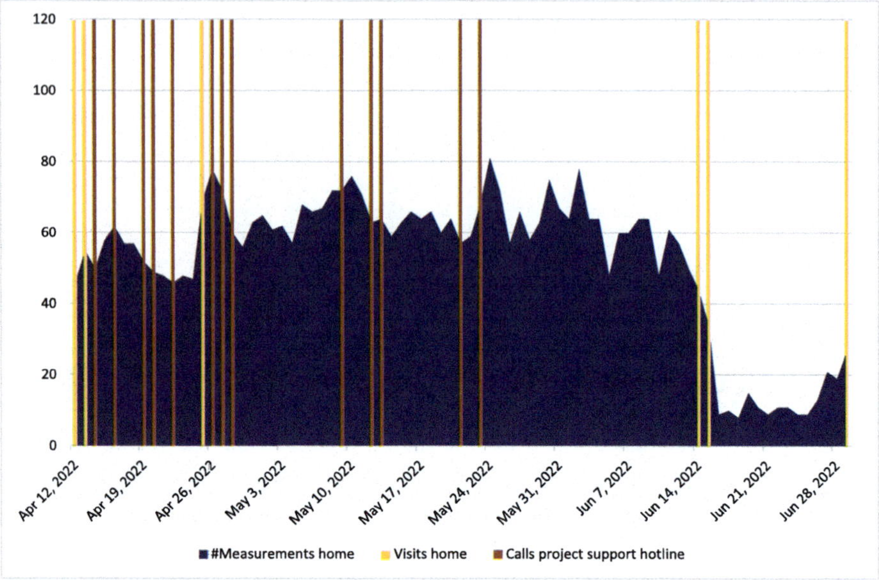

Fig. 6 Number of measurements at patient's home [36]

7.3 Acceptance by Physicians

The GPs involved saw great potential in data-based medicine, as it can provide better insight into the patient's state of health. Based on the vital data, several medications were also adjusted during the course of the project. The range of functions and the visual design of the web interface were optimized several times in dialog with the physicians.

However, in order to be used practically in future, it is not sufficient to have the vital data on a separate web interface. It must be integrated into the patient's electronic health record within the practice management system, as it is already done for example in Denmark [37], and practice management systems so that this can be easily accessed.

8 Conclusion

Data driven medicine is a key factor for high quality medical service that will remain affordable for societies struggling with more and more elderly people. The replacement of patient mobility by data mobility, e.g. in outpatient care using methods of telemonitoring with self-measurement of vital data by patients at home is a promising approach.

A model for a health care process based on remote patient monitoring has been introduces. Numerous vital signs like blood pressure, oxygen saturation, ECG and blood glucose can already be measured independently by the patient at home. These correspond to several common diseases like hypertension, diabetes mellitus or COPD in terms of prevention, treatment and to monitor possible consequences of the disease.

This vital data can easily be transferred to classical cloud environments, data spaces or the patients electronic health record and stored. To handle the massive amount of vital data AI analysis is necessary, to relieve physicians and detect emerging diseases at an early stage.

The legal requirements for this in Germany were shown as an example, with the great particularity of digital health applications. The results of a project study carried out in Germany proved the feasibility of this concept.

References

1. Statistisches Bundesamt: Anteil der Gesundheitsausgaben am Bruttoinlandsprodukt in Deutschland in den Jahren 1997 bis 2021 Statista. https://de.statista.com/statistik/daten/studie/76458/umfrage/deutschland-entwicklung-der-gesundheitsausgaben-seit-1997/ (Retrieved August 09, 2023)
2. Ulrich, R. E. (2005). Demographic change in Germany and implications for the health system. *Journal of Public Health, 13*, 10–15. https://doi.org/10.1007/s10389-004-0084-8
3. Denton, F. T., & Spencer, B. G. (2010). Chronic health conditions: Changing prevalence in an aging population and some implications for the delivery of health care services. *ICanadian Journal on Aging/La Revue canadienne du vieillissement, 29*(1), 11–21. https://doi.org/10.1017/S0714980809990390
4. El-Rashidy, N., El-Sappagh, S., Islam, S. R., El-Bakry, M., & H., & Abdelrazek, S. (2021). Mobile health in remote patient monitoring for chronic diseases: principles, trends, and challenges. *Diagnostics, 11*, 607. https://doi.org/10.3390/diagnostics11040607
5. Keil, A., Hahn, K., Farhan, N., Hillerkuss, D., Brück, R., & Gaus, O. (2022). Self-measurement of vital signs to prevent and treat common disease patterns. In *E-health and bioengineering conference (EHB), Iasi, Romania*. https://doi.org/10.1109/EHB55594.2022.9991709
6. Keil, A., Gaus, O., Brück, R., & Hahn, K. (2023). Concept of a new medical data-driven health care model based on remote patient monitoring. In *45th annual international conference of the IEEE engineering in medicine and biology society (EMBC), Sydney, Australia*, forthcoming
7. Qiao, D., Zulkernine, F., Masroor, R., Rasool, R., & Jaffar, N. (2021). Measuring heart rate and heart rate variability with smartphone camera. In *2021 22nd IEEE international conference on mobile data management (MDM)* (pp. 248–249). https://doi.org/10.1109/MDM52706.2021.00049
8. Bae, S., Borac, S., Emre, Y., et al. (2022). Prospective validation of smartphone-based heart rate and respiratory rate measurement algorithms. *Communications Medicine, 2*, 40. https://doi.org/10.1038/s43856-022-00102-x
9. Shang, T., Zhang, J. Y., Thomas, A., et al. (2022). Products for monitoring glucose levels in the human body with noninvasive optical, noninvasive fluid sampling, or minimally invasive technologies. *Journal of Diabetes Science and Technology, 16*(1), 168–214. https://doi.org/10.1177/19322968211007212
10. Apple Watch Series 8, apple.com. https://www.apple.com/de/apple-watch-series-8/ (Retrieved August 09, 2023)

11. Galaxy Watch5 Pro 45mm LTE. Samsung.com. https://www.samsung.com/de/watches/galaxy-watch/galaxy-watch5-pro-black-titanium-lte-sm-r925fzkadbt/ (Retrieved August 09, 2023)
12. Falter, M., Scherrenberg, M., Driesen, K., et al. (2022). Smartwatch-based blood pressure measurement demonstrates insufficient accuracy. *Front Cardiovascular Medicine, 9*. https://doi.org/10.3389/fcvm.2022.958212
13. Pätz, C., Michaelis, A., Markel, F., et al. (2022). Accuracy of the apple watch oxygen saturation measurement in adults and children with congenital heart disease. *Pediatr Cardiol*. https://doi.org/10.1007/s00246-022-02987-w
14. World Health Organization: The top 10 causes of death. (2020) https://www.who.int/news-room/fact-sheets/detail/the-top-10-causes-of-death (Retrieved August 09, 2023)
15. Kaduszkiewicz, H., Gerste, B., Eisele, M., Schäfer, I., & Scherer, M. (2014). Herzinsuffizienz: Epidemiologie und Versorgung. In J. Klauber & N. Schmacke (Eds.), *Versorgungs-Report 2013/2014* (pp. 209–229). Schattauer.
16. Hennersdorf, M., Perings, C., Kelm, M., et al. (2001). *Vorhofflimmern. Internist, 42*, 1631–1640. https://doi.org/10.1007/s001080170015
17. Neuhauser, H., Kuhnert, R., & Born, S. (2017). 12-Monats-Prävalenz von Bluthochdruck in Deutschland. *Journal of Health Monitoring, 2*(1), 57–63. https://doi.org/10.17886/RKI-GBE-2017-007
18. Heidemann, C., & Scheidt-Nave, C. (2017). Prävalenz, Inzidenz und Mortalität von Diabetes mellitus bei Erwachsenen in Deutschland - Bestandsaufnahme zur Diabetes-Surveillance. *Journal of Health Monitoring, 2*(3), 105–129. https://doi.org/10.17886/RKI-GBE-2017-050
19. Steppuhn, H., Kuhnert, R., & Scheidt-Nave, C. (2017). 12-Monats-Prävalenz der bekannten chronisch obstruktiven Lungenerkrankung (COPD) in Deutschland. *Journal of Health Monitoring, 2*(3), 46–54. https://doi.org/10.17886/RKI-GBE-2017-053
20. Herold, G. (2022). *Innere Medizin*. Köln.
21. Fischer, R., & Jarke, M. (2016). Medical Data Space – Digitale Souveränität über medizinische Daten. Fraunhofer-Gesellschaft zur Förderung der angewandten Forschung e.V.
22. Nagel, L., & Lycklama, D. (2021). Design principles for data spaces (1.0). *Zenodo*. https://doi.org/10.5281/zenodo.5244997
23. Otto, B., et al. *GAIA-X and IDS*. International Data Spaces. https://doi.org/10.5281/zenodo.5675897
24. Amazon Web Services: Compliance programs. https://aws.amazon.com/compliance/programs/ (Retrieved August 10, 2023)
25. Spieß, A. (2022). Bewertung und Klassifikation von Vitaldaten durch Methoden der künstlichen Intelligenz. Master thesis at University of Siegen, Department of Electrical Engineering and Computer Science. Supervised by Brück, R. and Keil, A.
26. Chen, T., & Guestrin, C. (2016). XGBoot: A scalable tree boosting system. *Association for Computing Machinery. doi, 10*(1145/2939672), 2939785.
27. Pedregosa, F., Varoquaux, G., Gramfort, A., Michel, V., Thirion, B., Grisel, O., Blondel, M., Prettenhofer, P., Weiss, R., Dubourg, V., Vanderplas, J., Passos, A., Cournapeau, D., Brucher, M., Perrot, M., & Duchesnay, E. (2011). Scikit-learn: Machine learning in python. *Journal of Machine Learning Research, 12*, 2825–2830.
28. Verordnung (EU) 2016/679 des Europäische Parlaments und des Rates vom 27. April 2016 zum Schutz natürlicher Personen bei der Verarbeitung personenbezogener Daten, zum freien Datenverkehr und zur Aufhebung der Richtlinie 95/46/EG (Datenschutz-Grundverordnung).
29. Bundesdatenschutzgesetz vom 30. Juni 2017 (BGBl. I S. 2097), das zuletzt durch Artikel 10 des Gesetzes vom 23. Juni 2021 (BGBl. I S. 1858; 2022 I 1045) geändert worden ist.
30. Das Fünfte Buch Sozialgesetzbuch - Gesetzliche Krankenversicherung - (Artikel 1 des Gesetzes vom 20. Dezember 1988, BGBl. I S. 2477, 2482), das zuletzt durch Artikel 2a des Gesetzes vom 19. Juli 2023 (BGBl. 2023 I Nr. 197) geändert worden ist.
31. gematik Gesellschaft für Telematikanwendungen der Gesundheitskarte mbH: Systemspezifisches Konzept ePA. Version 1.3.0, date 10-02-2019. https://www.ina.gematik.de/standard/formhandler/324/gemSysL_ePA_V1_3_0.pdf (Retrieved August 10, 2023)

32. Digitale Gesundheitsanwendungen-Verordnung vom 8. April 2020 (BGBl. I S. 768), die zuletzt durch Artikel 3 des Gesetzes vom 20. Dezember 2022 (BGBl. I S. 2793) geändert worden ist.
33. Weber, S., & Heitmann, K. U. (2021). Interoperabilität im Gesundheitswesen: auch für digitale Gesundheitsanwendungen (DiGA) verordnet. *Bundesgesundheitsbl, 64*, 1262–1268. https://doi.org/10.1007/s00103-021-03414-w
34. Keil, A., Hahn, K., Brück, R., Brombach, N., Farhan, N., & Gaus, O. (2022). Cloud-based system for vital data recording at Patients' home. In *Advances in intelligent systems and computing*. Springer. https://doi.org/10.1007/978-3-031-09135-3_2
35. Keil, A., Hahn, K., Brombach, N., Farhan, N., Brück, R., & Gaus, O. (2022). Acquisition and processing of biomedical data for outpatient care in rural areas. *Current Directions in Biomedical Engineering, 8*(2), 81–84. https://doi.org/10.1515/cdbme-2022-1022
36. Keil, A., Brombach, N., Gaus, O., Brück, R., & Hahn, K. (2023). Opportunities of data medicine: Telemonitoring of multimodal medical data in outpatient care. In *XXIII Polish conference on biocybernetics and biomedical engineering (PCBBE)*, forthcoming
37. Fischer, P. (2019). Digital Health: Untersuchung zur Akzeptanz der elektronischen Gesundheitsdatenspeicherung in Form der elektronischen Patientenakte (ePA) in Deutschland. ifgs Schriftenreihe der FOM 17, MA Akademie Verlags- und Druck-Gesellschaft mbH, Essen (2019). ISBN: 978-3-89275-116-8

Performance and Optimisation of Microfluidic Channels with Acute Angles and Serpentine Curvatures

Mozafar Saadat, Arran Hughes, Alexander Jungclaussen, Amir Hajiyavand, Gowsihan Poologasundarampillai, Amirpasha Moetazedian, and Majid Malboubi

Abstract This study presents the measurement of fluid fields of microfluidic channels with acute angle bends between 15° and 90° and examines the influence of both angle and geometric design on these bends through simulation and experimentation. Additionally, serpentine microchannels of different radii have been explored and their corresponding fluid fields. It was found that smaller angles cause significant destabilisation of the fluid field with rapid deceleration around bends with sharp internal and external points. Constant pressure regions were found at the bends which developed into rapid pressure changes in each of the inlet and outlet channels to the corners. 3D printed test structures revealed that a core issue of small angle bends was air bubble development during filling. Effective methods to eliminate this bubble were the rounding of inner and outer points, chamfering the outer point to reduce surface area and performing a much slower filling procedure to allow air to be pushed out prior to liquid filling. Rounding of the inner and outer corners led to the highest amount of flow continuity around 45° bends while the 15° corner still had significant flow instability. The serpentine channels were found to have no flow instability due to a constant pressure gradient through all the channels. A method of instigating higher fluid instability within serpentine microchannels was found by increasing the diameter at the bend apexes to distribute the fluid. Sharp angle bends and the redesigned serpentine channel have potential for highly compact and effective designs of micromixers.

Keywords Microfluidics · Microfabrication · Rapid prototyping · Fluid dynamics · Acute angle bends · Micromixers

---

M. Saadat · A. Hughes (✉) · A. Jungclaussen · A. Hajiyavand
School of Engineering, Department of Mechanical Engineering, University of Birmingham, Birmingham, UK
e-mail: ajh345@bham.ac.uk

G. Poologasundarampillai · A. Moetazedian
School of Dentistry, University of Birmingham, Birmingham, UK

M. Malboubi
School of Engineering, Department of Mechanical Engineering, University of Birmingham, Dubai, United Arab Emirates

M.-R. Alam and M. Fathi (eds.), *Integrated Systems: Data Driven Engineering*,
https://doi.org/10.1007/978-3-031-53652-6_5

1 Introduction

Microfluidics is a continually evolving field with far reaching applications into biomedical sensors, [14, 18, 40] organs on a chip [15, 21, 25], or micro-mixing of fluids for chemical reactors [19, 24, 27]. There has, however, been a relatively small amount of exploration into the effects of different geometric characteristics of channels especially with regards to the impact of varying curvature along with acute angle bends. As a result of this there has been little exploration of the impact these geometries have on microfluidic design and control for these applications. A key advantage of microfluidic devices is that the flow is laminar. Laminar flow produces steady and predictable streamlines allowing for accurate and precise manipulation of fluid flows and therefore forming predictable and repeatable flow patterns [3, 12].

Rapid prototyping is an attractive method of manufacture for microfluidics due to constant design and redesign of channel architecture requiring testing and retesting for each new design. Soft lithography has been utilised for manufacturing of channels for high pressure flows using epoxy resin to develop a rigid test structure [29]. Methods using laser micromachining have been developed for both directly developing channels from polymethyl methacrylate (PMMA) as well as with the laser ablation of steel moulds to develop a positive mould for polydimethylsiloxane (PDMS) casting for 'maskless' methods [23 , 33].

3D printing is a cheap and efficient method for developing rapid prototypes of many different designs through only generation of new CAD models and allows for complex shapes to be developed within channels with fast printing times [5, 13, 28]. Serpentine microchannels have previously been developed and analysed based on 3D printing manufacturing for DNA amplification and utilising this geometry for inertial focusing of particles [2, 16, 38]. Stereolithography (SLA) 3D printing is a useful method of microfluidic prototype manufacture due to good replication accuracy of approximately 10% and having the ability to produce square channels down to a resolution of 150 μm [37, 10].

One of the few reports on acute angles within microfluidics was through utilising the angles for zig-zag micromixers and their impact on the effectiveness of channels to mix the fluid at higher Reynold's numbers. It was found that as a general trend, increasing the acuteness of the angles improved the mixing ability of these zig-zag designs [36]. Previous exploration has been undertaken for examining the fluid velocity at average fluid speeds between 40 and 200 μm/s in acute angled channels. The angles were tested down to 60° with cross section areas of 90×10 μm where a deceleration effect was found for the channels with smaller angles [39]. Other studies have examined the fluid flow characteristics for sharp inner and outer corner channels of 90°, highlighting some vortex formation at different fluid speeds and examining velocity or temperature fluctuations around the bend [6, 32].

Serpentine channel geometries have many uses as micromixers and their effectiveness has been analysed previously but only a few geometric optimisations to the design have been developed. For example, Rhoades et al. examined different internal radius curved channels with the inclusion of slanted groove-ridge patterns and found this patterning as a useful tool to achieve better mixing [30]. López et al. also examined the impact of sharpening of the edges within microfluidic mixers and having semi-circular shaped channels were it was found this geometry may introduce Dean vortices for more efficient mixing [22]. Other studies have been performed to evaluate the mixing efficiency of different shaped serpentine mixers (square and round), but there are few studies on the optimisation of the fluid field for these types of channels [1, 17, 34]. Additionally, inertial focusing has been examined when combined with serpentine channels with periodic obstacles and highlights another application of these channel types for separation of cancer cells from beads [8].

This paper aims to contribute to a research gap within the literature surrounding the understanding and characterising of microfluidic flows particularly in serpentine and acute angle channels of which to our knowledge is one of the few studies available in this area. Finding the limitations and effects of simple corners and bends is vital for channel design, not only to avoid unwanted parameters such as bubble formation, but also to understand how to advantageously use them for microfluidic applications. This could be for example with space-saving for larger microfluidic networks by condensing the fluid channels and allowing for longer channel lengths within the same region. The aim of this paper is to characterise fluid flow behaviours through simple changes in channel geometry and offer a simulation model for understanding the behaviours of the fluid fields.

2 Materials & Methods

The microfluidic channels were designed in Solidworks (Dassaut Systèmes) as set of 3D CAD models with general channel dimensions of 300 μm width and 200 μm height. To ensure that all the tests were consistent, the channel length was set to 35 mm to allow for a better comparison between each channel geometry. This length is sufficient enough to allow for any entry disturbances experienced by the fluid flow in the channel to be dissipated prior to the region of interest. Additionally, this allows for sufficient amounts of the different types of geometries to be examined such as two bends to be fit with larger radius of curvature [35]. Inlet and outlet diameters were set to 1.5 mm to match the requirements of the tubing fixtures.

2.1 *Fluid Simulation*

The channel designs were simulated within COMSOL Multiphysics for which the flow was assumed to be laminar. To determine this the Reynolds number was calculated from:

$$Re = \frac{\rho u D_h}{\mu} \tag{1}$$

where ρ is the fluid density, u is the average fluid velocity, μ is the fluid dynamic viscosity and D_h is the hydraulic diameter which for square channels can be calculated by:

$$D_h = \frac{4A}{P} \tag{2}$$

where A is the channel area (height*width) and P is the wetted perimeter of the channel (2*height + 2*width). For the maximum flow rate within the device, the Reynolds number is expected to be 0.224 at a maximum, therefore satisfying the condition of Re << 1 for laminar Stokes flow to occur.

The governing equations within the simulation are firstly Navier–Stokes equation to evaluate the fluid field:

$$\rho\frac{\partial \vec{u}}{\partial t} + \rho(\vec{u} \cdot \nabla)\vec{u} = -\nabla P + \mu \cdot \nabla^2\vec{u} + \vec{F} \tag{3}$$

Due to the expected laminar flow (dictated by the low Reynolds number), the transient and inertial terms can be considered negligible, thus meaning the governing equation for the fluid field is the reduced version of the Navier–Stokes equation [20]:

$$\nabla P = \mu \cdot \nabla^2\vec{u} + \vec{F} \tag{4}$$

where P defines the pressure gradient across the channel and $\vec{F}$ is the sum of any external forces acting on the system. For this set of simulations, the external forces are considered to be zero due to no external driving forces acting on the fluid apart from pressure, and gravity is seen as negligible due to the chip being placed on a flat surface.

The other governing equations are those defined along the boundaries at the walls with the no-penetration and no-slip condition as defined by:

$$\vec{u} \cdot \hat{n} = u_n = 0 \tag{5}$$

$$\vec{u} \cdot \hat{t} = u_t = 0 \tag{6}$$

where $\hat{n}$ and $\hat{t}$ are the unity normal and tangential vectors to the wall respectively. Finally, due to the laminar state of the fluid flow, Stokes drag was utilised for the force calculated acting on the objects while travelling through the fluid as defined by [11]:

$$F_d = 6\pi\mu R u \tag{7}$$

where R is the radius of the object moving through the fluid. The simulation was then ran for each of the channels from fluid velocities of 50 μm/s up to 1000 μm/

s in steps of 10 μm/s to determine fluid speeds of interest from low velocities up to very high velocities and find any changes that occur at different magnitudes of fluid velocity.

2.2 Chip Design and Manufacture

Figure 1 highlights the different chips that were set out to be simulated and manufactured in order to analyse the effects of different kinds of geometries on the fluid flow, along with their effect on any transport objects within them [26].

The aim of the model in Fig. 1a is to explore the effect of sharp angle bends on the flow, flow rates and their effects on the flow with angles of 15°, 30°, 45°, 60° and 90° respectively with the amount of angles explored limited due to the chip size that was fabricated. Figure 1b shows different serpentine channels with curve radii of 1 mm, 2 mm, 4 mm and 8 mm.

CAD models were exported into the Preform (Formlabs) software to generate suitable supports during printing and to specify that the models were to be printed both flat for the angled channels and at a 45° angle for the curved channels. Positive moulds were 3D printed using a Form 3B + Stereolithography (SLA) printer (Formlabs, PKG-F2-REFURB) and was printed using a biocompatible resin (Formlabs, BioMed Clear Resin, RS-F2-BMCL-01). Once printed, it was then washed using a Form Wash L machine (Formlabs, PKG-FHL-WA) using isopropyl alcohol (IPA) to remove any leftover residue. Finally the chips were cured using a Form Cure machine (Formlabs, CH-CU-01) for 1 h at 60°C to harden them, improve the surface finish and improve the material properties [31].

To develop the chips, PDMS was casted onto the printed moulds. A box was 3D printed to hold the chip and contain the PDMS during the curing process while it was

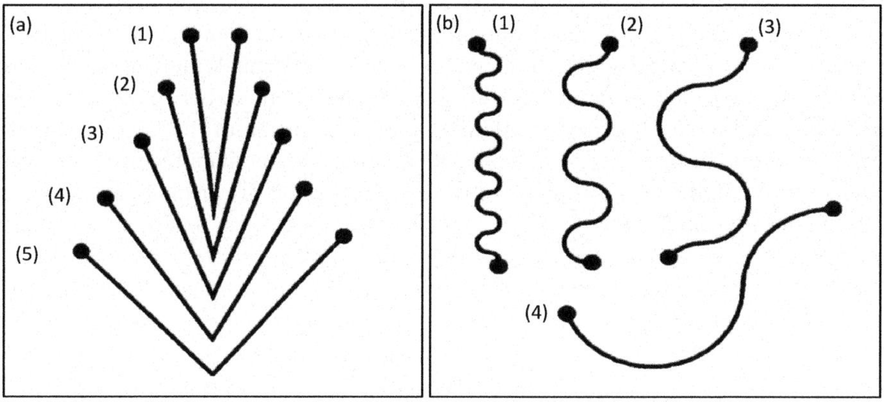

Fig. 1 CAD models of the test structures developed. **a** Highlights a single bend of varying degrees. **b** Highlights a series of bends of different amounts of curvature

still liquid. Sylgard 184 silicone elastomer kit (Dow Corning) was mixed at a ratio of 10:1 and 33 g of liquid was used total (combined elastomer and curing agent) and mixed thoroughly for 15 min prior to being poured on top of the positive mould. The mixture was then degassed in a vacuum oven at −0.8 bar for 20 min, after which it was re-pressurised and then re-vacuumed 2–3 times until almost all air bubbles had disappeared. Any remaining bubbles were punctured using a sharp needle or pushed to the edges and popped. The mixture was then baked at 100°C for 45 min to cure it. Once cured, it was allowed to cool and then cut and peeled out of the box with an EMS-core sampling tool used to puncture holes for the inlets and the outlets. The open side was then sealed using a thick piece of double sided, optically transparent adhesive tape and attached to a conventional microscope slide for structural support.

10 μm microbeads were used to evaluate the flow within the different channel geometries which were microparticles based on polystyrene in aqueous suspension (Merck, MFCD00243243) [7]. 1 ml of beads were diluted with de-ionised water at a ratio of 30:1 to allow for individual bead tracking while still offering a large enough sample to observe fluidic behaviours. Beads were only used for 1-h periods before being returned to the fridge for storage to prevent any melting/evaporation defects occurring within them [9]. Fluid was controlled using an AL-300 syringe pump (World Precision Instruments) with a 10 ml syringe and images were recorded through an inverted microscope (Bestscope, bs2090 inverted microscope) in conjunction with a microscopy camera (Basler) at 100 frames per second. Channels were initially filled at a flow rate of 10 μl/min and then the bead solution was ran after at flow rates of 1.5 μl/min and then 3.0 μl/min.

3 Results and Discussion

The channels were firstly analysed to determine the ability of the 3D printer to replicate the designed geometries as shown in Fig. 1.

Figure 2 highlights fairly good replication of the required channel dimensions with overshoot of the channel widths being under 10% for all channels along with the angle of the bend being closely replicated with only a slight difference between the intended and the actual angle. Figure 2a highlights good replication of the general channel geometries with relatively few defects with this design and manufacturing process. Figure 2b highlights a specific defect within sharp angle 3D printed geometries at the intersection point between the channels. The printer was unable to develop a sufficiently high channel wall as indicated by the blurred section in the image due to the printer not being able to reach this resolution [4]. This, however, only causes the channel floor to be higher, having little impact on the flow and could still be fully sealed. This meant that channels with very low bend angles (<15°) were unable to be manufactured with this method, however, the channels with higher bend angles were all able to be replicated well.

Next, the second set of channels were analysed when printed at 45° for the channels with curvature in them as shown in Fig. 3.

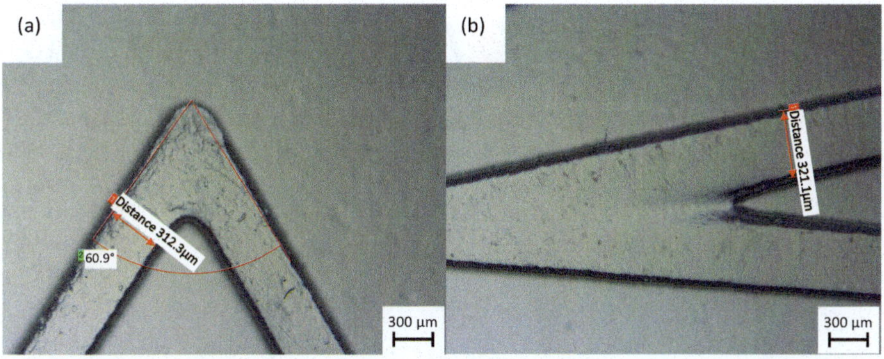

Fig. 2 Microscope images of the positive moulds post-cure for the **a** 60° and **b** 15° bends with distance and angle measurements

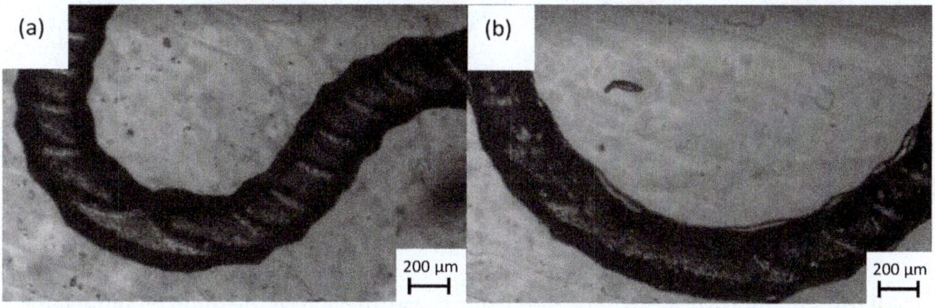

Fig. 3 Microscope images of the PDMS channels for the curved fluid channels. **a** shows the channel with 1 mm radius and **b** shows the channel with 4 mm radius

Figure 3 shows a defect that occurs we found when printing at an angle where a scaling effect occurs along the channel floor along with wavy walls. This was seen in all of the curved channels when the channel geometries were printed (and built up) at a 45° angle to the printing bed (rather than parallel to it). This however, did allow for easier removal of the chip from the printer bed to prevent any damage to the surface during extraction. The bumps on the edges of the channels were relatively small and due to the laminar flow again, was deemed to not have much of an effect on the overall bulk flow of the fluid through the system within the centre. A similar artifact can also be seen on the channel floors giving a ribbing effect where each printed layer is at the angle and was less pronounced at the bends in the channels. The distance between each point corresponds to the resolution of the printer which was approximately 100 μm.

To understand the fluid flow in each of these geometries, a COMSOL Multiphysics model was developed, the first of which is shown in Fig. 4 which shows the simulative results from chip 1 at an inlet speed of 500 μm/s.

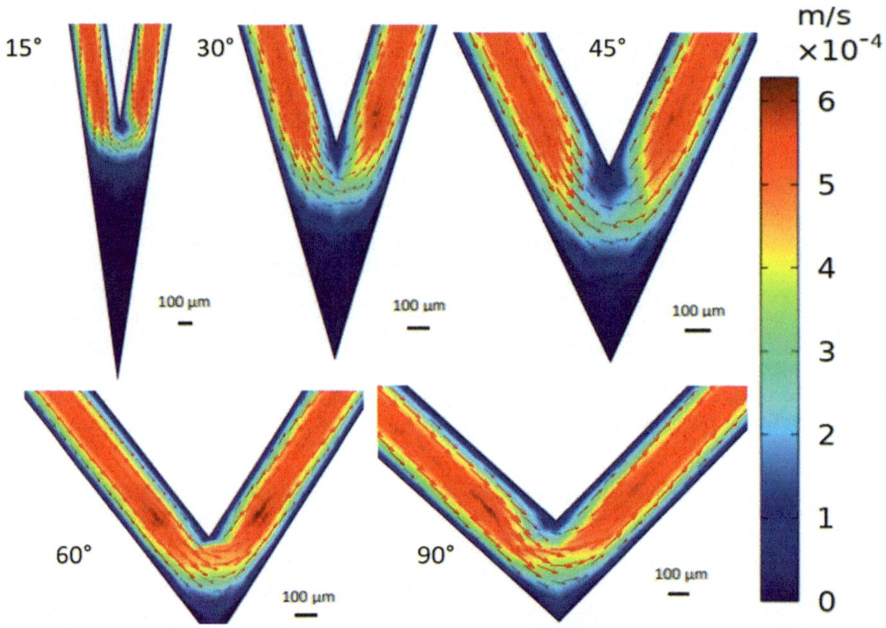

Fig. 4 Fluid field simulation results for each of the single angle bend channels. Angles are labelled to the left of or above each of the channels

All five channels show an expected velocity profile across the main channel lengths prior to the bend as Hagen–Poiseuille is shown due to the no-slip boundary condition. A clear change in velocities can be seen once the fluid reaches the bend. The fluid is seen to decelerate once the corner is reached and then accelerate upon exiting the corner. This effect is more prominent in the sharper angles. In the most acute channel of 15°, the peak fluid velocity in the middle of the bend is ~350 µm/s, whereas in the most obtuse channel of 90°, the peak fluid velocity is around 450 µm/s. These observations are supported by the pressure contours shown in Fig. 5.

Throughout the channels' length, the pressure decreases until it reaches 0 Pa at the outlet. The parallel pressure lines before and after the bends show that the pressure change is uniform (hence the Hagen-Poiseuille flow). However, at the bends, the pressure differential is much smaller at the apexes of the bends which is particularly clear in the 30° and 45° channels. This area of constant pressure leads to a push from the left side and a pull on the right side which distorts the flow field, effectively isolating each side of the channel. This is especially prominent at the lower angles where it can be seen in Fig. 4 that the fluid slows down dramatically as it reaches the corner until it is eventually pulled in again as it passes. This can be compared to the higher angle channels where there is a much lower decrease in fluid flow across the corner and the fluid field transitions much smoother around the bend.

A similar simulation was performed on the curved channels with 500 µm/s inlet speed as shown in Fig. 6.

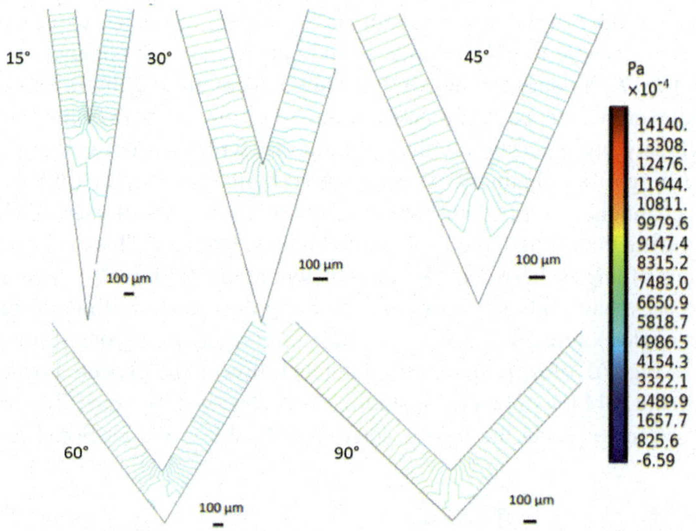

Fig. 5 Pressure contour simulation results for each of the single angle bend channels. Angles are labelled to the left of or above each of the channels

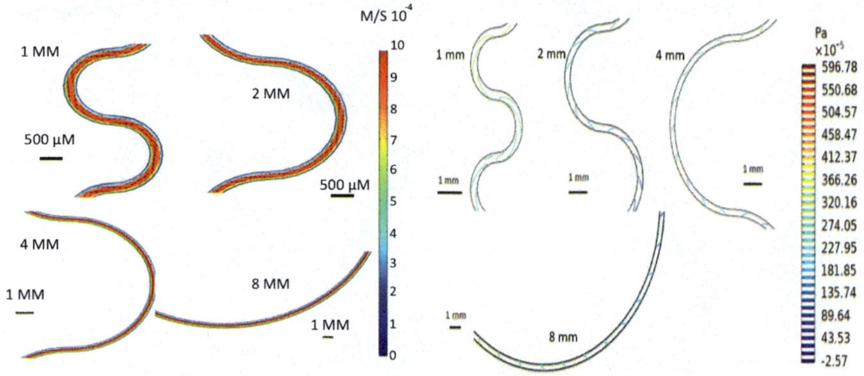

Fig. 6 **a** Fluid field simulation results for each of the curved channels. **b** Pressure contour simulation results for each of the curved channels. Curve radii are labelled to the left of or above each of the channels

Figure 6 shows that even with the varying curve radii, the effect on the characteristics of the flow is largely unchanged. The expected velocity profile was seen with the highest velocity being in the centre channel, reducing to 0 again caused by no slip condition. Differences in maximum velocities throughout the corners are negligible, regardless of curve radii which is shared by the pressure contours being relatively linear in their reduction as shown in Fig. 6b and the only difference between each of the channels being the angle of the contours which is dictated by the radius curvature.

To validate the flow in the angled channels, they were evaluated experimentally as shown in Fig. 7 where the filling process and operation of the channels is shown.

Within Fig. 7, the darker areas in the channel were air and the light areas within the channel were water. The beads were found to follow the path of the fluid that had been previously predicted by the fluid simulation, with the inlet velocity set to 500 μm/s for filling of all channels. As highlighted by the blue circles in Fig. 7a, when filling the channels, a trapped air bubble forms in the corners of varying sizes for all channels during the filling which is to be considered a defect. This occurred for all of the channels of angles between 15° and 60° where the bubble size decreases as the angle increases, with only the geometry changing during each filling procedure. For the largest angle examined of 90°, no bubble was formed during the filling procedure as shown in Fig. 7d. Flow in the main channels followed the expected path but at the corner with the bubbles some flow instability was observed. Beads still moved around the corner effectively, but the bubble acted as a flexible wall, accelerating the beads

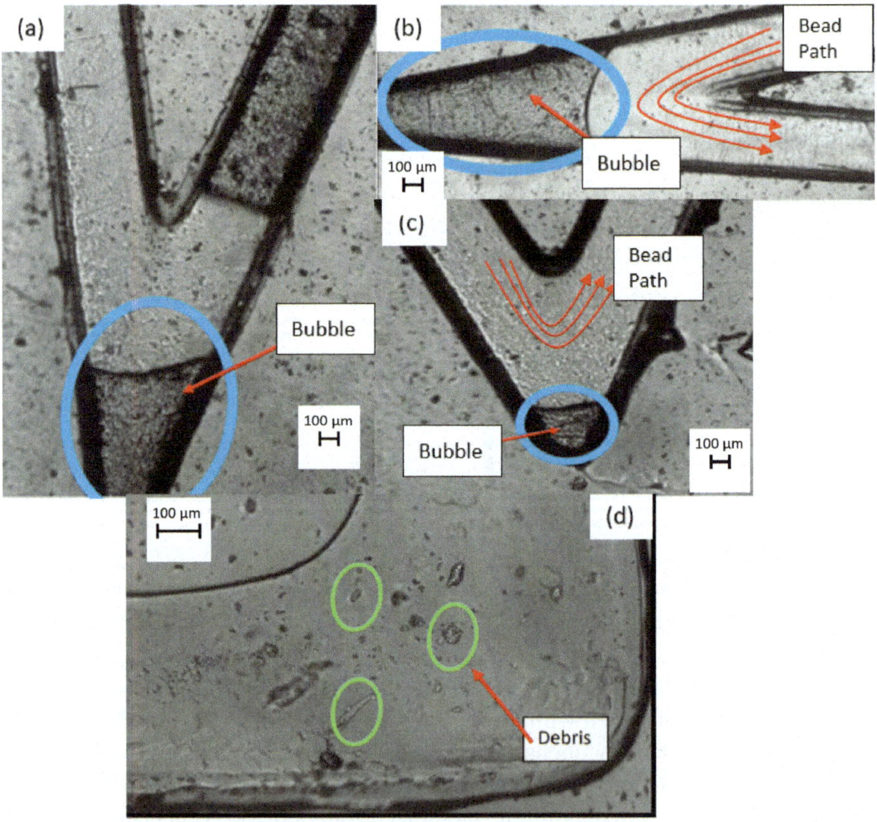

Fig. 7 Fluid movement of the microbeads in the sharp angle channels with fluid inlet speed of 500 μm/s. **a** Filling of the 30° channel. **b** Operation of the 15° channel. **c** Operation of the 45° channel. **d** Operation of the 90° channel

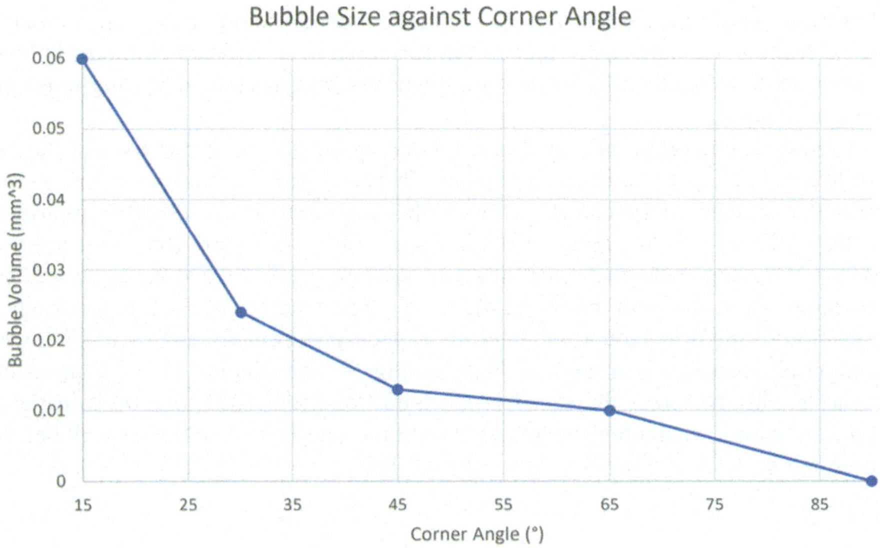

Fig. 8 Bubble size for each of the angles explored for the single corner microchannels

around the corner as if the channel was thinner and causing some flow instability which does have a potential for a better mixing effect. The effects of bubble size based on the angle is shown in Fig. 8 with the bubble volume being estimated as the area of the triangle along with the depth of the channel.

Figure 8 shows a clear relationship between the volume of the bubble in each of the channels and the angle of the channel with the bubble completely disappearing at 90°. The bubble size seems to fall relatively exponentially from the sharper angle due to two factors. One is the angle itself of the corner but also due to the shape of the channels, the point was much longer in the lower angle channels leading to a larger volume that the air could fill. Therefore, an assumption can be made that there is a relationship between surface area and bubble formation. To further explain this phenomenon, Young's equation (Eq. (8)) could be applied. Young's equation defines the relationship between contact angle, surface tension, surface free energy as well as the interfacial tension between the solid and liquid surfaces.

$$\sigma_{sg} = \sigma_{sl} + \sigma_{lg} \cdot \cos(\theta) \tag{8}$$

where σ is the surface tension and sg, sl and lg defined solid–gas, solid–liquid and liquid–gas surface tension respectively with θ being the contact angle between the liquid–gas interface and the solid. Developing these equations would be a useful endeavour to understand exactly why bubbles are forming in the channels but is out of the scope of this paper.

To combat the effects seen in both of the original designs, a set of redesigns were developed where firstly, the curved paths were altered to have some effect on the fluid flow rather than acting effectively as a straight channel. Additionally, the geometry of the corners with angles 30° and 15° were altered in an effort to eliminate bubble formation and disturbances to the flow field with the simulation for the redesign shown in Fig. 9.

Figure 9a shows how the velocity accelerates in the straighter lengths of channels, while decelerating when in channel bends. This is a geometric response due to the altering diameters of the channel along with some inertia of the fluid being pushed to the outer walls at the bends. This leads to flow field disturbances that did not exist in the original channel where the flow field was uniform. For the sharper bend redesigns a few observations can be made, in general there is now a much smoother transition of the flow around the bend in all redesign scenarios with a much lower amount of deceleration around the bend. In Fig. 9b, the effect of the smooth inside corner greatly increases the speed of the fluid flow from 1000 to 1300 μm/s as it turns the corner whereafter the higher speed fluid moves against the far wall before resettling in the centre further along the channel.

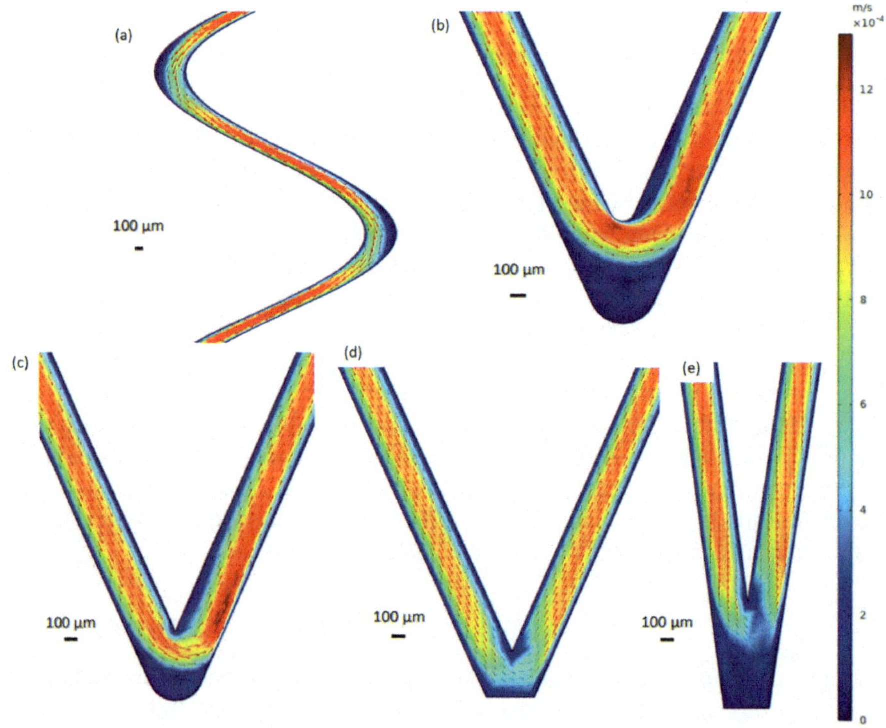

Fig. 9 Fluid fields for the redesigns of the geometries **a** Redesign of the curved channel with a serpentine shape. **b** Redesign of the 45° corner channel with inner and outer corner rounding. **c** Redesign of the 45° corner channel with outer corner rounding. **d** Redesign of the 45° corner channel with the sharp point chamfered. **e** Redesign of the 15° corner channel with the sharp point chamfered

Comparatively, the sharp inner edge in Fig. 9c and 9d cause a significant decrease in fluid velocity around the bend and highlights similar characteristics seen in the flow field of the first design with significant deceleration around the bend. The outer edge also has an effect on the flow characteristics as seen by the maximum velocity in 9c being approximately 1000 μm/s when the fluid moves around the bend. Comparatively, the maximum velocity in 9d is much lower at approximately 500 μm/s at the bend due to the chamfer design. This effect is further extrapolated in the sharper channel in Fig. 9e where the maximum velocity at the corner is approximately 400 μm/s. The pressure contours can again be utilised to explain the effects seen in the fluid field as shown in Fig. 10.

The general trend is relatively similar between the original and the redesigned channels for the pressure contours, however there are some noticeable differences. Figure 10a shows the curved pressure contours now not being as linear as seen previously by altering the diameter of the channel. It can be seen that, at the curve apex, a similar gradient to the other single corner designs is shown without having rapid pressure changes between each side of the apex. Additionally, there is now an area of constant pressure as with the other designs at the outer edge which is reflected in the fluid field being very low speeds.

A comparison shows that in Fig. 10b, the pressure contours for the inner and outer rounded walls are much denser around the bend, the density decreases slightly in Fig. 10c with just the outer wall being rounded and then decreases significantly with only a chamfered outer wall. This is reflected in the flow field again where Fig. 10b has a smoother transition and accelerates the fluid while turning the sharp corner. However, channel (d) decelerates the flow and destabilises the flow streamlines where this effect is likely due to the push and pull between each sides of the bend as with the previous designs. Additionally, Fig. 10d shows much more jagged pressure contours compared to the rounded designs which leads to the magnitude of flow instability when compared to the other designs. The designs were then remanufactured through 3D printing as previously as shown and evaluated experimentally again in Fig. 11 with beads in the flow.

Figure 11a highlights the new path of the beads where they were seen to move much further out at the curves distributing themselves across the channel through advection as predicted by the field flow. This setup could now be utilised as a design for a micromixer due to the squeezing and expansion of the channels forcing the fluid to disperse outwards by geometric changes and small amounts of inertia. Figure 11b shows a significantly improved bubble formation when compared to the original design, where the bubble is now much smaller and hardly disturbs the flow as it is very deep in the area of low fluid velocity. The vast majority of beads moved around the corner without touching or getting near the bubble as indicated by the streamlines. Some beads however still touched the bubble and its presence would still cause some flow instability from what would be expected from the simulation.

Figure 11c shows a similar defect from the manufacturing technique at the inner corner as seen in previous designs along with some defects at the wall due to the peeling process. For this test however, the inlet speed was set to much lower at 1.5 μl/min where due to the much slower fluid velocity, along with the chamfered design, a

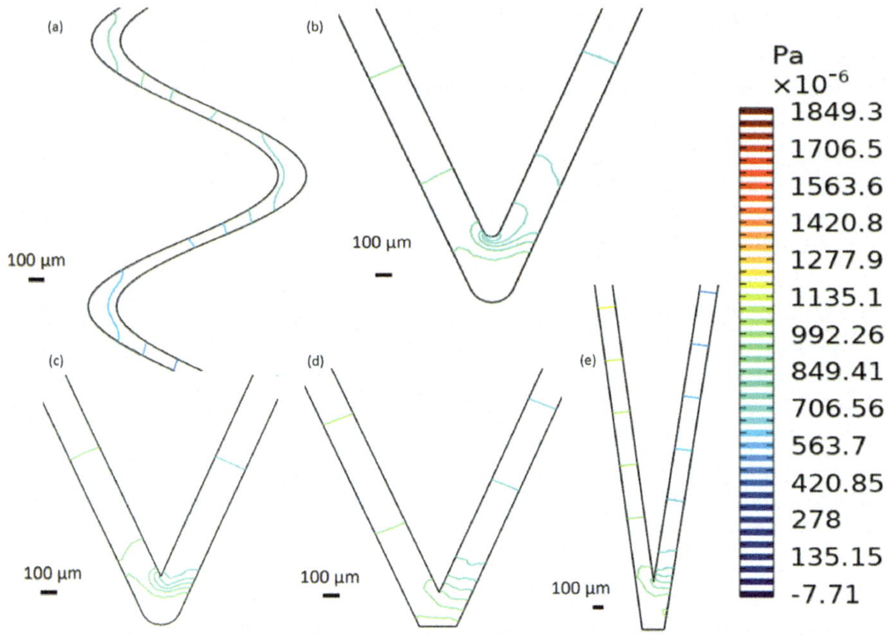

Fig. 10 Pressure contours for the redesigned geometries redesigns of the geometries **a** Redesign of the curved channel with a serpentine shape. **b** Redesign of the 45° corner channel with inner and outer corner rounding. **c** Redesign of the 45° corner channel with outer corner rounding. **d** Redesign of the 45° corner channel with the sharp point chamfered. **e** Redesign of the 15° corner channel with the sharp point chamfered

visible bubble did not form. Therefore, to utilise acute angles in microfluidic designs, a combination of altering the geometries to avoid sharp, high angle points along with a slower filling speed will lead to ideal flow conditions and avoid bubble formation.

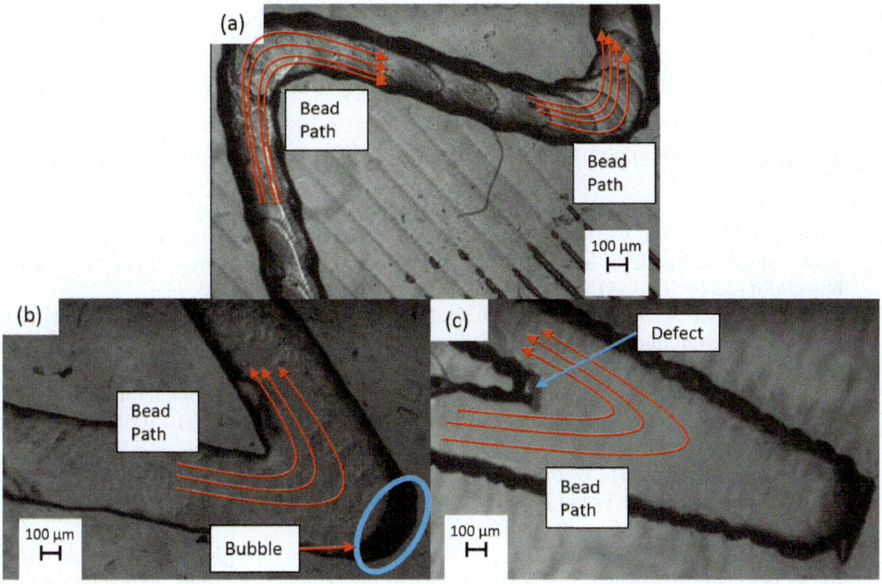

Fig. 11 Filled PDMS channels of the redesigned geometries. **a** Redesign of the curved channel with a serpentine shape. **b** Redesign of the 45° corner channel with outer corner rounding. **c** Redesign of the 15° channel with the sharp point chamfered

4 Conclusion

Within this paper we have shown the development of 3D printed channels for the utilisation of acute angled bends and curved channels. This was achieved through initial testing of design concepts, and then further redesign to optimise the geometric characteristics of the channels. Simulations were performed for each of the designs where pressure contours have been utilised to explain the flow fields within the channels. Sharp angled channels initially were found to have almost no pressure differences at the bends, however, the redesigns had higher pressure contours densities. It was found that having the inner and outer wall rounded caused rapid flow acceleration around the bend along with flow stability. Initially curved channels were seen to have very little effect on the flow, so they were redesigned into a serpentine shape with a wider diameter at curves, and a thinner diameter in the straight areas of the channel. It was found that this causes flow distribution along the lengths of the channel as characterised by advection of the beads in the channel. This indicates that this could be utilised as part of a micromixer design or to distribute particles within the flow. Bubbles were found in the initial acute angle designs where several redesign attempts were shown to eliminate this problem. We have highlighted design methods for acute angles within microfluidics where rounding the corners, or chamfering sharp points along with utilising lower fluid filling speeds were able to eliminate bubble formation within the channels. The geometries shown may have potential applications into micromixer designs along with condensing of microfluidic networks into smaller spaces.

References

1. Amini, Y., Ghazanfari, V., Heydari, M., et al. (2023). Computational fluid dynamics simulation of two-phase flow patterns in a serpentine microfluidic device. *Science and Reports, 13*, 9483. https://doi.org/10.1038/s41598-023-36672-6
2. Arcot, Y., Samuel, G. L., & Kong, L. (2022). Manufacturability and surface characterisation of polymeric microfluidic devices for biomedical applications. *International Journal of Advanced Manufacturing Technology, 121*, 3093–3110. https://doi.org/10.1007/s00170-022-09505-5
3. Ayuso, J. M., Virumbrales-Muñoz, M., Lang, J. M., et al. (2022). A role for microfluidic systems in precision medicine. *Nature Communications, 13*, 3086. https://doi.org/10.1038/s41467-022-30384-7
4. Beauchamp, M., Gong, H., Woolley, A., & Nordin, G. (2018). 3D printed microfluidic features using dose control in X, Y, and Z dimensions. *Micromachines [online], 9*(7), 326. https://doi.org/10.3390/mi9070326
5. Belmonte, I., & White, R. J. (2022). 3-D printed microfluidics for rapid prototyping and testing of electrochemical, aptamer-based sensor devices under flow conditions. *Analytica Chimica Acta, 1192*, 339377. https://doi.org/10.1016/j.aca.2021.339377
6. Huang, C.-Y. & others. (2014). The study of temperature rise in a 90-degree sharp bend microchannel flow under constant wall temperature condition. *Journal of Mechanics, 30*(6), 661–666. https://doi.org/10.1017/jmech.2014.60
7. Campo-Deaño, L. (2017). Fluid-flow characterization in microfluidics. *Complex Fluid-Flows in Microfluidics*, 53–71. https://doi.org/10.1007/978-3-319-59593-1_3
8. Cha, H., Amiri, H. A., Moshafi, S., Karimi, A., Nikkhah, A., Chen, X., Ta, H. T., Nguyen, N. T., & Zhang, J. (2023). Effects of obstacles on inertial focusing and separation in sinusoidal channels: An experimental and numerical study. *Chemical Engineering Science, 276*. https://doi.org/10.1016/j.ces.2023.118826
9. Chen, C.-W., & Chen, C.-Y. (2009). Preparation of monodisperse polystyrene microspheres: Effect of reaction parameters on particle formation, and optical performances of its diffusive agent application. *Colloid and Polymer Science, 287*(12), 1377–1389. https://doi.org/10.1007/s00396-009-2098-9
10. de Costa, B. M. C., Griveau, S., Bedioui, F., d' Orlye, F., & da Silva, J. A. F., & Varenne, A. (2022). Stereolithography based 3D-printed microfluidic device with integrated electrochemical detection. *Electrochimica Acta, 407*. https://doi.org/10.1016/j.electacta.2022.139888
11. Dey, S., Zeeshan Ali, S., & Padhi, E. (2019). Terminal fall velocity: The legacy of stokes from the perspective of fluvial hydraulics. *Proceedings of the Royal Society A: Mathematical, Physical and Engineering Sciences, 475*(2228), 20190277. https://doi.org/10.1098/rspa.2019.0277
12. Duncombe, T. A., Tentori, A. M., & Herr, A. E. (2015). Microfluidics: Reframing biological enquiry. *Nature Reviews Molecular Cell Biology, 16*(9), 554–567. https://doi.org/10.1038/nrm4041
13. Griffin, K., & Pappas, D. (2023). 3D printed microfluidics for bioanalysis: A review of recent advancements and applications. *TrAC Trends in Analytical Chemistry, 158*. https://doi.org/10.1016/j.trac.2022.116892
14. He, X., Fan, C., Luo, Y., et al. (2022). Flexible microfluidic nanoplasmonic sensors for refreshable and portable recognition of sweat biochemical fingerprint. npj Flexible Electronics, *6*, 60. https://doi.org/10.1038/s41528-022-00192-6

15. Hong, S., Kang, E. Y., Byeon, J., Jung, S.-H., & Hwang, C. (2019). Embossed membranes with vascular patterns guide vascularization in a 3D tissue model. *Polymers, 11*, 792. https://doi.org/10.3390/polym11050792
16. Park, J., & Park, H. (2017). Thermal cycling characteristics of a 3D-printed serpentine microchannel for DNA amplification by polymerase chain reaction. *Sensors and Actuators A: Physical, 268*, 183–187. https://doi.org/10.1016/j.sna.2017.10.044
17. Tsai, J.-H., & Lin, L. (2002). Active microfluidic mixer and gas bubble filter driven by thermal bubble micropump. *Sensors and Actuators A: Physical, 97–98*, 665–671. https://doi.org/10.1016/S0924-4247(02)00031-6
18. Karimi, S., Farré-Lladós, J., Mir, E., Escolar, G., & Casals-Terré, J. (2019). Hemostasis-on-a-chip: Impedance spectroscopy meets microfluidics for hemostasis evaluation. *Micromachines (Basel), 10*(8), 534. https://doi.org/10.3390/mi10080534.PMID:31416133;PMCID:PMC6722990
19. Kawanami, H., Matsushima, K., Sato, M., & Ikushima, Y. (2007). Rapid and highly selective copper-free sonogashira coupling in high-pressure, high-temperature water in a microfluidic system. *Angewandte Chemie International Edition, 46*, 5129–5132. https://doi.org/10.1002/anie.200700611
20. Kilic, M. S., Jacobs, G. B., Hesthaven, J. S., & Haller, G. (2006). Reduced Navier–stokes equations near a flow boundary. *Physica D: Nonlinear Phenomena [online], 217*(2), 161–185. https://doi.org/10.1016/j.physd.2006.04.008
21. Lee, P. J., Hung, P. J., & Lee, L. P. (2007). An artificial liver sinusoid with a microfluidic endothelial-like barrier for primary hepatocyte culture. *Biotechnology and Bioengineering, 97*, 1340–1346. https://doi.org/10.1002/bit.21360
22. López, R. R., Ocampo, I., Sánchez, L.-M., Alazzam, A., Bergeron, K.-F., Camacho-León, S., Mounier, C., Stiharu, I., & Nerguizian, V. (2020). Surface response based modeling of liposome characteristics in a periodic disturbance mixer. *Micromachines, 11*, 235. https://doi.org/10.3390/mi11030235
23. Matellan, C., & del Río Hernández, A. E. (2018). Cost-effective rapid prototyping and assembly of poly(methyl methacrylate) microfluidic devices. *Scientific Reports, 8*, 6971. https://doi.org/10.1038/s41598-018-25202-4
24. Moetazedian, A., Candeo, A., Liu, S., Hughes, A., Nasrollahi, V., Saadat, M., Bassi, A., Grover, L. M., Cox, L. R., & Poologasundarampillai, G. (2023). Versatile microfluidics for biofabrication platforms enabled by an agile and inexpensive fabrication pipeline. *Advanced healthcare materials*, 2300636. https://doi.org/10.1002/adhm.202300636
25. Dabaghi, M., Fusch, G., Saraei, N., Rochow, N., Brash, J. L., Fusch, C., & Ravi Selvaganapathy, P. (2018). An artificial placenta type microfluidic blood oxygenator with double-sided gas transfer microchannels and its integration as a neonatal lung assist device. *Biomicrofluidics,12*(4), 044101. https://doi.org/10.1063/1.5034791
26. Neeraj, M. N., Bundele, M., & Suri, P.K. (2020). Pressure and velocity measure in the pipeline for leak detection using COMSOL multiphysics. *International Journal of Recent Technology and Engineering (IJRTE) [online], 9*(1), 1812–1816. https://doi.org/10.35940/ijrte.a2727.059120
27. Plouffe, P., Roberge, D. M., Sieber, J., Bittel, M., & Macchi, A. (2016). Liquid–liquid mass transfer in a serpentine micro-reactor using various solvents. *Chemical Engineering Journal, 285*, 605–615. https://doi.org/10.1016/j.cej.2015.09.115
28. Razavi Bazaz, S., Rouhi, O., Raoufi, M. A., et al. (2020). 3D Printing of inertial microfluidic devices. *Science and Reports, 10*, 5929. https://doi.org/10.1038/s41598-020-62569-9
29. Rein, C., Toner, M., & Sevenler, D. (2023). Rapid prototyping for high-pressure microfluidics. *Science and Reports, 13*, 1232. https://doi.org/10.1038/s41598-023-28495-2
30. Rhoades, T., Kothapalli, C. R., & Fodor, P. S. (2020). Mixing optimization in grooved serpentine microchannels. *Micromachines, 11*, 61. https://doi.org/10.3390/mi11010061
31. Riccio, C., Civera, M., Grimaldo Ruiz, O., Pedullà, P., Rodriguez Reinoso, M., Tommasi, G., Vollaro, M., Burgio, V., & Surace, C. (2021). Effects of curing on photosensitive resins in SLA additive manufacturing. *Applied Mechanics, 2*, 942–955. https://doi.org/10.3390/applmech2040055

32. Gulati, S., Dutcher, C. S., Liepmann, D., & Muller, S. J. (2010). Elastic secondary flows in sharp 90 degree micro-bends: A comparison of PEO and DNA solutions. *Journal of Rheology, 54*(2), 375–392. https://doi.org/10.1122/1.3308643
33. Saadat, M., Taylor, M., Hughes, A., & Hajiyavand, A. M. (2020). Rapid prototyping method for 3D PDMS microfluidic devices using a red femtosecond laser. *Advances in Mechanical Engineering, 12*(12). https://doi.org/10.1177/1687814020982713
34. Arockiam, S., Cheng, Y. H., Armenante, P. M., & Basuray, S. (2021). Experimental determination and computational prediction of the mixing efficiency of a simple, continuous, serpentine-channel microdevice. *Chemical Engineering Research and Design, 167*, 303−317. https://doi.org/10.1016/j.cherd.2021.01.022
35. Tasoglu, S., Safaee, H., Zhang, X., Kingsley, J. L., Catalano, P. N., Gurkan, U. A., Nureddin, A., Kayaalp, E., Anchan, R. M., Maas, R. L., Tüzel, E., & Demirci, U. (2013). Exhaustion of racing sperm in naturemimicking microfluidic channels during sorting. *Small (Weinheim an Der Bergstrasse, Germany) [online], 9*(20), 3374–3384. https://pubmed.ncbi.nlm.nih.gov/23677651/
36. Tsai, C. D., & Lin, X. Y. (2019). Experimental study on microfluidic mixing with different zigzag angles. *Micromachines (Basel), 10*(9), 583. https://doi.org/10.3390/mi10090583.PMID:31480452;PMCID:PMC6780503
37. Waheed, S., Cabot, J. M., Niall M. P., Lewis, T., Guijt, R. M., Paull, B., & Breadmore, M. C. (2016). 3D printed microfluidic devices: Enablers and barriers. *Lab Chip, 16*(11), 1993–2013.The Royal Society of Chemistry. https://doi.org/10.1039/C6LC00284F
38. Yin, P., Zhao, L., Chen, Z., Jiao, Z., Shi, H., Hu, B., Yuan, S., & Tian, J. (2020). Simulation and practice of particle inertial focusing in 3D-printed serpentine microfluidic chips via commercial 3D-printers. *Soft Matter, 16*(12), 3096–3105. https://doi.org/10.1039/D0SM00084A
39. You, J., Flores-Herrera, L., Packirisamy, M., Stiharu, I. (2005). Modeling the effect of channel bends on microfluidic flow. *IASME Trans, 1.*
40. Zarifi, M. H., Sadabadi, H., Hejazi, S. H., Daneshmand, M., & Sanati-Nezhad, A. (2018). Noncontact and nonintrusive microwave-microfluidic flow sensor for energy and biomedical engineering. *Scientific Reports, 8*(1), 139. https://doi.org/10.1038/s41598-017-18621-2.PMID:29317767;PMCID:PMC5760583

Swimming Robots for Targeted Chaotic Mixing in Viscous Environments

Mehdi Mirzakhanloo, Mohsen Saadat, Mir Abbas Jalali, and Mohammad-Reza Alam

Abstract Here we demonstrate that employing the controlled motion of a highly maneuverable swimming micro-robot (called Quadroar) enables a highly-efficient mixing in viscous environment, specifically in the Stokes regime of a Newtonian fluid. Detailed description of the design, modeling, and experimental realization of such versatile swimming micro-robots are discussed. The work presented here provides a roadmap toward realizing a controlled flock of robotic micro-mixers capable of inducing targeted chaotic mixing in viscous fluid environments.

1 Introduction

Mixing plays a crucial role in supporting life on Earth across various scales. It leaves its trails in nature, such as the distribution of spatial energy in aquatic ecosystems [1], as well as in our industrial world, including fields like petroleum engineering [2], food preparation [3], and pharmaceuticals [4]. Mixing is also essential in biology, particularly for lab-on-a-chip operations where rapid mixing of different reagents is necessary within a biochemical micro-reactor [5].

Finding an effective mixing strategy has long been a sought-after goal in many applications, and with the increasing popularity of microfluidic devices, there has been a growing focus [6] on achieving optimal mixing performance at the micron scale typically characterized by the low-Reynolds-number (Stokes) regime.

Fundamentally, mixing can occur via diffusion, convection, or turbulence, and can be achieved through either active or passive approaches [7]. The relative importance of convective over diffusive mixing is characterized by the Péclet number. For micro-scale fluid flows, the Péclet number typically reaches to tens of thousands [8].

M. Mirzakhanloo · M. Saadat · M. A. Jalali · M.-R. Alam (✉)
University of California, Berkeley, CA 94720, USA
e-mail: reza.alam@berkeley.edu

M. Mirzakhanloo
e-mail: m.mirzakhanloo@berkeley.edu

M. Saadat
e-mail: msaadat@berkeley.edu

M.-R. Alam and M. Fathi (eds.), *Integrated Systems: Data Driven Engineering*,
https://doi.org/10.1007/978-3-031-53652-6_6

This simply points out the inefficiency of diffusion for mixing in most microfluidic operations and lab-on-a-chip devices. On the other hand, linearity of the governing equations in Stokes (i.e., the low-Reynolds-number) regime, results in a highly laminar flow which suffers from the absence of natural turbulence. Therefore, active stirring and vortex generation is preferred in many of such applications [9–11].

Recent experiments have revealed the important role of microorganisms in active biomixing [12], which tacitly suggests the use of artificial micro-swimmers (e.g., autonomous swimming micro-robots) to induce active mixing. The concept of stirring by self-propelled micro-swimmers is different from the traditional mixing through moving forces/torques, and this is well studied in the literature both theoretically and numerically [1, 13, 14]. In fact, the force-free and torque-free characteristic of autonomous (i.e., self-propelled) micro-swimmers results in a dipolar flow field (to the leading order) which drops sharply with distance [15, 16]. This further weakens the so-called wall-effect, eliminating extra waste of energy due to viscous dissipation near solid boundaries of the container.

Here, we disucss the induction of targeted chaotic mixing through quasi-periodic motion of a swimming micro-robot, called *Quadroar* (Quadru+oar), introduced in our earlier studies [17–19]. By combining the rotations of four independently controlled rotary disks (see Fig. 1) along with the expansion-contraction of its chassis, Quadroar can perform full three-dimensional (3D) maneuvers in viscous environments. It can also carry out in-place reorientation maneuvers by rotating about any of its body-fixed axes. Therefore, this robotic micro-swimmer can follow any prescribed path in 3D space using stepwise control strategies. An end-to-end design, theoretical modeling, and experimental realization of a mm-scale Quadroar is also presented here from our earlier investigations [19].

The targeted nature of mixing induced by such autonomous swimming micro-robots is particularly useful for applications in which only a local mixing is required. For instance, in cases where there are biological molecules in the host medium that cannot be strained arbitrarily and must be treated carefully [20]. To mix a large volume of fluid, on the other hand, one can either use a single swimmer sweeping the entire volume, accessing remote regions and containers with complex geometries [18], or a controlled flock of such robotic micro-mixers [21].

2 The Swimmer and Kinematics

The schematic representation of the swimming robot, i.e., Quadroar, is shown in Fig. 1 [21]. It consists of three links with an I-shape arrangement and four independently controlled rotary disks of radius a which can rotate stepwise or continuously about their axes. Here, we refer to the set of swimmer links as the *I-frame*, which includes two axles with length of $2b$ and a chassis with variable length of $2l + 2s$, where s is the contribution of middle linear actuator. The rotation of nth disk ($n = 1, \ldots, 4$), rotating about its supporting flange link, is measured by the angle $-\pi \leq \vartheta_n \leq \pi$ that the plane of disk makes with the plane of I-frame.

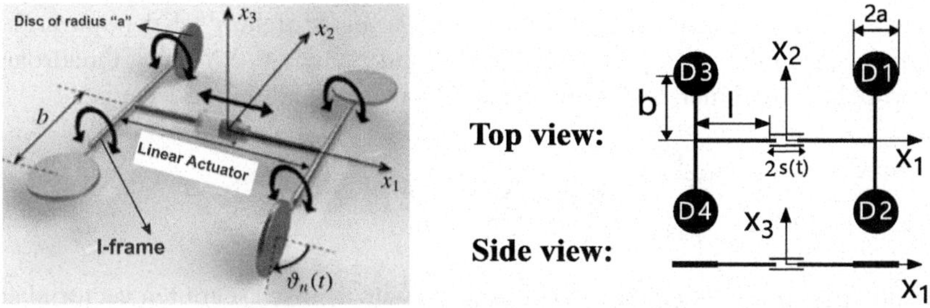

Fig. 1 Schematics of the Quadroar [17]: The origin of body-fixed Cartesian coordinate system (x_1, x_2, x_3), is at the geometrical center of the frame, x_1, x_2-axes lie in the plane of swimmer and x_1-axis is along the chassis

The global frame of reference is denoted by X_i, and we also consider a body fixed coordinate system x_i, $i = \{1, 2, 3\}$, such that its origin is at the geometric center of Quadroar and $x_1 - x_2$ plane is fixed to the I-frame (see Fig. 1). Vectors expressed in any of these two frames can be mapped to the other one using the rotation matrix **R** [18]. Here we assume each disc exerts a point force and a point torque to the surrounding fluid due to its translation and rotation, respectively:

$$\boldsymbol{f}_n = \mu K_n \cdot (\boldsymbol{v}_n - \boldsymbol{u}_n), \tag{1}$$

$$\boldsymbol{\tau}_n = \mu G \cdot (\boldsymbol{\omega}_n - \boldsymbol{\Omega}_n), \tag{2}$$

where μ is dynamic viscosity; u_n, Ω_n are the velocity and vorticity field at the center of each disc in the surrounding fluid; K_n and G are the translation and isotropic tensors defined in the body-fixed frame of reference [17]; v_n and ω_n are the absolute linear and angular velocity of each disk at the center point, which are determined by the linear velocity of swimmer's geometric center, i.e., $\boldsymbol{v}_c$, and its angular velocity $\boldsymbol{\omega}_{body}$. The resultant contribution of each propeller (i.e., disk) to the streaming in background fluid is then the combination of Stokeslet and rotlet [22], from the point-force ($\boldsymbol{f}$) and point-torque ($\boldsymbol{\tau}$) presented in Eqs. (1) and (2), respectively. Velocity and vorticity fields, at the position of each disk, are then calculated as the sum of contributions from all other disks:

$$\boldsymbol{u}_n = \sum_{k=1 k\neq n}^{4} \left[\frac{1}{8\pi\mu} \left(\frac{\boldsymbol{f}_k}{z_{kn}} + \frac{\boldsymbol{f}_k \cdot \boldsymbol{X}_{kn}}{z_{kn}^3} \boldsymbol{X}_{kn} + \frac{\boldsymbol{\tau}_k \times \boldsymbol{X}_{kn}}{z_{kn}^3} \right) \right], \tag{3}$$

$$2\boldsymbol{\Omega}_n = \sum_{k=1 k\neq n}^{4} \left[\frac{1}{8\pi\mu} \left(\frac{2\boldsymbol{f}_k \times \boldsymbol{X}_{kn}}{z_{kn}^3} + \frac{3(\boldsymbol{\tau}_k \cdot \boldsymbol{X}_{kn})\boldsymbol{X}_{kn} - z_{kn}^2 \boldsymbol{\tau}_k}{z_{kn}^5} \right) \right], \tag{4}$$

where $\boldsymbol{X}_{kn} = \boldsymbol{r}_n - \boldsymbol{r}_k$, in which $\boldsymbol{r}_n$ is the position vector of the nth disk ($n = 1, \ldots, 4$) with respect to the local frame of reference, and $z_{kn} = (\boldsymbol{X}_{kn}.\boldsymbol{X}_{kn})^{0.5}$. Quadroar is a self-propelled swimmer, which means it is force-free and torque-free, i.e.,

$$\sum_{n=1}^{4} \boldsymbol{f}_n = 0, \quad \sum_{n=1}^{4} \left(\boldsymbol{r}_n \times \boldsymbol{f}_n + \boldsymbol{\tau}_n\right) = 0. \tag{5}$$

The equations presented in (1)–(5) provide us with a closed set of ten vectorial equations and ten vectorial unknowns, namely: $\boldsymbol{v_c}$, $\boldsymbol{\omega}_{body}$, $\boldsymbol{u}_n$, and $\boldsymbol{\Omega}_n$ for $n = 1, \ldots, 4$. By solving this system of equations at each time step, one can then find the trajectory and orientation of the swimmer over time by integrating:

$$\dot{\boldsymbol{X}}_c = R^T \cdot \boldsymbol{v}_c, \quad \dot{\boldsymbol{\alpha}}_c = T^{-1} \cdot \boldsymbol{\omega}_{body}, \tag{6}$$

where $\boldsymbol{X}_c$ is the position vector of geometric center in the global frame of reference; $\boldsymbol{\alpha}_c$ is the swimmer's yaw-pitch-roll sequence of Euler's angles $\alpha = (\phi, \theta, \psi)$, and T is the transformation matrix [17].

The flow field induced by a Quadroar at each arbitrary point is the superposition of all Stokeslets and rotlets generated by the propellers (i.e., disks). We study the flow properties tracking Lagrangian path-lines of passive tracers by integrating:

$$\frac{d\boldsymbol{X}}{dt} = \boldsymbol{u}(\boldsymbol{X}, t). \tag{7}$$

Detailed explanation of governing equations and different numerical techniques which have to be used for solving the resultant system of equations are discussed with details in our previous studies [17, 18, 23, 24].

3 Trajectory and Flow-Field

Tuning geometrical constants (i.e., a, b, l) and dynamical parameters, i.e., $s\ (t)$ and $\vartheta_n\ (t)$, which are the control inputs, results in various behaviors for the Quadroar swimmer [17]. Here we assume harmonic expansion-contraction of the linear actuator, $s\ (t) = s_0\,[1 - \cos\,(\omega_s t)]\,/2$, where $s_0/l = 1/2$, and $\omega_s = 1$. The disk D_n ($n = 1, \ldots, 4$) is spinning with angular speed $\dot{\vartheta}_1 = \dot{\vartheta}_2 = c_0\ \omega_s$ and $\dot{\vartheta}_3 = \dot{\vartheta}_4 = -c_0\ \omega_s\ + \delta\nu$, where constants c_0 and $\delta\nu$ (detuning parameter) are set to different values depending on the specific application. Two characteristic time scales then emerge in the problem: $T_s = 2\pi/\delta\nu$ and $T_f = 2\pi/\omega_s$. For simplicity, we also set $a = 1$, and $b/a = l/a = 4$ throughout our simulations.

For each specific c_0, non-zero values of $\delta\nu$ results in quasi-periodic single-loop (cardioid) or multi-loop orbits (see Fig. 2). A swimmer traveling on Quasi-periodic orbits, experiences two different phases during the motion: (i) lingering phase, when it

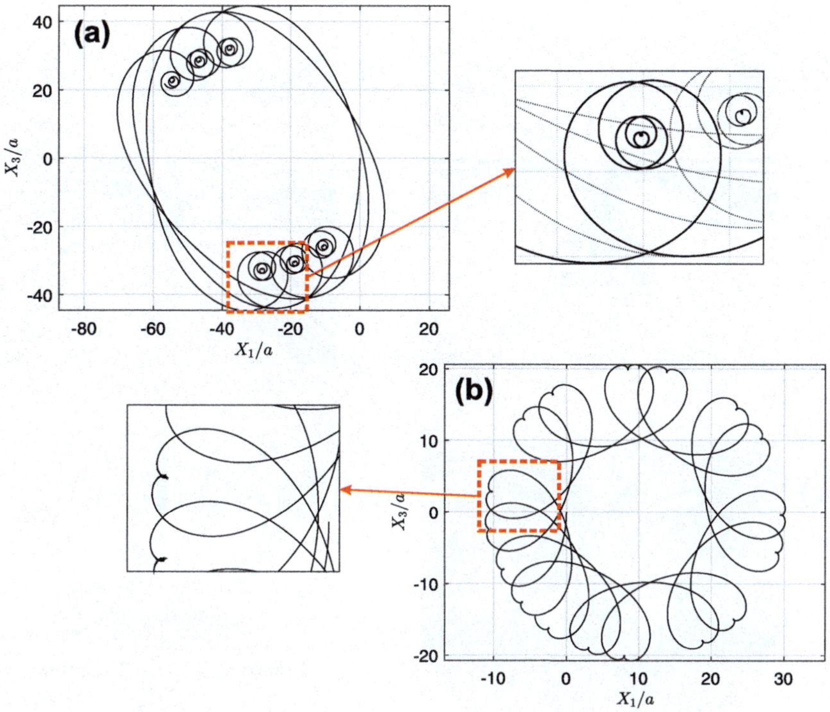

Fig. 2 Quadroar represent various maneuvers, including multi-loop **a**, and cardioid single-loop **b** orbits. The swimmer has $c_0 = 1/2$ in both cases and $\delta\nu$ is equal to 0.001 and 0.00625 for **a** and **b**, respectively. Trajectories are shown during $0 \leq t \leq 2000T_f$ for the cardioid orbit and $0 \leq t \leq 5000T_f$ for the multi-loop turns

undergoes a single- or multi-loop turn, and (ii) propulsive phase, when it travels a long distance in space. There exist non-linear relations between the detuning parameter, i.e., $\delta\nu$, with (i) the number of loops in each turn, (ii) the bundles being arranged toward or away from the center, and (iii) the mean radius of orbits.

On the other hand, for a specific parameter setup with $\delta\nu = 0$, the Quadroar moves on a straight line inducing an oscillatory (pusher-puller) flow field with posterior, side, and anterior vortices to the surrounding environment (Fig. 3). Our previous investigations reveal the resembelence of such flow characteristics to that observed for real flagellar microorganisms [24]. This makes the Quadroar a unique candidate to study hydrodynamic interactions between swimming microorganisms [23] as well as their effects on the surrounding environments [24].

Moreover, Quadroar can represent a full three-dimensional maneuver when $\dot{\vartheta}_1 \neq \dot{\vartheta}_2$, and $|\dot{\vartheta}_1| \neq |\dot{\vartheta}_3|$, i.e., when there is no symmetry or mirroring between disks' angular speeds. Thus, the number of control inputs increase to five.

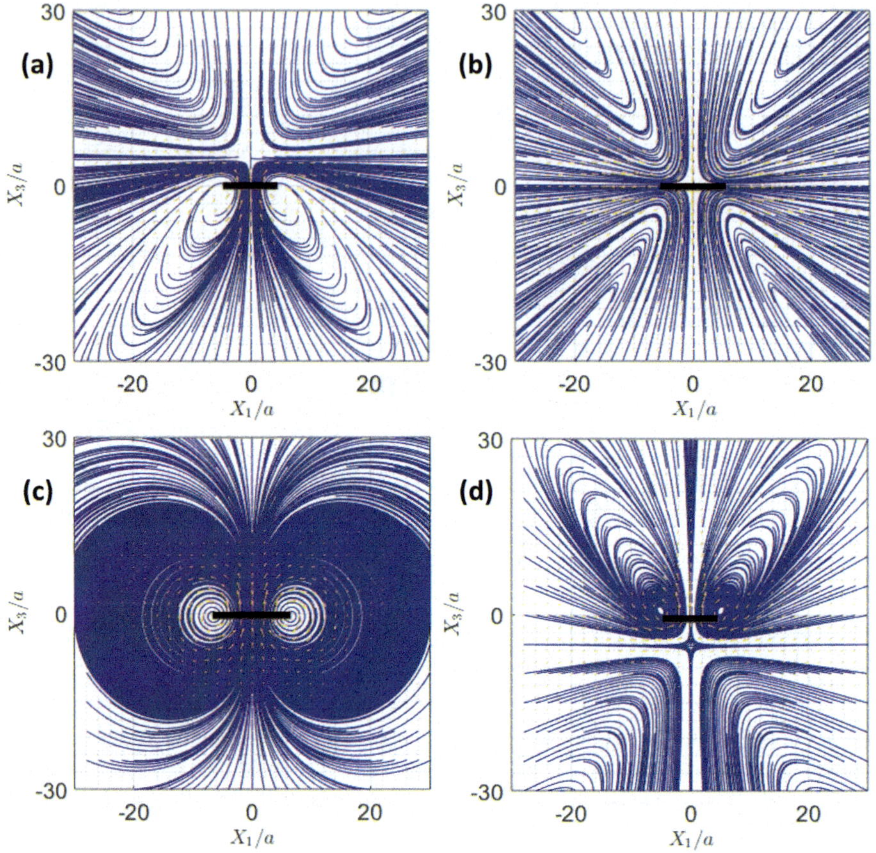

Fig. 3 Streamlines and the flow field induced by a single swimmer when moving on a transversed straight trajectory along X_3-axis ($c_0 = 1/2$ and $\delta\nu = 0$). The oscillatory nature of the induced flow field with posterior **a**, side **c**, and anterior **d** vortices closely resembles to that observed for Chlamydomonas reinhardtii [18, 24]. Snapshots are given at: **a** $1/64\ T_f$ **b** $16/64\ T_f$ **c** $32/64\ T_f$ **d** $63/64\ T_f$. The thick solid black bar demonstrates the swimmer's chassis at the moment

4 Chaotic Mixing

Swimmer moving on a straight line, only makes distant fluid particles move on a loop-like paths [1], which is not desired for mixing. Therefore, here we focus on quasi-periodic orbit trajectories of the swimmer, corresponding to the non-zero values of $\delta\nu$. For instance, setting $\delta\nu = 0.00125$ results in a quasi-periodic trajectory with multi-loop turns (Fig. 4a), which is the projection of an *invariant torus* in the (X_1, X_3, θ)-space on the (X_1, X_3)-plane [18].

To demonstrate mixing process, we track Lagrangian trajectories for a packet of 2500 tracers which are uniformly distributed in the square region $\mathscr{R}_0 = \{(X_1, X_3)| -$

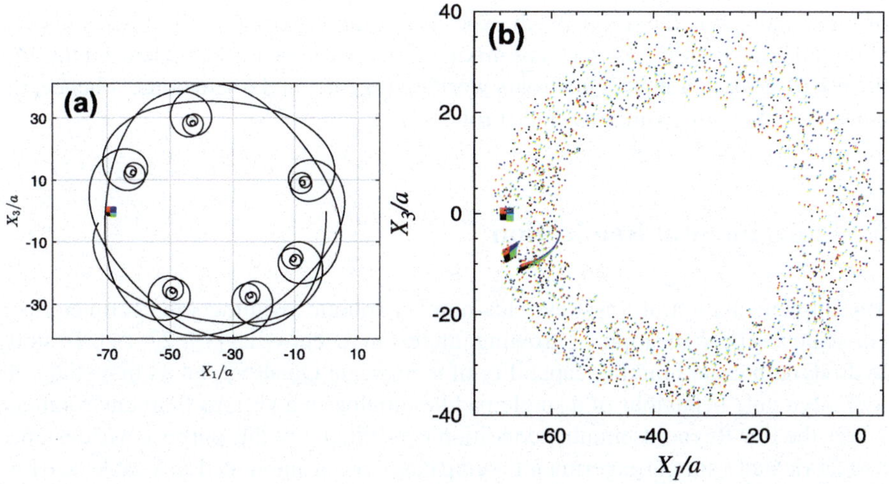

Fig. 4 Visualization of mixing process using a Quadroar swimmer: **a** The swimmer moves on quasi-periodic orbits with multi-loop turns, corresponding to $\delta\nu = 0.00125$. Black solid line represents the trajectory of swimmer. A packet of 2500 tracers, which has been divided into four squares with different colors, is uniformly distributed in the square region $\mathscr{R}_0 = \{(X_1, X_3)| - 70 \leq X_1 \leq -67.5, -1.25 \leq 1.25\}$. **b** Mixing process is visualized through snapshots of the packet of tracer at $t = 0,\ 3.75T_s,\ 6.25T_s$, and the final state, i.e., $t = 125\ T_s$

$70 \leq X_1 \leq -67.5, -1.25 \leq 1.25\}$, as shown in Fig. 4a. For better visualization of the process, we have also divided the region to for equal sub-squares with different colors. The swimmer is released from $(X_1, X_3) = (0, 0)$, and we simultaneously integrate Eq. (6) for the swimmer and Eq. (7) for the tracers.

Snapshots of the mixing process, including tracers at the initial and final states are presented in Fig. 4b. As time elapses, the packet of tracers is highly stretched and folded, $\mathscr{R}_0$ is mapped to some $\mathscr{R}(t)$, and eventually, as $t \to \infty$, it has been dispersed uniformly in an annular region, where the projection of invariant torus becomes dense with time. Snapshots of the tracer packet presented in Fig. 4b respectively correspond to the following states: $t = 0$, $t = 3.75\ T_s$, $t = 6.25\ T_s$, and the final state at $t = 125\ T_s$ in which tracers have been perfectly mixed up and uniformly distributed over the annular region of the swimmers pathlines.

The presented mixing process can be quantified using various methods [25]. In our previous study [18] we used the finite-time *Lyapunov exponent* method to investigate the presence of chaos. A positive Lyapunov exponent was reported [18] for the induced mixing, suggesting an exponential divergence for the adjacent tracers on a line of markers, and that is an indication of *chaotic* mixing. Our repeated experiments reveal that the mixing process remains chaotic for different variations of quasi-periodic orbits. The thickness of final annular mixing region, however, is proportional to the number of loops in each turn of the swimmer's trajectory. In fact, as the number of loops increases, the tracer patch experiences more foldings, and this results in a wider affected region. The speed of mixing process can be further enhanced through the use of several or many Quadroars working together at the same time—hence, the controlled flock of swimming micro-robots [21].

5 Experimental Realization

The implementation of low-Reynolds-number robotic swimmers in centimeter and cm-scale has been the topic of growing interest in recent years [19, 27–29]. To study hydrodynamics and mixing capability of a cm-scale Quadroar, let us first study the hydrodynamic behaviour of a single paddle moving in a viscous fluid environment, where the low Reynolds number condition is fulfilled. For this purpose, we designed and fabricated a simple experimental setup based on an automated X-Y table (see Fig. 5). Two stepper motors are used to move the paddle (attached to the X-Y table end effector) in horizontal plane, while a third motor is utilized to rotate the paddle around the vertical axis. The paddle is a thin circular aluminum plate which moves inside a container filled with corn syrup (to achieve the low Reynolds number condition). A digital camera is placed under the container (facing up) to record the fluid motion when the paddle moves/rotates inside the viscous fluid environment.

The starting state snapshot of the experiment is shown in Fig. 5a, where several blobs of dye are added to the liquid surface for visualizing the induced mixing process as well as tracking the fluid motion in the vicinity of the moving paddle (by applying image processing techniques). The paddle undergoes a linear motion in the Y^+ direction (see Fig. 5b) with the constant angle of attack θ, then stops at point B and rotates by 2θ in the counter-clockwise direction. It then moves in the Y^- direction until it stops at point A, makes a clockwise turn of 2θ, and repeats the sequence. Note that the linear velocity of the paddle moving between point A and B remains constant during each test. To quantify the propulsion (or pumping) speed of the dye wave, the linear motion of a point on the frontier has been tracked over time (see Fig. 5b). While this point has an oscillatory back-and-forth motion in short time scales (due to up-and-down motion of the paddle), its average velocity in long run would be toward the left-hand-side in Fig. 5.

The effect of paddle angle and its linear velocity on propulsion are also studied on this test-bed. A set of experiments with different combinations of blade angle and velocity is conducted. Results corresponding to one set of these case studies are shown in Fig. 5c–d, where the linear velocity of the paddle and its total stroke length are 4.87 cm/s and 7.3 cm, respectively. The experiment results indicate that for a given paddle angle and a given linear velocity, the pumping speed at a given location is a function of the distance from the paddle (denoted by X). Therefore, as the front borderline moves away from the paddle (to the left), its linear expansion rate decreases. Additionally, for a constant paddle linear velocity, the maximum expansion rate occurs at an optimal blade angle. For example, for the case shown in Fig. 5, the maximum expansion rate is achieved at around 40 degrees. This is while

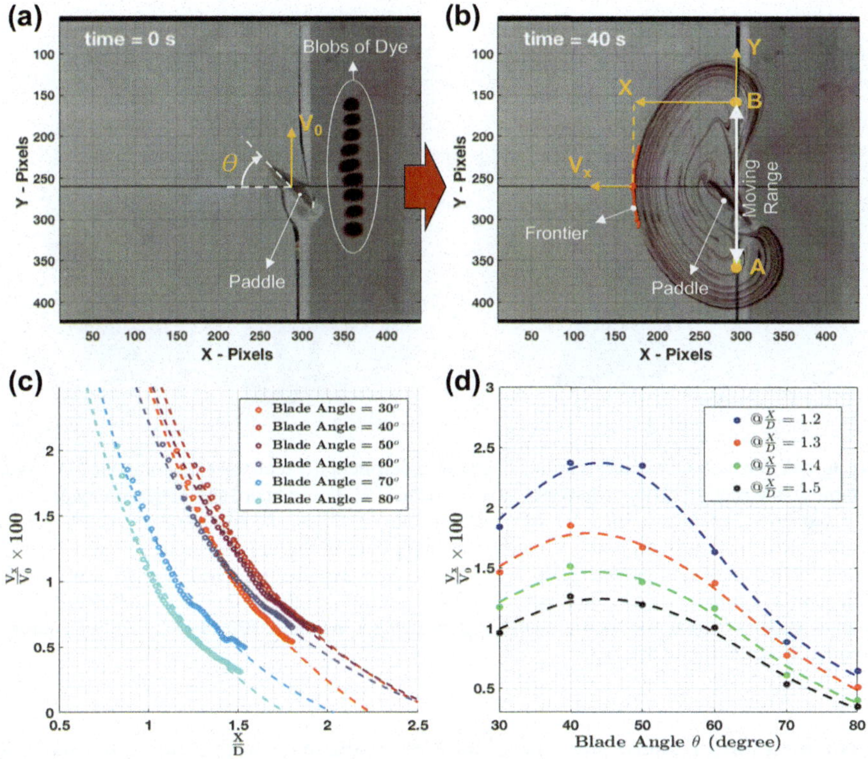

Fig. 5 Experimental study for the hydrodynamic behavior of a single paddle in the low Reynolds number condition. **a** Snapshot of the liquid tank at the beginning of the experiment. The tank is filled with corn syrup where several blobs of dye are placed on the liquid surface. Paddle diameter is 3 cm; **b** Snapshot of the liquid tank at $t = 40$ s. Rocking motion of the paddle has mixed liquid. The propulsion and pumping performance of the paddle has been evaluated by the average velocity of the frontier curve to the left (V_x); **c–d** The effect of paddle angle θ on the expansion rate (i.e., frontier speed) of the dye mixture in different distances from the paddle. D and V_0 are the paddle diameter and linear speed, respectively

if the linear speed increases to 7.3 cm/s, the maximum expansion rate happens at around 50 degrees. The mixing effect of the paddle movement (with different speeds and angles) is also visualized using separate dye colors and inspectinet the dye mixture downstream of the paddle.

With the valuable insight gained from the single paddle hydrodynamic study, we then engineered and experimentally tested a cm-scale Quadroar swimmer based on the proposed idea (see Fig. 6). The system is composed of four rotating disks connected to the tips of front and rear axles. The axles are located on a variable length chassis. The chassis and axles make an I-shape planar frame. The maximum required torque τ for rotating a circular disk for the given geometry and the Reynolds number condition is calculated to be about 2 N.mm. Considering this maximum torque as well as other required properties such as complexity, cost and availability,

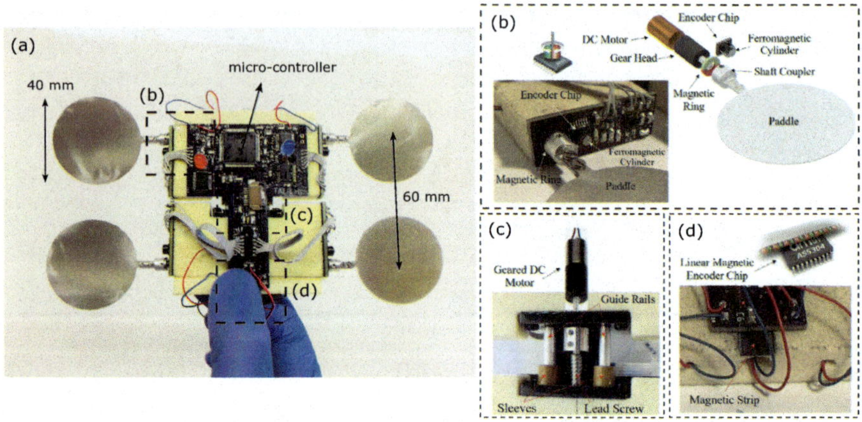

Fig. 6 **a** A cm-scale Quadroar swimmer has been experimentally realized with complete 3D maneuverability in viscous fluid environments, as reported in [19]. The onboard angle measurement system uses a magnetic encoder to measure the angles of rotary paddles, accomplished by attaching a ring magnet to the paddle shaft and inducing a magnetic field in the ferromagnetic cylinder installed on top of the magnetic encoder chip **b** A lead-screw mechanism is used to realize the expansion/contraction motion between the front and rear sections of the swimmer **c** Additionally, there is an expansion/contraction measurement system using a magnetic strip **d** Detailed description is reported in our earlier work [19]

a combination of micro geared DC-motors is selected. Additionally, a small ball screw system is utilized to generate a rapid expansion-retraction between the front and rear axles. Since accurate angular phase shift between four disks are critical to achieve different behaviors for the Quadroar swimmer, a sensory system is designed to measure the angle between each disk and the Quadroar planar I-frame. Here, we used $AS5050$ magnetic encoder due to its compactness (4 mm $\times$ 4 mm $\times$ 1 mm), accuracy ($\sim 0.18^o$ resolution) and its low price. All the sensors and actuators are connected to an on-board micro-controller (Atmega-2560) which is attached to the Quadroar main frame. While the low level controller algorithms are performed by the on-board micro-controller, the higher level (i.e., supervisory) control commands are communicated between a desktop computer and micro-controller via a USB connection (using a flex cable). Finally, to reduce the density of the Quadroar in order to plunge it inside the liquid, two "T" shape blocks made out of low-density foam are added to the system.

For a complete report on the mechatronic design, fabrication, and experimental study of this cm-scale Quadroar swimmer the reader is referred to our earlier work [19]. We have described the design methodology and component selection of the system based on the required performance. A supervisory control scheme is also presented to achieve an accurate trajectory tracking for all the actuators used in the swimmer. Finally, we have also conducted experiments in silicone oil (with 5000 cP viscosity) where two primary modes of swimming–forward translation and planar reorientation–have been tested and compared with the theoretical model.

In the end, it is worth noting that owing to the recent achievements in design and synthesizing molecular motors with photochemical or electrochemical actuators [30, 31], implementation of Quadroar seems also feasible in micro- and nano-scales. Light-driven monodirectional molecular motors [32] or their second generation [33] can be suitable candidates for propellers in a Quadroar swimmer, as they have already been used in light-driven nanocars [34]. The realization of linear actuator at the middle of swimmer's I-frame can also be done through expansion-contraction of an azobenzene chassis by *cis-trans* photo-isomerization. A similar idea has been used in synthesis of molecular scissors [35], and photo-active nanoworms [36].

6 Conclusion

Here we introduced swimming micro-robots as autonomous robotic micro-mixers capable of inducing targeted chaotic mixing in viscous fluid environments. Detailed description of the design, modeling, and experimental realization of such swimming micro-robots (named Quadroar) are also discussed. Through computational and experimental experimentation, we then showed how the controlled motion of a Quadroar can induce chaotic mixing in the surrounding fluid. To visualize the process, we accurately followed a packet of tracers as it was stretched and folded through the motion of a Quadroar on quasi-periodic orbits. At the final state, tracers were completely mixed and uniformly distributed over an annular region where trajectories of the swimmer become dense. The mixing process can be further sped up by using a controlled flock of such robotic micro-mixers.

Acknowledgements The authors acknowledge the support provided by the American Bureau of Shipping and the National Science Foundation under grant CPS-1932595.

References

1. Pushkin, D. O., & Yeomans, J. M. (2013). Fluid mixing by curved trajectories of microswimmers. *Physical Review Letters, 111*(18), 188101.
2. Ennis-King, J. P., Paterson, L., et al. (2005). Role of convective mixing in the long-term storage of carbon dioxide in deep saline formations. *Spe Journal, 10*(03), 349–356.
3. Cullen, P.J. (2009). *Food mixing: Principles and applications*. John Wiley & Sons.
4. Hersey, J. A. (1975). Ordered mixing: a new concept in powder mixing practice. *Powder Technology, 11*(1), 41–44.
5. Dittrich, P. S., & Manz, A. (2006). Lab-on-a-chip: microfluidics in drug discovery. *Nature Reviews Drug Discovery, 5*(3), 210–218.
6. Lee, C.-Y., Chang, C.-L., Wang, Y.-N., & Lung-Ming, F. (2011). Microfluidic mixing: a review. *International Journal of Molecular Sciences, 12*(5), 3263–3287.
7. Volker, H., Holger, L., & Schönfeld, F. (2005). Micromixers-a review on passive and active mixing principles. *Chemical Engineering Science, 60*(8), 2479–2501.

8. Ottino, J. M., & Wiggins, S. (2004). *Introduction: mixing in microfluidics* (pp. 923–935). Philosophical Transactions: Mathematical, Physical and Engineering Sciences.
9. Mathew, G., Mezić, I., Grivopoulos, S., Vaidya, U., & Petzold, L. (2007). Optimal control of mixing in stokes fluid flows. *Journal of Fluid Mechanics, 580*, 261–281.
10. Couchman, I. J., & Kerrigan, E. C. (2010). Control of mixing in a stokes' fluid flow. *Journal of Process Control, 20*(10), 1103–1115.
11. Shamloo, A., Mirzakhanloo, M., & Reza Dabirzadeh, M. (2016). Numerical simulation for efficient mixing of newtonian and non-newtonian fluids in an electro-osmotic micro-mixer. *Chemical Engineering and Processing: Process Intensification, 107*, 11–20.
12. Kurtuldu, H., Guasto, J. S., Johnson, K. A., & Gollub, J. P. (2011). Enhancement of biomixing by swimming algal cells in two-dimensional films. *Proceedings of the National Academy of Sciences, 108*(26), 10391–10395.
13. Eckhardt, B., & Zammert, S. (2012). Non-normal tracer diffusion from stirring by swimming microorganisms. *The European Physical Journal E: Soft Matter and Biological Physics, 35*(9), 1–2.
14. Wagner, G. L., Young, W. R., & Lauga, E. (2014). Mixing by microorganisms in stratified fluids. *Journal of Marine Research, 72*(2), 47–72.
15. Mirzakhanloo, M., & Alam, M.-R. (2020). Stealthy movements and concealed swarms of swimming micro-robots. *Physics of Fluids, 32*(7), 071901.
16. Mirzakhanloo, M., Esmaeilzadeh, S., & Alam, M.-R. (2020). Active cloaking in stokes flows via reinforcement learning. *Journal of Fluid Mechanics, 903*, A34.
17. Jalali, M. A., Alam, M.-R., & Mousavi, S. (2014). Versatile low-reynolds-number swimmer with three-dimensional maneuverability. *Physical Review E, 90*(5), 053006.
18. Jalali, M. A., Khoshnood, A., & Alam, M.-R. (2015). Microswimmer-induced chaotic mixing. *Journal of Fluid Mechanics, 779*, 669–683.
19. Saadat, M., Mirzakhanloo, M., Shen, J., Tomizuka, M., & Alam, M.-R. (2019). The experimental realization of an artificial low-reynolds-number swimmer with three-dimensional maneuverability. In *2019 American Control Conference (ACC)* (pp. 4478–4484). IEEE.
20. Aref, H. (1991). Stochastic particle motion in laminar flows. *Physics of Fluids A: Fluid Dynamics, 3*(5), 1009–1016.
21. Mirzakhanloo, M. (2020). *Flocks of Artificially Intelligent Swimming micro-Robots with Long-range Hydrodynamic Interaction and Objectives*. Berkeley: University of California.
22. Lopez, D., & Lauga, E. (2014). Dynamics of swimming bacteria at complex interfaces. *Physics of Fluids, 26*(7), 400–412.
23. Mirzakhanloo, M., Mir Abbas, J., & Mohammad-Reza, A. (2018). Hydrodynamic choreographies of microswimmers. *Scientific Reports, 8*(1), 3670.
24. Mirzakhanloo, M., & Alam, M.-R. (2018). Flow characteristics of chlamydomonas result in purely hydrodynamic scattering. *Physical Review E, 98*(1), 012603.
25. Mathew, G., Mezić, I., & Petzold, L. (2005). A multiscale measure for mixing. *Physica D: Nonlinear Phenomena, 211*(1), 23–46.
26. Chabreyrie, R., Chandre, C., & Aubry, N. (2011). Complete chaotic mixing in an electro-osmotic flow by destabilization of key periodic pathlines. *Physics of Fluids, 23*(7), 072002.
27. Herr, H., & Dennis, R. G. (2004). A swimming robot actuated by living muscle tissue. *Journal of Neuroengineering and Rehabilitation, 1*(1), 6.
28. Ye, X., Su, Y., Guo, S., & Wang, L. (2008). Design and realization of a remote control centimeter-scale robotic fish. In *Advanced Intelligent Mechatronics, 2008. AIM 2008. IEEE/ASME International Conference on* (pp. 25–30). IEEE.
29. Silverberg, O., Demir, E., Mishler, G., Hosoume, B., Trivedi, N., Tisch, C., Plascencia, D., Pak, O. S., & Araci, I. M. (2020). Realization of a push-me-pull-you swimmer at low reynolds numbers. *Bioinspiration & Biomimetics, 15*(6), 064001.
30. Kottas, G. S., Clarke, L. I., Horinek, D., & Michl, J. (2005). Artificial molecular rotors. *Chemical Reviews, 105*(4), 1281–1376.
31. Michl, J., & Charles H Sykes, E. (2009). *Molecular rotors and motors: recent advances and future challenges*.

32. Koumura, N., Robert, W. J. Z., van Delden, A., & R., Harada, N., & Feringa, B.L. (1999). Light-driven monodirectional molecular rotor. *Nature, 401*(6749), 152.
33. Koumura, N., Geertsema, E.M., van Gelder, M.B., Meetsma, A., & Feringa, B.L. (2002). Second generation light-driven molecular motors. unidirectional rotation controlled by a single stereogenic center with near-perfect photoequilibria and acceleration of the speed of rotation by structural modification. *Journal of the American Chemical Society, 124*(18), 5037–5051.
34. Pinn-Tsong, C., Johannes, M., Jazmin, G., Guerrero, J. M., Alemany, L. B., Villagomez, C. J., Saywell, A., Grill, L., & Tour, J. M. (2011). Toward a light-driven motorized nanocar: Synthesis and initial imaging of single molecules. *ACS Nano, 6*(1), 592–597.
35. Muraoka, T., Kinbara, K., Kobayashi, Y., & Aida, T. (2003). Light-driven open- close motion of chiral molecular scissors. *Journal of the American Chemical Society, 125*(19), 5612–5613.
36. Sasaki, T., & Tour, J. M. (2008). Synthesis of a new photoactive nanovehicle: a nanoworm. *Organic Letters, 10*(5), 897–900.

Intelligent Robotics and Embedded Controls

Adaptive Hierarchical Meta-Scheduling for Distributed Time-Triggered Systems with Diverse Internal and System Architectures

Pascal Muoka, Tom Kiesow, Daniel Onwuchekwa, and Roman Obermaisser

Abstract Time-triggered (TT) systems are widely adopted in safety-critical applications for their predictability and fault containment. With the increasing deployment of TT nodes in distributed environments, adaptation becomes essential to address changing system conditions as they impact nodes differently. This paper presents a hierarchical approach to multi-node/multi-core adaptation in distributed TT systems with diverse internal architectures. Our hierarchical meta-scheduler leverages the system's hierarchical topology to compute node-level and system-level adapted schedules. We introduce a superset reconvergence technique to minimise the size of the Multi-Schedule Graph (MSG) generated by the traditional meta-scheduler, minimising memory requirements in nodes with limited storage capacity. The scheduler employed by the hierarchical meta-scheduler is evaluated for performance and the superset reconvergence technique's MSG size reduction rate. The results demonstrate the benefits of hierarchical adaptation and minimised memory consumption.

1 Introduction

Time-triggered (TT) nodes have gained significant attention and adoption in real-time applications. Their temporal predictability and inherent support for fault containment contribute to their deterministic properties providing reliable platforms for systems operating in environments where safety and performance are critical [1]. However, as the demand for distributed applications increases, there is a growing trend of

P. Muoka (✉) · T. Kiesow · D. Onwuchekwa · R. Obermaisser
University of Siegen, Hölderlinstrasse 3, Siegen, Germany
e-mail: Pascal.Muoka@uni-siegen.de

T. Kiesow
e-mail: Tom.Kiesow@uni-siegen.de

D. Onwuchekwa
e-mail: Daniel.Onwuchekwa@uni-siegen.de

R. Obermaisser
e-mail: Roman.Obermaisser@uni-siegen.de

M.-R. Alam and M. Fathi (eds.), *Integrated Systems: Data Driven Engineering*,
https://doi.org/10.1007/978-3-031-53652-6_7

deploying TT nodes with different internal architectures, introducing new challenges and opportunities for system adaptation.

In distributed applications, where TT nodes with diverse internal architectures are interconnected, changing system parameters and conditions impact different nodes across the system, such as a faulty core or an application job completed earlier than its scheduled completion time [2]. Under these conditions, adaption is desirable to enhance dependability and performance by dynamically adjusting the system behaviour in response to evolving conditions.

Prior work has addressed node-level adaptation using meta-scheduling to preserve the system's temporal predictability and implicit synchronisation [2, 3], limiting their effectiveness in distributed environments. However, leveraging the hierarchical topology of distributed nodes with different internal architectures presents an opportunity for multi-node adaptation. By utilising the hierarchical structure, adaptation mechanisms are leveraged to provide an optimised adaptation to mitigate the impact of events at different system levels.

In addition to the challenge of multi-node adaptation, limited internal memory in specific nodes further complicates the runtime switching of schedules required for adaptation [4]. Storing many schedules for runtime switching becomes impractical in nodes with constrained resources. Therefore, innovative approaches are needed to optimise memory usage and facilitate efficient adaptation in such scenarios.

This paper proposes a hierarchical meta-scheduler for multi-level adaptation in distributed TT systems. Our approach takes advantage of the hierarchical topology of the system, allowing for optimised adaptation at different levels. Considering the unique characteristics of nodes with diverse internal architectures, our meta-scheduling technique preserves the TT properties while delivering adaptation across multiple nodes, enhancing system performance and dependability.

Furthermore, we introduce a superset reconvergence technique to address the challenge of node memory limitations. This technique minimises the size of the multi-schedule graphs generated by the meta-scheduler, enabling efficient storage and utilisation of schedules needed for runtime switching. By reducing the memory overhead, our proposed technique improves the feasibility of adaptation in resource-constrained nodes, expanding the applicability of this adaptation technique to distributed TT systems.

Through comprehensive experiments and evaluations, we demonstrate the effectiveness and advantages of our hierarchical meta-scheduler and superset reconvergence technique. The results highlight the potential of our approach in addressing the evolving needs of distributed TT systems with different internal architectures.

The rest of the paper is structured as follows-Sect. 2 the related work. The hierarchical time-triggered system is presented in Sect. 3. Section 4 addresses the hierarchical scheduling and meta-scheduling technique. The evaluation of the hierarchical meta-scheduler is presented in Sect. 5, and the results are discussed. Finally, Sect. 6 summarises the main contributions and findings.

2 Related Work

Several research efforts have investigated hierarchical scheduling techniques for distributed systems. Oliver and Craciunas [5] introduce an abstraction of a distributed hierarchical time-triggered system composed of networked multi-core nodes. Application tasks are executed based on a table-driven schedule where communication between tasks is in the form of time-triggered (TT) messages. Two levels in the system are differentiated: an on-chip network connecting the tiles within nodes and an off-chip network connecting the multiple nodes. In the event of failure, tasks are migrated to available tiles or nodes. Instead of reconfiguring the system at runtime, task sets are merged, and an adapted schedule with a longer makespan is computed to accommodate the system's adaptation to a finite number of events.

Other researchers [6] focus on scheduling at the off-chip network layer for safety-critical applications employing deterministic networks such as FlexRay. Control applications consisting of sensor, controller, and actuator tasks are scheduled on distributed Electronic Control Unit (ECU) nodes, while messages carrying sensor data are mapped onto specific FlexRay schedules. A co-optimisation problem is formulated where a MILP scheduler is employed to find scheduling solutions for each control application while optimising the total FlexRay bus resource utilisation and overall system performance. On the other hand, Craciunas et al. in [7] address the scheduling problem arising from the IEEE 802.1Qbv extension on multi-hop fully switched Time-Sensitive Networking (TSN) networks, presenting a method for computing static schedules for 802.1Qbv-capable network devices using Satisfiability Modulo Theories (SMT) and, alternatively, Optimization Modulo Theories (OMT) solvers.

Furthermore, an adaptive sample-path algorithm is proposed to compute schedules considering the convergence and stability of the adaptive scheme when an event is reported. Derasevic et al. [8] addresses fault tolerance of Flexible Time-Triggered (FTT)-based systems, proposing a node replication scheme with voting to improve system reliability. A proposal for fault-tolerant hierarchical systems is presented in [9], where the authors suggest a mechanism for handling faults at the task, partition, and distributed levels to increase system reliability. Mouka et al. in [10] propose an adaptive scheduling method for energy saving in NoC-based multi-core systems. Their approach uses a meta-scheduling approach to adapt the system schedule to slack events by periodically sampling the system. This approach preserves the predictability of the system while providing system adaptation for energy saving.

Although recent research has explored various aspects of hierarchical architectures and system scheduling and adaptation to runtime context events, our work contributes to the field by focusing on achieving hierarchical adaption and optimisation while preserving the time-triggered properties of the system. We also investigate the impact of failures at the different levels of the hierarchy, including core, router, switch and link failures and propose an effective adaptation strategy to minimise the impact of these failures while optimising system performance.

3 Hierarchical Time-Triggered System

A hierarchical time-triggered system describes a system that incorporates a hierarchical architecture, time-triggered scheduling, and efficient communication infrastructure. Decomposing the system into levels of hierarchy allows for scalability and resource allocation based on the specific requirements of each level. In addition, it ensures a reliable and coordinated operation across the system through synchronized global time and inter-core communication via a Time-triggered Network-on-Chip (TTNoC) and inter-node communication through a Time-triggered Off-chip Communication Network (TTOCN) such as Time Sensitive Networking (TSN), Time-triggered Ethernet (TTEthernet), Time-triggered Controller Area Network (TTCAN), suitable for real-time applications with stringent timing requirements. In this work, nodes are computational units within the networked system that perform jobs, communicating and coordinating with each other. On the other hand, cores are processing subunits within the node capable of executing tasks of a job concurrently.

3.1 Hierachical Platform

The Hierarchical Time-triggered Platform (HTP) is a multi-core and multi-node composite system where each node is described by its topology. The topology of the overall system emerges from the interconnections of the nodes. At the node level, communication between cores is through the TTNoC, and the TTOCN ensures communication between the nodes via a gateway within the nodes that provides a bridge for data sharing between the TTNoC and TTOCN.

Figure 1 illustrates two levels of HTP: the system and node levels. At the system level, four sets of four nodes $\{N_1 - N_4, N_5 - N_8, N_9 - N_{12}, N_{13} - N_{16}\}$ are clustered around switches $\{S_1, S_2, S_3, S_4\}$, respectively, and the switches are connected in a ring. Each core, C_n or gateway, G_n, is interconnected at the node level through routers, R_n, in different topologies forming the TTNoC.

The HTP system level is a tuple represented as an undirected graph, $\mathbb{P}_s = \{V_s, L_s\}$ where $V_s = \{N \cup S\}$ is the set of nodes N and switches S. The time-triggered edges $L_s = \{(u, v)|u, v \in V_s\}$ represent the communication links between nodes and switches and are unordered pairs of vertices. In Fig. 1, the nodes are clustered around a switch, connected in a star topology, where switches are connected in a ring.

The node level is a tuple represented as an undirected graph, $\mathbb{P}_n = \{V_n, L_n\}$, where $V_n = \{C \cup R \cup G\}$ is the set of vertices containing cores C, routers R and a gateway G. The edges $L_n = \{(u, v)|u, v \in V_n\}$ are half-duplex links connecting unordered pairs of vertices. The delay factor, D_f of a given link, $l \in L_s$ and $l \in L_n$, is a positive weight describing the communication cost for message traffic through the link.

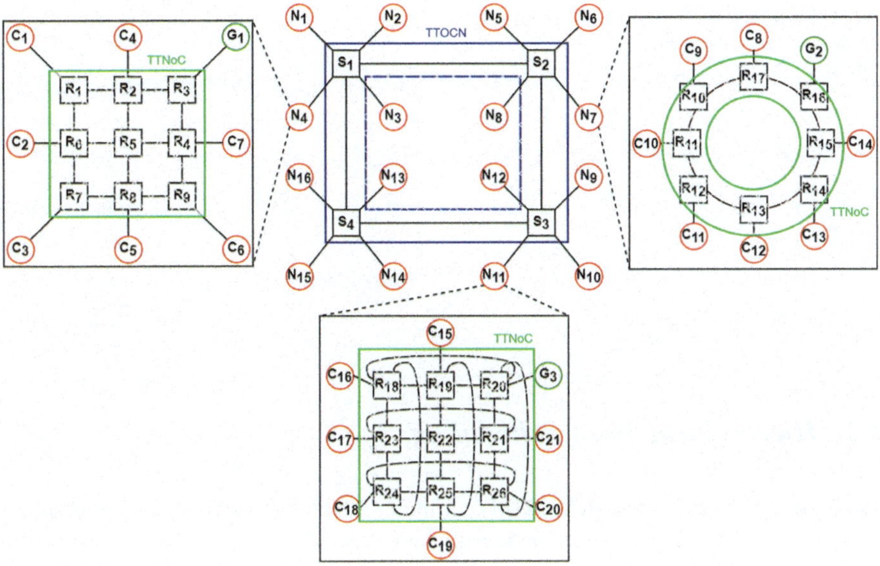

Fig. 1 Hierarchical NoC-based multi-core, multi-node system model

Each node contains a gateway, G, which facilitates data sharing between the TTNoC and TTOCN, represented as a bridge in the node that connects the two levels in the hierarchy. In addition, the TTOCN provides real-time guarantees for inter-level communication, ensuring the communication between the nodes or cores does not delay the execution of time-triggered schedules.

Dynamically adapting the hierarchical time-triggered system to system events to save energy or isolate faulty resources is possible by switching the system schedule. Each system schedule is a complete temporal and spatial allocation of tasks to system resources for a given system event. These time-triggered schedules ensure implicit synchronisation and timing constraints and avoid resource contention, eliminating dynamic resource arbitration [2]. In addition, the system event or context information is consistent across the system hierarchy and is used to switch the system schedule. We refer to the hierarchical time-triggered system capable of runtime adaptation through the switching of system schedules as the Hierarchical Adaptive Time-triggered Multi-core/Multi-node Architecture (HATMA). HATMA is extended from the Adaptive Time-triggered Multi-core Architecture (ATMA) [2] to allow for hierarchical adaptation of schedules.

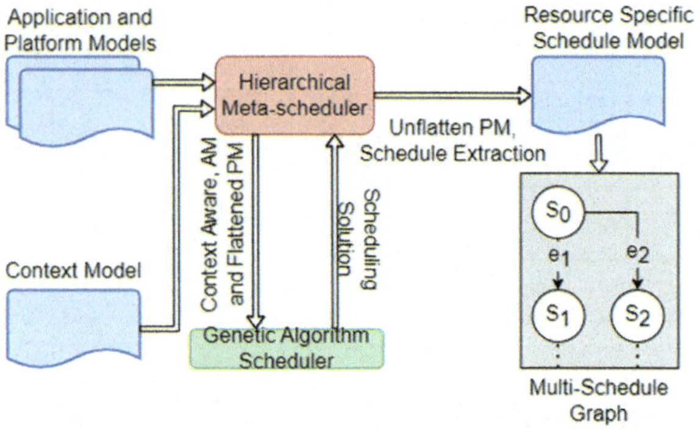

Fig. 2 Overview of the hierarchical meta-scheduler (HM)

3.2 *Hierarchical Meta-Scheduling*

The hierarchical meta-scheduler (HM) computes adapted static time-triggered schedules for hierarchical platforms considering hierarchical constraints for a given system context event. Based on the system context monitored for at runtime and specified at design time, adapted system schedules are computed at design time and switched to at runtime. The HM repeatedly constructs scheduling problems from the systems application model (AM) and platform model (PM) using the context model (CM), and a scheduler is invoked to find scheduling solutions to these problems, as illustrated in Fig. 2. Scheduling solutions are represented in a schedule model (SM) and stored in a multi-schedule graph (MSG). The MSG is a Directed Acyclic Graph (DAG) where the nodes are the schedules and edges the context event for which child schedules are computed from parent schedules. HATMA uses the MSG to orchestrate runtime adaptation for monitored context events.

3.2.1 Application Model (AM)

The runtime system's application is described in an Application Model (AM) as a DAG, $\mathbb{A} = \{J, M\}$, where a job $j \in J$ is a node representing the computational tasks of the application. Each job has a worst-case execution time $wcet$, deadline d and node allocation constraint can_run_on, which constrains the execution of a job to a node in the PM or a group of nodes. Messages $M = \{(j, k) | j, k \in J\}$ are directed edges between two jobs representing data exchange from a parent job to a child job and set a precedence constraint between the jobs. Each message is characterised by a sender b, receiver z and size s_z.

3.2.2 Platform Model (PM)

The Platform Model (PM) is a representation of HATMA and is an undirected graph $\mathbb{P} = \{\mathbb{P}_s \cup \mathbb{P}_n\}$, describing the hierarchical structure of the system.

3.2.3 Context Model (CM)

The context model (CM) is the set of runtime events $\mathbb{E}$ monitored at runtime for which the HM computes adapted system schedules. For example, the set of slack events $e_s \in \mathbb{E}$ is a tuple (j, e_t) and indicates that a job j finishes earlier e_t than its associated $wcet$. A failure event $e_f \in \mathbb{E}$ is a tuple (p, t) describing the failure of a hardware resource $p \in \mathbb{P}$ at a time t.

3.2.4 Schedule Model (SM)

The system scheduling solution is defined in the schedule model (SM), $\mathbb{S} = \{A_j, A_m\}$ as an allocation of the AM to the PM. A job allocation is a tuple $(j_i, c_i, t_i) \in A_j$, where a job $j \in J$ is spatially allocated to a core $c \in C$ and temporally allocated at time t_i. A message allocation is a tuple $(m_i, c_i, t_i, t_a, r_n, c_j) \in A_m$, where a message m_i is injected at a core $c \in C$ at a time t_i and travels across route r_n, arriving at time t_a at its destination c_j. A route r_n is an array of hops, and every hop is a tuple of a source node n_i, destination node n_j and link l_k such that $r_n = [(n_i, l_k, n_j)]$.

3.2.5 Multi-Schedule Graph (MSG)

The nodes of the Multi-Schedule Graph, $MSG = \{\mathbb{S}, \mathbb{E}\}$ is a time-triggered schedule $s \in S$, and the directed edge between two nodes is a context event $e \in E$. A child node is an adapted schedule of the parent schedule for which a context event is monitored at time t. Adaptation involves starting jobs earlier and reallocating jobs to other cores, message routes, or injection times. Figure 3 is an example MSG where S_n is an adapted schedule based on context events $e_{sn}, e_{fn} \in E$. In this example, during the execution of schedule S_0, the system schedule is switched to S_4 if e_{f1} is reported, where S_4 is a schedule adapted for e_{f1}. Schedule S_0 has no incoming edges and is thus the initial system schedule for which every execution of the system application begins at this node.

4 Hierarchical Scheduling

In order to simplify the scheduling process, we transform the PM into a flat single-level representation of the system, abstracting away the hierarchical structure. All

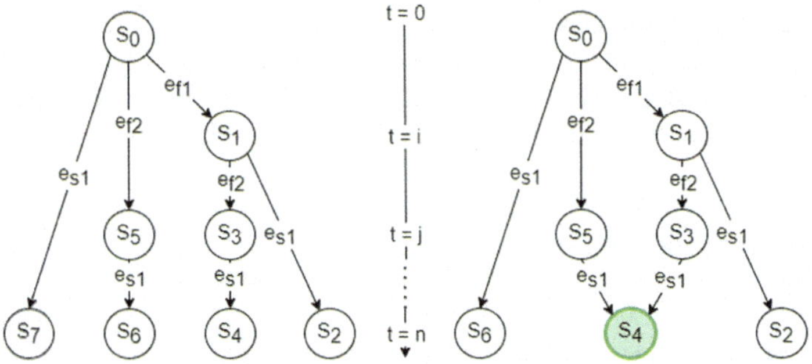

Fig. 3 Multi-schedule graph (MSG) with and without reconvergence of path (green)

vertices at the node-level $\mathbb{P}_n$ are enumerated for all nodes at the system-level $\mathbb{P}_s$, assigning unique identifiers to each core, router and gateway. The links between vertices are assigned a delay factor D_f, determining the communication cost of one hop across the link. The value of the delay factor is expressed as,

$$Df = h_{nt} * \frac{link\_speed}{on - chip\ link\_speed}, \tag{1}$$

where h_{nt} is the on-chip hop time between two vertices, $link\_speed$ is a physical quantity and is the communication speed of the link, and $on - chip\ link\_speed$ is the communication speed of the on-chip link.

4.1 Scheduler

We use a Genetic Algorithm (GA) to find scheduling solutions for HATMA from problems constructed by the HM. GAs are search algorithms that employ the principles of natural evolution to optimise a population of genomes [11]. First, the scheduling problem is mapped to the prototype genome such that every job in the AM is assigned a set of possible allocations to cores in the system. Then, an initial population of genomes is randomly generated from the prototype representing a solution to the scheduling problem.

This set of possible allocations can_run_on is mapped to a gene in the prototype genome. Each job is also assigned a gene, a positive weight range, and is used to determine the priority of job allocation. Each message is also assigned a range of route indexes. Message routes are precomputed for all possible routes between vertices in the platform model using a k-shortest path algorithm [12], where each route is assigned an index. Weights of messages are also assigned a gene used to determine the priority of message allocation.

Each genome in a population is evaluated based on a fitness function. The fitness of a genome is determined by its resulting schedule makespan. The GA evolves a population constantly evaluating the fitness of its genomes until no further better solution to the scheduling problem is found. The $wcet$ $\forall$ jobs $j \in J$ is used to allocate time slots of jobs in the resulting schedule as detailed in Algorithm 1. The genome with the least makespan is returned as the best solution to the scheduling problem.

```
function GenomeToSchedule(AM, PM, genome)
    initialize result schedule;
    initialize set of locks;
    initialize in-degree indegree;
    queue ← all jobs of AM that do not receive a message;
    while queue is not empty do
        x ← highest weight item in queue according to genome;
        if x is a job then
            AllocateJob(x, genome, locks, schedule);
            for message m sent by x do
                add m to queue;
            end
        end
        if x is a message then
            AllocateMessage(x, genome, locks, schedule);
            indegree (receiver(x)) ← decremented by 1;
            if indegree (receiver(x)) is 0 then
                add receiver(x) to queue;
            end
        end
    end
function AllocateJob(job, genome, locks, schedule)
    n ← node on which job is allocated according to g;
    t ← earliest time job can be allocated according to locks and received messages;
    locks ← new lock at t for duration wcet(job) on node n;
    schedule ← new job allocation for job at t for duration wcet(job) on node n;
function AllocateMessage(message, genome, locks, schedule)
    n1 ← node on which sender(message) is allocated;
    n2 ← node on which receiver(message) will be allocated;
    i ← route index from genome;
    r ← ith route from n1 to n2;
    t ← finish time of sender job;
    t ← max(t, earliest time message can be allocated according to locks);
    locks ← add cascading lock for every node on r;
    schedule ← new message allocation for message at t on route r;
```

Algorithm 1: Converting a genome into a schedule

```
function Metaschedule(MSG, CM, AM, PM, ParentNode)
    if CM is empty then
        return;
    end
    e ← earliest occurring event from CM;
    CM′ ← CM without e;
    Metaschedule(MSG, CM', AM, PM, ParentNode);
    (AM′, HPM′) ← apply e to (AM, PM);
    s ← ComputeSchedule(AM', PM');
    if s ∈ MSG then
        Merge node and path;
        return;
    end
    else
        add s to MSG as NewNode;
        add edge from ParentNode over e to NewNode in MSG;
        CM″ ← CM without e and events excluded by e;
        Metaschedule(MSG, CM", AM', PM', s);
    end
function ComputeMSG(AM, PM, CM)
    initialize empty MSG;
    s ← ComputeSchedule(AM, PM);
    add s to MSG as RootNode;
    Metaschedule(MSG, CM, AM, PM, RootNode);
    return MSG;
```

Algorithm 2: Meta-scheduler Algorithm

4.2 Hierarchical Meta-Scheduler (HM)

The HM is a recursive tool that creates scheduling problems for each context event in the CM. It calls the GA repeatedly to compute a scheduling solution represented in the SM and added to the Multi-Schedule Graph (MSG) as nodes and the context event as an edge from the parent node to the child node. The HM first generates a root node that indicates the absence of a context event.

For each generated node, an equivalence comparison is made with other nodes in the MSG. Two nodes $\mathbb{S}_i$ and $\mathbb{S}_j$ are returned as equivalent if,

$$\{\mathbb{S}_i(t) \cap \{A_j, A_m\} = \mathbb{S}_j(t) \cap \{A_j, A_m\}\} \cap \{\mathbb{E}_i(t) \cap \{e_s, e_f\} = \mathbb{E}_j(t) \cap \{e_s, e_f\}\}. \tag{2}$$

Nodes returned as having an equivalent job and message allocation are referred to as *deja vu* nodes. *deja vu* nodes are merged, and an edge is created between the parent node and the merged node. If returned as not having an equivalence, the node is added to the MSG and an edge is created. In each case, HM recursively calls itself with the created node and the updated CM without the previous event as detailed in Algorithm 2.

The HM uses the events in the CM to update either the PM in the case of failure events or the AM in the case of slack events. Applying context events to update the

AM or PM involves creating a new application (sAM) and platform model (sPM), a sub-problem of the parent application and platform models. Sub-problems are solved separately by the scheduler and reintegrated into the parent schedule. This approach enables the uniform handling of all context event types. In the case of slack events, the sPM represents the node for which the slack was reported. sAM represents jobs and messages within a reconvergence horizon, where an adaptation of the parent schedule is only permitted within a limited time interval after the reported slack event [13]. The reconvergence horizon is a limited period after the occurrence of the context event [2]. For failure events, the sPM contains all hardware resources without the failed resources, and the sAM contains all jobs and messages after the event time.

4.2.1 Handling Slack Events

To ensure that sAM is consistent with its parent AM, the allocation of jobs added to sAM is restricted to the set of cores associated with the node for which the slack event is reported. Jobs in sAM are constrained to a start and finish time within the reconvergence horizon,

$$t < [t_i, (t_i + wcet_j)] < (t + h), \tag{3}$$

where t_i is the start time of the job, h is the length of the reconvergence horizon, and t is the context report time. Jobs not added to sAM are locked, remaining unchanged from the parent schedule in the child schedule, as illustrated in Fig. 4 where rescheduled jobs are marked in green and fixed jobs in red. In Fig. 4, C_1 and C_2 are cores in a node for which a slack event is reported, and C_3 is a core in another node in the system. Jobs j_6, j_{10} and j_{11} satisfy the adaptation constraint and are rescheduled, while other jobs are either outside the horizon, cut across it or allocated to other nodes.

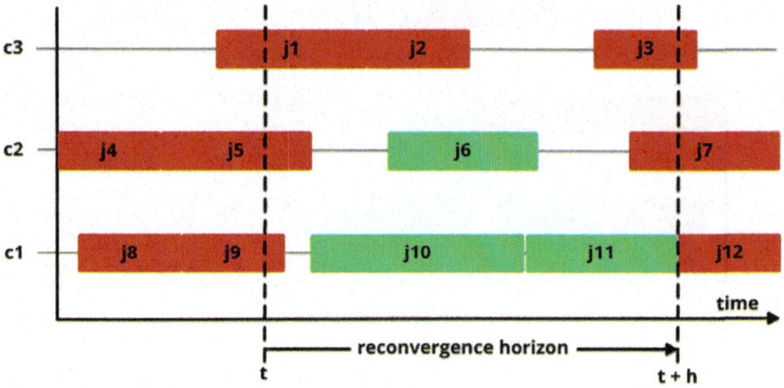

Fig. 4 Node level adaptable jobs due to a slack event

A message in the horizon is included in sAM if both its sender and receiver jobs are in the sAM. The message is fixed, and its receiver job is constrained to the allocated core but can be rescheduled later. The solution to the scheduling problem of sAM and sPM is combined with the parent schedule and the child schedule created. The child schedule differs from the parent schedule only within the node for which the slack event is reported and within the reconvergence horizon. We refer to this as a node-level schedule adaptation, and in this work, slack events are adapted to only node level.

4.2.2 Handling Failure Events

For failure events, sPM represents all hardware resources in the system, excluding the failed resource e_f such that $sPM = PM \setminus e_f$. In addition, the sAM is a representation of all jobs and messages in the parent schedule during and after the time of the reported failure event. For all jobs in sAM, *can_run_on* is updated with the available cores while message routes are recomputed. Jobs after the event time are then rescheduled. Jobs and messages before the event are locked as they were in the past and remain unchanged in the child schedule, illustrated in Fig. 5.

In Fig. 5, n_1, n_2 and n_3 are nodes in the system. A failure event e_f in node n_1 is reported at time t when jobs in green are rescheduled and those in red are fixed. Jobs j_8 and j_4 are in the past, while j_5 and j_1 are executed on different nodes. Due to the rescheduled messages exchanged between the nodes, all other jobs on n_1 are rescheduled on available resources and those on n_2 and n_3. Similar to handling slack events, the solution to the subproblem is integrated with the parent schedule, resulting in the child schedule. We refer to the process of handling failure events as a system-level schedule adaptation, and in this work, failure events are adapted to at the system level.

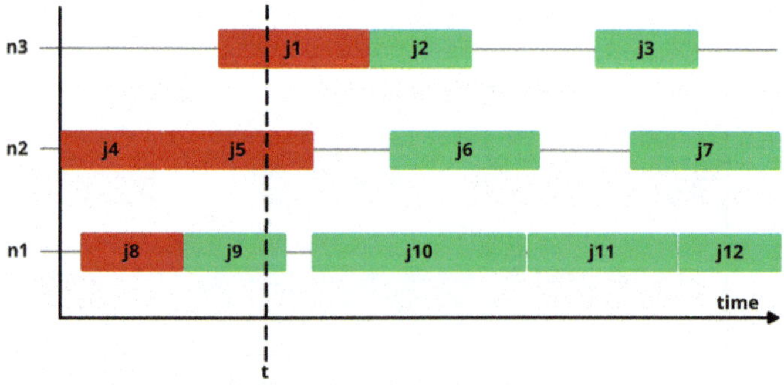

Fig. 5 System level adaptable jobs due to a failure event

4.3 *Platform Model Unflattening and Resource Schedule Extraction*

The flat model of the platform is reconstructed into the hierarchical platform, and resource-specific schedules are extracted. All vertices and links are grouped for each node in the system based on their unique identifiers and links as they align with the communication patterns and dependencies of the original hierarchical platform. A schedule table is created for each node and subdivided by cores, routers and gateway. For each job allocated to a core, its start and end times are added to the schedule table. The injection time and route for messages across routers are added to the schedule table of routers. For gateway schedule tables, arrival time, injection time and routing are extracted. The process of extracting the schedule table of the off-chip switches is similar to the gateways. Finally, each schedule table is validated for synchronization, collision avoidance and real-time requirements.

5 Evaluation and Results

The AM contains 100 jobs and 120 messages where the *wcet* of jobs range from 8–20 ns and deadlines of 100 ns. Messages are between 3 and 7 packets. The PM is a hierarchy of two nodes interconnected through a switch. The link between nodes and the switch has a delay factor of 5 compared with the on-chip communication speed. Each node in the system contains three cores and a gateway interconnected through a 3X3 NoC. Experiments are carried out on the OMNI Cluster of the University of Siegen [14].

5.1 *Results for Impact of GA Hyperparameters on Schedule Makespan*

Hyperparameters of the genetic algorithm (GA) are predefined settings responsible for achieving optimal or near-optimal performance and convergence in solving complex optimisation problems. These hyperparameters, such as the population size, crossover and mutation probability rates, genome selection mechanisms, and termination criteria, influence the exploration and exploitation capabilities of the algorithm, requiring careful tuning.

The results of the GA are a distribution of values not as a function of its inputs and hyperparameters, for which traditional methods for optimising black box functions like Bayesian optimisation and grid search are not ideal. In this work, we evaluate different hyperparameter values by establishing a baseline of values. Next, hyperparameter values are varied, where changes in the distribution of results are plotted against the baseline. In [15], mutation probabilities in the range of 0.001 to 0.01,

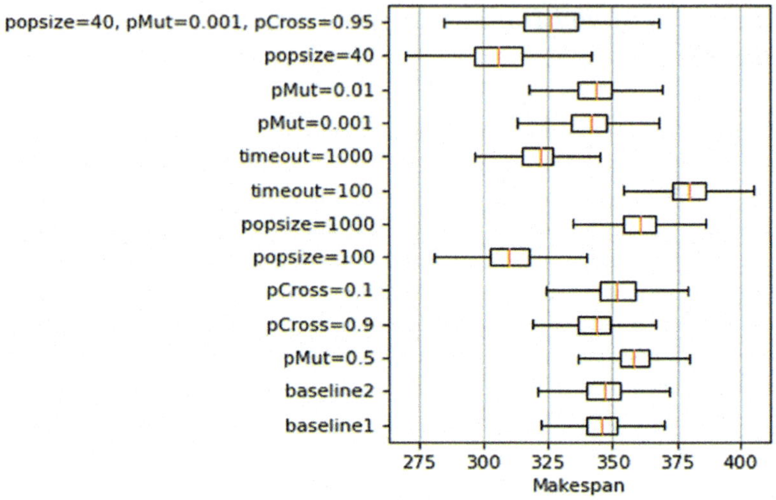

Fig. 6 Schedule makespans for GA hyperparameter selection

population sizes varying from 2 to 4000 and crossover probabilities from 0 to 0.95 are reported as optimal. For our baseline, a mutation probability of 0.004, a population size of 500 and a crossover rate of 0.5 are used, and the GA terminated after 500 ms for every schedule generation. Consistency in the result distribution is ensured by executing the experiment 10000 times. Figure 6 is a box plot of the results of various GA hyperparameters. $baseline_1$ and $baseline_2$ are the resulting makespans of the schedules generated by the GA for the hyperparameters, AM and PM described previously. They serve as reference points for which other parameters are compared. $pMut$ results from varying mutation rates. For example, a mutation rate of 0.5 yielded a makespan of 360. In contrast, for reduced mutation rates of 0.01 and 0.001, the makespan improved to 340 and 338, respectively, indicating lower mutation rates leading to better performance within the scope of minimising the makespan.

The impact of varying crossover rates, denoted as $pCross$, shows a crossover rate of 0.9, resulting in a makespan of 338. In contrast, a lower crossover rate of 0.1 increases the makespan to 352, suggesting a higher crossover rate contributes to improved performance, as the reduced makespan indicates. On the other hand, a population size "popsize" of 100 achieved a makespan of 315, while a smaller population size of 40 yielded a reduced makespan of 310. Conversely, a larger population size of 1000 resulted in a higher makespan of 360. These findings indicate that smaller population sizes tend to produce better results due to the GA further optimisation of the genomes in a smaller population within the given timeout value as opposed to a larger population. Furthermore, a larger population leads to more diversity, which lowers the risk of solutions that are local maxima of the solution space.

Finally, a longer timeout $timeout$ improves performance, as reflected by the lower makespan. Lowering the timeout to 100 ms leads to the worst performance of all runs. Larger timeouts improve makespans over the baseline, but low population size leads to better results. In this work, the scheduler is repeatedly called thousands of times to generate schedules to fill an MSG, rendering larger timeouts infeasible. Therefore, the most significant factor contributing to the overall performance of the HM is the number of context events for which adapted schedules are computed, resulting in an exponential increase in the size of the MSG.

5.2 Superset Reconvergence in the MSG

Reconvergence allows merging paths in the MSG which reduces the MSG size. A path in the MSG refers to a sequence of nodes connected by directed edges, where each edge in the sequence represents a context event. A path traverses multiple vertices and edges in the MSG and contains no cycles. For paths in the MSG to be merged, two nodes $n1, n2 \in MSG$ must have an identical sequence of context events in their context model ordered in a calender such that,

$$CM_1 = (e_{i,t}, e_{i,(t+k)}, ..., e_{i,(t+n)}) \supseteq CM_2 = (e_1, e_2, ..., e_n), \tag{4}$$

where $e_{i,t} < e_{i,(t+k)} < ... < e_{i,(t+n)}$ are context events reported in increasing order of discrete times, and,

$$n_{1,ti} = n_{2,tk} \forall t_i \geq t_k, \tag{5}$$

where t_i and t_k are the reported times of the incoming edges of n_1 and n_2, respectively. Figure 7 shows the results of reconvergence for a context model of 6, 10 and 14 context events, reconvergence horizon size of 20 ns and 50 ns and the MSG computation times, using the AM and PM described earlier.

The results indicate compression of the MSG size compared to MSG sizes without reconvergence. For a CM of 6 events, this approach of merging paths results in a 65% average reduction in the MSG size for a reconvergence horizon of 20 ns and 46% average reduction for reconvergence horizon of 50 ns. Similarly, for 10 context events, the average MSG size reduction is 82% for a reconvergence horizon of 20 ns and 47% for a reconvergence horizon of 50 ns. A smaller horizon leads to less adaptable jobs within the horizon and a higher probability of identical sequences of context events, increasing the reconvergence rate as observed from the results of the reduction ratio of horizon sizes of 20 ns and 50 ns. There is also a direct relationship between the MSG size and the computation time of the MSG.

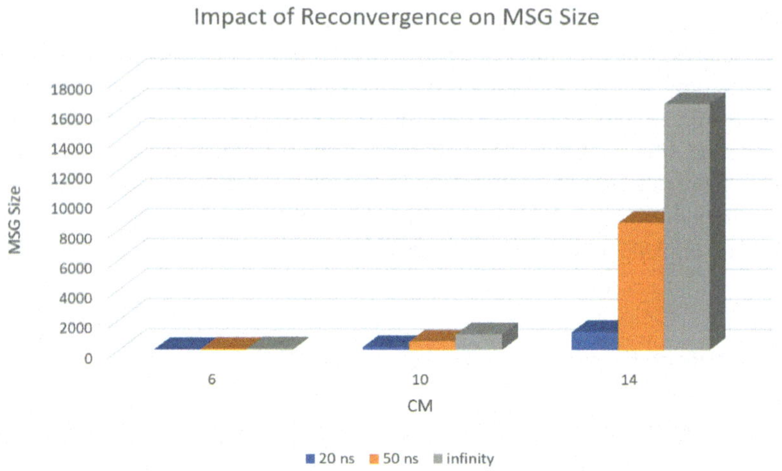

Fig. 7 MSG size reduction as a result of reconvergence of paths

6 Conclusion

This work introduces the Hierarchical Meta-scheduler (HM) as an offline tool to compute adapted hierarchical schedules for multi-level runtime adaptation in distributed Time-triggered (TT) systems. The hierarchical platform is simplified into a flat system where constraints such as the system's architectural topology and urgency of context events are considered in adapting the system schedules at different levels in the system hierarchy. Jobs and messages are adapted at the node level for slack events and system-wide for failure events. Schedules are adapted to minimise the makespan, and the number of generated schedules is also minimised through an equivalence comparison to identify *deja vu* nodes. *deja vu* nodes are merged in the multi-schedule graph (MSG) through a superset reconvergence technique to compress the MSG size, reducing the runtime memory needed to store the adapted schedules. Resource-specific schedule tables are extracted and validated for synchronisation, collision avoidance and real-time requirements for each schedule in the MSG.

Acknowledgements This work has received funding from the ECSEL Joint Undertaking(JU) under grant agreement No 877056. The JU receives support from the European Union's Horizon 2020 research and innovation programme and Spain, Italy, Austria, Germany, France, Finland, Switzerland.

References

1. Buttazzo, G.C. (2011). *Hard real-time computing systems: Predictable scheduling algorithms and applications* (3rd ed.). Springer Publishing Company.
2. Obermaisser, R., Ahmadian, H., Maleki A., Bebaway, Y., Lenz, A., Sorkhpour, B. (2019). Adaptive time-triggered multi-core architecture. *Designs, 3*.
3. Sorkhpour, B., Murshed, A., Obermaisser, R. (2017). Meta-scheduling techniques for energy-efficient robust and adaptive time-triggered systems. In *IEEE 4th International Conference on Knowledge-Based Engineering and Innovation (KBEI)* (pp. 0143–0150). Tehran, Iran.
4. Fohler, G., Gala, G., Pérez, D. G., Pagetti, C. (2018). Evaluation of DREAMS resource management solutions on a mixed-critical demonstrator. In *9th European Congress on Embedded Real Time Software and Systems (ERTS)*. Toulouse, France.
5. Oliver, R. S., & Craciunas, S. S. (2016). Hierarchical scheduling over off- and on-chip deterministic networks. *SIGBED, 13*, 14–19.
6. Roy, D., Zhang, L., Chang, W., Goswami, D., Chakraborty, S. (2016). Multi-objective co-optimization of FlexRay-based distributed control systems. In *IEEE Real-Time and Embedded Technology and Applications Symposium (RTAS),* (pp. 1–12). Vienna, Austria.
7. Craciunas, S.S., Oliver, R.S., Chmelík, M., Steiner, S. (2016). Scheduling real-time communication in IEEE 802.1Qbv time sensitive networks. In *24th International Conference on Real-Time Networks and Systems (RTNS '16)* (pp. 183–192). New York, USA.
8. Derasevic, S., Proenza, J., Barranco, M. (2014). Using FTT-ethernet for the coordinated dispatching of tasks and messages for node replication. In *IEEE Emerging Technology and Factory Automation (ETFA)* (pp. 1–4). Barcelona, Spain.
9. Wijiutomo, C., Trilaksono, B., Kistijantoro, A. (2018). Fault tolerant dynamic scheduling on real time hierarchical system: Proposals for fault tolerant mechanism on safety-critical system. *IJET, 7*.
10. Muoka, P., Umuomo, O., Onwuchekwa, D., & Obermaisser, R. (2023). Adaptation for energy saving in time-triggered systems using meta-scheduling with sample points. In S. Henkler, M. Kreutz, M. A. Wehrmeister, M. Götz, & A. Rettberg (Eds.), *Designing Modern Embedded Systems: Software, Hardware, and Applications*. Cham: Springer.
11. Yan, X. (2021). *Nature-inspired optimization algorithms* (2nd ed.). London: Academic Press.
12. Wang, Y., Liu, Q., Ren, H., Ma, X., Liu, L., Wang, W., Zhang, J. (2018). Optimizing multi-criteria k-shortest paths in graph by a natural routing genotype-based genetic algorithm. In *13th IEEE Conference on Industrial Electronics and Applications (ICIEA)* (pp. 341–345). Wuhan, China.
13. Muoka, P., Onwuchekwa, D., Obermaisser, R. (2022). Adaptive scheduling for time-triggered network-on-chip-based multi-core architecture using genetic algorithm. *Electronics, 11*.
14. Steiner, J. (2023). *OMNI-system information*. Available via DIALOG. https://cluster.uni-siegen.de/omni/system-information/?lang=en. Cited 5 June 2023.
15. Hassanat, A., Almohammadi, K., Alkafaween, E., Abunawas, E., Hammouri, A., & Prasath, V. S. (2019). Choosing mutation and crossover Ratios for genetic algorithms-a review with a new dynamic approach. *Information, 10*, 390.

Cognitive Controllable Local System Improving Blackstart Resilience in Smart Distribution Grids

Kai Daniel, Daniel Henn, and Daniel Kurzer

Abstract Blackouts are unlikely to occur in most countries of the world. For such kind of critical grid situations both transmission and distribution system operators have defined a very detailed processes to overcome a blackout and manage a blackstart. By means of pumped-storage power plants voltage and frequency recovery are initiated before conventional power plants working with flywheel mass can be switched to the transmission grid. This simple but effective procedure is reliably working for traditional electricity grids with minor decentralized feed-in of renewable energy resources. However, with the energy turnaround in Germany, the percentage of solar and wind energy is constantly increasing while conventional power plants have a minor part with respect to the connected power plant capacity. By 2045 more than 80% of the power mix will be contributed by renewables. However, solar and wind energy are volatile and hence not reliably available to manage the European system frequency of 50 Hz. Since solar inverters mostly have no off-grid capability and wind energy is not available anytime, a blackstart can fail, if too much load depresses the grid frequency. To overcome this deficiency, this article proposes a central and a decentral approach. Both novel algorithms improve the blackstart capability of electricity grids after a blackout by measuring the grid frequency and voltage as switching conditions for decentral controllable local systems like heat pumps, night heatings and charging points of electric vehicles.

Keywords Blackstart · Blackout · Energy turnaround · Controllable local system (CLS) · Coordination function (COF) · Distribution grid · Transmission grid · Resilience · Smart grid · Off-grid functions · Decentral energy management system · Power system restoration

---

K. Daniel (✉) · D. Henn · D. Kurzer
Chair for Simulation Techniques and Scientific Computing, University Siegen, Siegen, Germany
e-mail: kai.daniel@uni-siegen.de

M.-R. Alam and M. Fathi (eds.), *Integrated Systems: Data Driven Engineering*,
https://doi.org/10.1007/978-3-031-53652-6_8

1 Introduction

Blackouts–power supply breakdowns or large-scale power failures–in the electricity grid are unlikely to occur in most countries of the world. However, in times of energy and gas shortage a blackout becomes more likely [1], impacting not only private households, but also critical infrastructure [2]. Recent developments due to the energy turnaround need to be considered in order to prevent a blackout from happening or to enable for a proper blackstart in case of a crisis. Against that background pumped-storage power plants (PSPP) have a quite high impact since elevation energy of water of PSPPs can not be exploited repeatedly for blackstart. Thus, the reliability and resilience of interdependent grid entities needs to be ensured.

1.1 Demand for Improving Resilience in Electricity Grids

Due to climate change, the energy mix is globally targeting for an energy mix, relying on a high share of volatile renewable energy sources (RES). In order to make the quantitative structure of the power mix and the resulting demand for more resilient power sources more transparent, the German power mix will be explained subsequently as a prominent example.

In Germany the mean power outage in 2021 was 12.7 min per household [2] but a large-scale blackout never occurred. However, the German power mix of today/the future is defined by political goals, such as nuclear phase-out and climate neutrality, facing a complete turnaround and increasing the probability of a blackout due to congestion or overload situations in the grid structure. This entails the need to rethink grid expansion and to control the connected loads and feeders.

In 2023 the last German nuclear power plant was taken off grid, concluding the phase-out stated in Atomic Energy Act (AtG). The present base load power is mainly delivered by coal (lignite and hard coal), with an installed power of 37.8 GW or 16.1% in the German grid system [3, 4]. But with the necessity to reduce greenhouse gases (GHG) the German parliament passed the Coal Phase-Out Act in 2020, ending coal power production by 2038 at the latest [5].

Today RES are more and more utilized in power production–thermal power plants incinerating biomass or using geothermal heat, hydropower, wind power on- and offshore or photovoltaics. The main focus is on wind energy with an installed power of 67.4 GW and photovoltaics with 72.5 GW [4], disintegrating the previously existing top-down operation structure of the grid.

Offshore wind energy has to be transported a long distance, breaking the structure of locally distributed centralized power plants and bringing DC back to power transmission in a high voltage range, where onshore wind power is locally dispersed and feeding into high, medium and low voltage grids. Unidirectional power supply in the electricity grid is more and more evolving into a bidirectionally power flow over all voltage levels, making curtailment due to grid congestions necessary.

"The energy sector, currently causing more than 80% of total emissions in Germany, has a key role to play for reducing greenhouse gas emissions (GHG emissions). Electricity supply is responsible for about 40% of the energy sector's CO_2 emissions" [6]. The German Government gave out the target for the energy sector to be completely sustainable by 2050, interim targets being 35% RES by 2020 and 65% RES by 2030. This correlates with a reduction of 40% GHG by 2020 and 80–95% by 2050. Opposed to this, there are other political goals like 500,000 new electrical heat generators per year–summing up to 3 million by 2030–and 15 million electrical cars by 2030, increasing the energy demand further.

Studies [7, 8] show that, aside from technical solutions, public acceptance is key to sufficient system change, depending on the changeover of private homes and the grid expansion to carry high loads. Predictions place the installed power of RES needed to reach sustainability by 2050 at 700 to 962 GW in Germany alone and the power demand at around 1450 TWh, going up from now almost 550 TWh [8].

Figure 1 shows the predicted installed power in Germany from 2025 to 2045, based on the premise of 95% GHG reduction by 2045.

Depicted are four scenarios; the first (Reference) focuses on cost optimal transformation of the energy system without any distinct behavior of the population. The other three scenarios are based on the first, but describe different behaviors. The

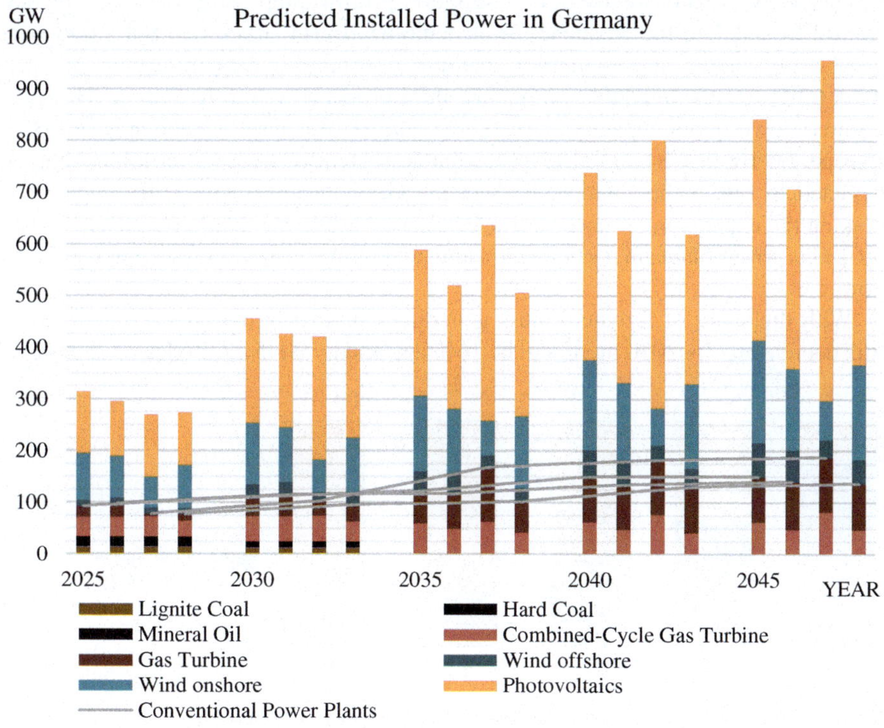

Fig. 1 Predicted installed power in Germany

second scenario (Insistence) takes a resistance against new technology in the private area, such as electrical heat pumps or electrical cars, into consideration, whereas the third (Inacceptance) assumes less acceptance of big infrastructural measures, such as wind parks or overhead cables for trucks on highways. Finally the fourth scenario (Sufficiency) anticipates a complete change in values, decreasing energy demand drastically by 45%.

All four scenarios try to predict the German power mix in a few decades, getting more and more blurry with the range of prediction and highly depending on behavior and acceptance of the population.

But in all scenarios the share of RES, mainly consisting of wind and solar power, is assumed to be around 80% of the overall power mix by 2045, being the main source power for the energy transition in Germany. Premise of the steady expansion being an intelligent and reasonable use of the volatile energy, making use of storage systems such as batteries and pumped storage power plants (PSPPs) [8, 9].

Note a big share of the installed power in all four simulated scenarios goes to gas turbines and combined-cycle gas turbines–around 20% or 138 to 188 GW. With the volatility of solar and wind energy, the option of flexibility from gas power plants is needed. To reduce the emission of GHG, they can be fueled with synthetic methane and hydrogen. The expansion of gas power can be scaled down with a higher integration of the German into the European grid [8].

Generally, RES currently do not support "gridforming" with power–frequency-control to stabilize the frequency in decentralized grid structures: Solar and wind energy are volatile and hence not available constantly ('dark doldrums' or 'dark wind lull') reliable available to manage the European system frequency. However, ensured power plant capacity is crucial to stabilize the system frequency (50 Hz in Europe, 60 Hz in US) and also for a reliable system recovery after a blackout. As depicted in Fig. 2 the share of RES in the German power mix will steadily rise, making the expansion of gas power necessary to deliver a reliable frequency control [10]. Gas power plants in particular are able to provide the required secondary control power.

With the share of flywheel mass generators supplying that power in the power mix decreasing–by 2030 there will be 16% less and by 2045 22.5% less than today–a

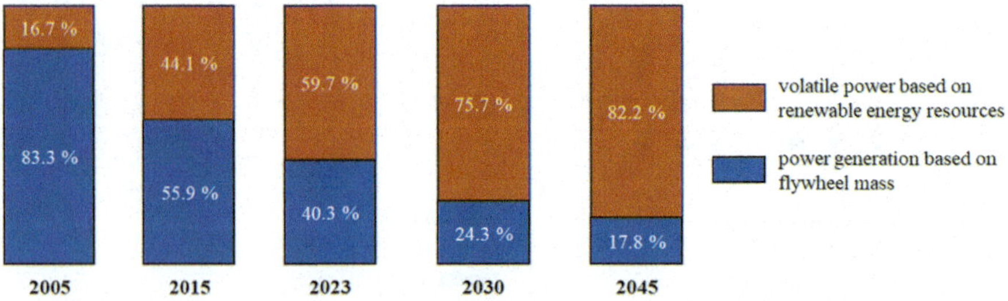

Fig. 2 Devolution of ensured power plant capacity (reference scenario)

compensation by agent-based algorithms in low voltage grid areas is necessary and useful.

1.2 State of the Art: Blackstart Procedure

Thermal power plants, such as nuclear, lignite or combined heat and power plants, cannot be started without an external power source: A large number of pumps, turbines and fuel production systems need to be supplied with external electricity before the electricity production of the power plant can start. This is challenging in case of a blackout situation:

In case of a widespread blackout thermal power plants must be bridged with PSPPs for voltage and frequency recovery. In a first step the grid operator ensures the communication capability between the Network control center (NCC) and the substation in the electricity grid (cf. (I) Fig. 3). Afterwards, a grid segmentation to be prepared for island operation of the established grid segments is conducted (II). This enables for successive load increasement and frequency synchronization in the subsequent step. The PSPP comes from island operation, is turned on, i.e. "water runs down the hill" (III) and then synchronized with the power grid segment, where also thermal power plants are connected to and started up (IV). The feed-in and recovery process begins. The PSPP is not only connected to a single thermal power plant but also to consumers depending on the grid structure and geographical location of the PSPP. Hence, it is crucial to successively connect the consumers to the electricity grid in order adjust the actual grid frequency to a small target frequency band, which is 60 Hz in the US and 50 Hz in Europe (V). Note, that electrical load leads to lower frequencies, since flywheel mass of thermal power plants are affected by inertia. Thus, from load perspective feed-in of renewable energy resources (RER) increases the grid frequency in general–if no further control takes place. If two or more grid segments are running, they are (semi-manually) synchronized on the same frequency and phase and then connected to each other (VI). Switch-disconnectors and circuit breakers, which can be remotely switched, are (in most electricity grids) usually available in (extra) large voltage grids, i.e. >110 kV. As shown in Fig. 3 steps (III) and (IV) are crucial, since it is not possible to exploit the elevation energy of water of PSPPs repeatedly for a blackstart.

1.3 Evaluation of Latest Research on Decentral Grid Cognition

Possible power system restoration dynamics issues are presented in [1], However, the use of decentral energy management and controllable local systems during a blackstart are not considered. An inverter-based resource (IBR) driven black-start

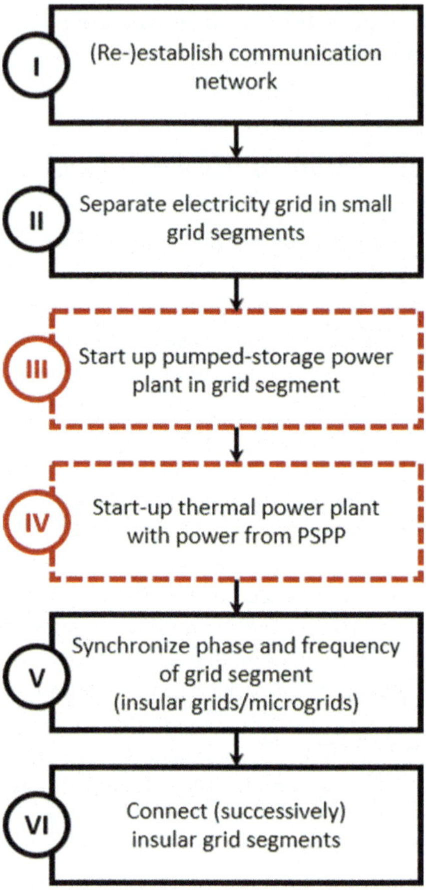

Fig. 3 Blackstart process schema

of electric grids is presented in [11, 12], which shows, that it is in general possible to perform a power system restoration with decentral located inverters with off-grid function. This can be combined with micro segmentation of the electricity grid as proposed in [13, 14]. Smart circuit breakers are introduced in [12], which integrate autonomous functions based on a phased-locked-loop. This algorithm enables for decentralized phase synchronization within a highly partitioned power system. However, algorithms to overcome overload and peak situations during a blackstart by using frequency measurements and backoff intervals are not yet included.

2 Coordination Function and Communication Architecture

To maintain frequency stability when connecting individual grid areas to the already established microgrid during a blackstart, intelligent coordination of non-frequency-generating generators is necessary, compensating for the low share of thermal power plants and thus a reduced amount of rotating mass. For this purpose, a system architecture known from the field of load coordination is proposed to avoid overload situations in the transmission and distribution grid [15]. In the event of a blackout in a grid with many decentralized power generation facilities, it is crucial for various stakeholders such as transmission grid operators, distribution grid operators, and power plant operators to have coordinated and unified communication channels to include producers and consumers that have not been considered in black start processes before. To address different interests and information density requirements, it is recommended to establish a coordination and measurement infrastructure [16–18]. Such an arrangement is illustrated in Fig. 4. During a black start, the higher-level global Coordination Function (COF) is responsible for coordinating the connection of the underlying grid areas and establishing an appropriate order depending on the situation [19]. The subordinate COFs then autonomously decide how to proceed with the grid areas under their administration–High Voltage (HV)/Medium Voltage (MV)/Low Voltage (LV)–to fulfill the requirements set by the higher-level entity. These requirements may include connecting the grid area, increasing or reducing net injection, or increasing or reducing net consumption.

Despite the different information bases, the task of the COF is similar across different hierarchical levels. The structure and functioning are explained in Fig. 5. The

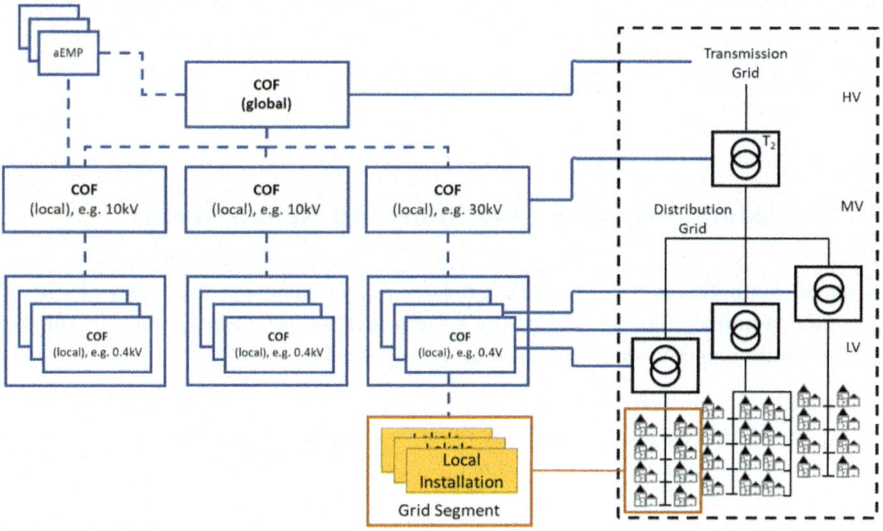

Fig. 4 Voltage levels cascaded structure of the COF

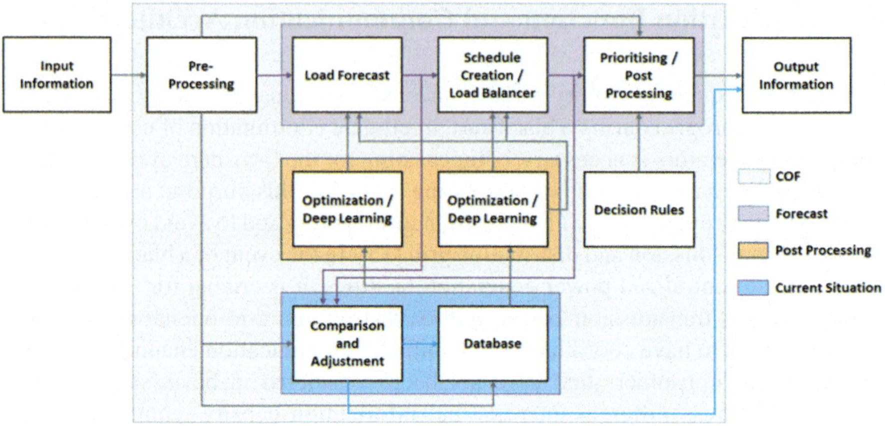

Fig. 5 Architecture of the COF

appropriate actions are taken, after deciding whether the data should be sent as direct commands to the next level or processed beforehand. In the load forecast, a prediction is made regarding the expected impact of individual regions on the grid. Historical data and current information such as weather data and time of day are relevant to obtain realistic expected values [20]. Based on these forecasts, a schedule is created that ensures a balanced and grid-supportive connection of the individual grid areas and the optimal utilization of flexibility [21, 22]. This data is then communicated via the output unit. The effectiveness of the schedules and the accuracy of the forecast are continuously monitored to enable immediate and targeted actions in case of misjudgment. During a blackstart, the limitation of power consumption (LPC) and limitation of power production (LPP) are relevant for regulation; if the time for connecting the grid areas can be estimated well, using a power envelope (POEN) is also conceivable.

3 Blackstart Boundary Conditions and Requirements

A major requirement for the blackstart is to recover voltage and frequency in the presence of fault conditions. Following system-level requirements need to be fulfilled for a reliable power system restoration:

- Amount of gridforming (decentral) inverters with decentral synchronization capability for phase and frequency must be significantly increased.
- Start-up of power in a microgrid needs to be limited in order to mainly transfer power from the PSPP to the connected thermal power plant.
- Power peaks must be avoided to prevent from undesired overload situations, which trigger circuit breakers or protection circuits respectively.

- Consumer load must be switched to the separated microgrid as slow and as iteratively as possible
- In order to ensure a stable grid frequency, a minimum of power plants working with flywheel mass must be available during the transition period, e.g. thermal backup power plants.

3.1 Decentral Algorithm for Gridforming Controllable Local Systems

Since the status of the electricity grid is unknown and communication between the devices might be limited after a blackout, autonomous behavior on demand side can mitigate fault risk like overload situations, which might be caused by simultaneous return peaks of controllable local systems (CLS) such as heat pumps, electric vehicles, night storages and other major consumers in the separated microgrids. Based on the introduction of advanced metering devices, e.g. smart meter gateways, CLS devices on demand side have the ability to for "grid forming", i.e. providing/influencing frequency, voltage and reactive power.

The proposed decentral and agent-based algorithm is based on a context sensitive autonomous agent [23] implemented on a CLS device at demand side, which measures the system frequency and voltage in order to determine its own decision for connecting to the microgrid. The algorithm as shown in Fig. 6 relies on two switching conditions: In a first step the grid status (frequency, voltage) is measured on the CLS device in order to determine its own switching target: Depending on frequency and voltage.

(a) Controllable load or
(b) Controllable feeders like solar and wind power

are switched on/off. In cases of under-frequency feeders are switched to the microgrid and CLS are switched off. Vice versa in cases of over-frequency consumer load is switched on and feed-in is switched off in order to support the rotation of fly-wheel

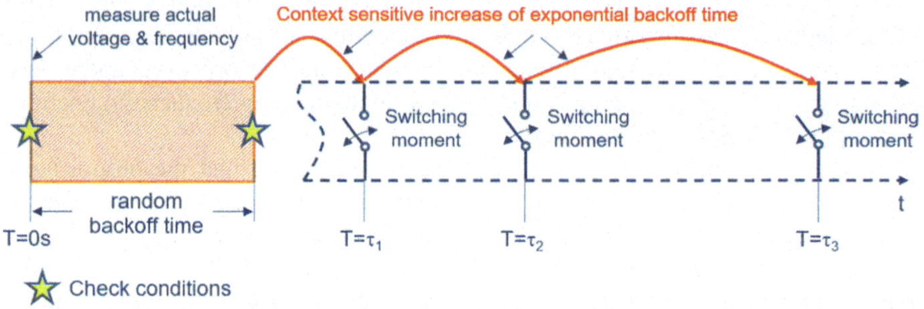

Fig. 6 Random backoff for autonomous agents on CLS devices prevent from peak

mass in the connected thermal power plant. If the switch condition is not fulfilled, a backoff timer is starting again in order to distribute the load in the grid over time.

After the end of the backoff timer, the switching condition is again reviewed. This process is repeated until all CLS devices are connected to the system. On this basis the power–frequency-control can be supported by decentral CLS on low voltage level. However, a full system recovery needs additional frequency, analysis and synchronization functionality.

3.2 Coordination Function and Communication Architecture

The proposed approach of autonomous agents functions as a supportive tool, even in cases of restricted communication. Additionally, this approach can also be advantageous in a black start scenario where communication pathways have been restored.

The increasing number of small generation units and flexible consumers in private households elevates their importance in ensuring supply security. In the distribution grid, the flexibilization of high-capacity consumer devices aims to prevent overload during periods of high simultaneity. It also mitigates overloads caused by generation, by reducing active power injection. During a blackstart, these functions can be coordinated to support the restart process.

The Energy Management System (EMS) manages consumption and production within the available flexibilities in a household. In the event of a voltage or communication interruption, it enters a failsafe mode. It exits this mode either after 48 h or once communication with the control center (COF) has been restored and current restrictions or plans have been transmitted.

In the case of a blackstart and the connection of a sub-region, all available flexibilities are activated in the failsafe mode, reducing consumer power consumption and limiting power injection by generators. Once communication with the COF is established, the distribution grid operator can control the flexibilities in a grid-supportive manner. This enables a proactive response to the connection of regions. If a net power generation is expected for the region being connected, it can be compensated by reducing energy generation or increasing consumption. If a net power consumption is expected, it can be supported by increasing generation and reducing consumption in the already connected regions. This equalizes the connection processes and reduces the impact of frequency changes due to load fluctuations in frequency-generating power plants, despite reduced inertia.

As described in the Introduction chapter, it is expected that the proportion of generators dependent on environmental conditions such as time of day and weather will increase in the future. This also means that the optimal blackstart strategy can be adjusted based on environmental conditions. To determine the best blackstart plan, the Coordination Function can provide information on consumption and production forecasts dependent on time of day and weather, as well as flexibility availability, and make suggestions for the connection order and utilization of flexibility.

4 Conclusions and Future Work

In this paper the concept for an autonomous agent for controllable local systems (CLS) was presented, which in contrast to other research work is focusing on stabilizing the rotation of flywheel mass in thermal power plants, which are mainly forming the system frequency within an electricity grid. Additionally, critical power peaks can be avoided based on decentralized random backoff interval, which leads to a distribution of CLS devices.

The authors claim, that political boundary conditions need to be changed in a way, that reliable balancing power is sufficiently available. Furthermore it must be mandatory, that solar inverters are gridforming and able to operate in a decentral synchronized insular mode.

Furthermore, the possible role of a Coordination Function was introduced, which can contribute to compensate the lack of rotational inertia during a blackstart through an interface to central and decentralized control mechanisms. In the future, a simulation and an experimental laboratory environment consisting of an agents-based controllable local system (CLS) and novel Coordination Function (COF) will be implemented to validate the proposed algorithm.

References

1. Adibi, M. M., & Martins, N. (2008). Power system restoration dynamics issues. In *2008 IEEE power and energy society general meeting-conversion and delivery of electrical energy in the 21st century* (pp. 1–8). Pittsburgh. https://doi.org/10.1109/PES.2008.4596495
2. Bundesamt für Bevölkerungsschutz und Katastrophenhilfe. https://www.bbk.bund.de/DE/Themen/Kritische-Infrastrukturen/KRITIS-Gefahrenlagen/Stromausfall/stromausfall_node. Stand 24.07.2023.
3. Bundesamt für Justiz, Gesetz über die friedliche Verwendung der Kernenergie und den Schutz gegen ihre Gefahren (Atomgesetz).
4. Bundesministerium für Wirtschaft und Energie, Statistisches Bundesamt, Arbeitsgruppe Erneuerbare Energien-Statistik (AGEE-Stat), ab 2011 Bundesnetzagentur (BNetzA).
5. Bundesamt für Justiz, Gesetz zur Reduzierung und zur Beendigung der Kohleverstromung.
6. Umweltbundesamt Germany, Energy target 2050: 100% renewable electricity supply, 2010.
7. eXtremOS, Summary Report, Modeling Kit and Scenarios for Pathways Towards a Climate Neutral Europe.
8. Fraunhofer ISE, Wege zu einem klimaneutralen Energiesystem, Die deutsche Energiewende im Kontext gesellschaftlicher Verhaltensweisen, Freiburg, Februar 2020.
9. Fraunhofer ISE, https://www.energy-charts.info/. Stand 18.07.2023.
10. Deutsche Energie-Agentur (dena): Bedarf und Erbringung von Momentanreserve im Jahr 2030. Berlin, January 2016.
11. Jain, H., Seo, G.-S., Lockhart, E., Gevorgian, V., & Kroposki, B. (2020). Blackstart of power grids with inverter-based resources. In *2020 IEEE power & energy society general meeting (PESGM)* (pp. 1–5). https://doi.org/10.1109/PESGM41954.2020.9281851
12. Banerjee, A., et al. (2022). Autonomous microgrid restoration using grid-forming inverters and smart circuit breakers. In *2022 IEEE power & energy society general meeting (PESGM)* (pp. 1–5). https://doi.org/10.1109/PESGM48719.2022.9916679

13. Demetriou, P., Asprou, M., & Kyriakides, E. (2019). A real-time controlled Islanding and restoration scheme based on estimated states. *IEEE Transactions on Power Systems, 34*(1), 606–615. https://doi.org/10.1109/TPWRS.2018.2866900
14. Jiang, Y., & Ortmeyer, T. H. (2021). Propagation-based network partitioning strategies for parallel power system restoration with variable renewable generation resources. *IEEE Access, 9*, 144965–144975. https://doi.org/10.1109/ACCESS.2021.3122799
15. Henn, D., Daniel, K., & Weber, T. (2023). Functional architecture for holistic grid and market oriented power management. In *2023 IEEE International Systems Conference (SysCon)* (pp. 1–6). https://doi.org/10.1109/SysCon53073.2023.10131226
16. Palaniappan, R., Bauernschmitt, B., Hilbrich, D., & Rehtanz, C. (2020). An intelligent measurement and control device for active distribution grids. In *2020 IEEE PES innovative smart grid technologies Europe (ISGT-Europe)* (pp. 975–979). https://doi.org/10.1109/ISGT-Europe47291.2020.9248862
17. Daniel, K. (2014). Verfahren zum Schalten von dezentralen Energiewandlern, patent *DE102014010019A1*.
18. Daniel, K., Englert, H., Kerber, G., & Kramny, J. (2014). *Das intelligente messsystem*. VDE Kongress, Smart Cities, Frankfurt/M.
19. Srikantha, P., & Kundur, D. (2019). Intelligent signal processing and coordination for the adaptive smart grid: An overview of data-driven grid management. *IEEE Signal Processing Magazine, 36*(3), 82–102.
20. Almalaq, A., & Edwards, G. (2017). A review of deep learning methods applied on load forecasting. In *2017 16th IEEE international conference on machine learning and applications (ICMLA)* (pp. 511–516). IEEE.
21. Zishan, A. A., Haji, M. M., & Ardakanian. (2020). Reputation-based fair power allocation to plug-in electric vehicles in the smart grid. In *2020 ACM/IEEE 11th international conference on cyber-physical systems (ICCPS)* (pp. 63–74). IEEE.
22. He, Q., Duan, J., Min, J., Hao, W., Ye, J., Xu, J., Wu, M., & Zhao, M. Task scheduling algorithm based on time and cost in smart grid cloud computing. In *2022 7th international conference on computer*.
23. Daniel, K. (2021). Regelvorrichtung für eine elektrische Anlage zum Wiederanfahren der elektrischen Anlage nach einem Stromausfall. patent *EP 2 899 828 B1*.

Real-Time Anomaly Detection in Connected Autonomous Vehicles: A Data-Driven Approach

Mahsa Jalali, Milad Moradi, and Mehdrad Saif

Abstract The impressive advancements in intelligent and interconnected transportation systems are largely attributed to the emergence of smart vehicles. However, the rise of vehicle connectivity, while offering clear benefits, also brings about significant safety issues that can jeopardize the lives of people. One of the challenging but essential tasks in this field is Anomaly Detection, Identification, and Recovery (ADIR) on connected autonomous vehicles (CAVs). The emergence of recent technologies that enable real-time and onboard processing of large data has made it possible to implement novel approaches in Anomaly Detection (AD) for complex systems. With the increased use of Artificial Intelligence (AI) in solving complex engineering problems in the face of anomaly detection, there has been significant increased use of these methods for systems that experience external disruptive intruding attacks. This chapter outlines some preliminary *experimental and proofs-of-concept* research work by using a deep autoencoder approach to verify the feasibility of AD task on a platoon of vehicles that are dynamically coupled through communication and feedback control moving in a rectilinear formation in a leader-follower architecture. The AD strategy was implemented on a set of Quanser Cars (QCars) within the Laboratory for Autonomous Resilient Systems (LARS) at the University of Windsor.

1 Introduction

The future of advanced transportation systems relies on CAV, which offer improved road safety, reliability, and comfort. Autonomous driving has greatly improved with advancements in AI and Internet of Things (IoT) technologies.

M. Jalali · M. Moradi · M. Saif (✉)
University of Windsor, 401 Sunset Avenue, Windsor, ON, Canada
e-mail: msaif@uwindsor.ca

M. Jalali
e-mail: jalali3@uwindsor.ca

M. Moradi
e-mail: moradih@uwindsor.ca

M.-R. Alam and M. Fathi (eds.), *Integrated Systems: Data Driven Engineering*,
https://doi.org/10.1007/978-3-031-53652-6_9

They are expected to become a common part of our daily lives in the near future [1]. As CAV technologies continue to evolve, the data transferred between sensors, actuators, and other integrated components can be intentionally disrupted in a shared communication or wireless setting. New threats and attacks are likely to emerge, posing risks to the CAV network. Any attacks on CAV can lead to a wide range of devastating consequences that could even endanger people's lives. Therefore, timely and efficient detection of such attacks is crucial as it can significantly mitigate risks, save lives, and prevent widespread chaos and harm.

Model-based approaches to ADI rely on mathematical models of the systems under consideration, along with known inputs, to detect deviations from expected behavior [2]. However, in recent years, there has been a growing trend in utilizing data-driven or Machine Learning (ML) in ADI, particularly in poorly modeled networked systems that are complex to be handled in real-time using traditional model-based strategies [3]. The fact that the data-driven approach is beneficial when the system dynamics are not completely understood or system behavior is highly nonlinear or complex, along with the advancement of computing technologies and signal processing, has made real-time ML-based ADI possible and increasingly popular.

Artificial Intelligence (AI) is a specialized field that focuses on creating computer systems that can perform tasks like human thinking. These tasks include learning from experience, solving complex problems, understanding language, and making decisions. AI represents a rapidly advancing field, focusing on exploring and advancing concepts, methodologies, tactics, and practical implementations aimed at emulating and amplifying human intelligence. AI systems can process and analyze data, make choices, adapt to different situations, and improve performance over time. They use techniques like machine learning, neural networks, deep learning, and reinforcement learning. The main goal of AI is to make machines smart enough to do tasks on their own as humans do [4].

Deep Learning (DL) techniques which utilize neural networks with multiple hidden layers to identify abnormal patterns in data, have recently been used in attack diagnosis applications [4, 5]. Various deep-learning methods have been proposed for anomaly detection, including variational autoencoders and generative adversarial networks [4]. These techniques have shown promising results in various domains, such as healthcare, manufacturing, and cybersecurity [6]. For example, in the attack diagnosis method proposed in [7], deep autoencoders are used to identify deviations from normal behavior in a complex system. Autoencoders are unsupervised neural networks including input encoder and decoder parts to reconstruct input data to learn useful features of data [8]. Indeed, using autoencoder neural network leads to identify hidden and compressed features relevant to normal condition [9]. A comprehensive dataset of normal behavior is used to train these neural networks. Once trained, the neural network is deployed to detect instances of abnormal behavior. By detecting deviations from the normal mode, the autoencoder can indicate the presence of an attack.

Vehicles in CAV, communicate with each other through wireless technology. The network topology explains how cars share information and form groups. The problem of organizing vehicles into a platoon can be approached using the consensus method. To achieve agreement on position, speed or acceleration, distributed control protocols allow vehicles to exchange information with their neighbors. Researchers have extensively studied the concept of consensus in systems with multiple agents over the past decade [10–13]. By sharing information about their own interests with neighboring vehicles, each vehicle can reach a common understanding . An information graph theory is used to understand how the vehicles interact with each other [14]. In this study, a consensus algorithm is designed for a group of vehicles traveling in a platoon. Furthermore, the CAVs in this formation can be subjected to adveserial attacks. The objective of the chapter is to experimentally and in a laboratory setting verify the effectiveness of the control designed control strategy along with the related ADI strategy, hence confirming that ML strategies could be an effective tool in ADI in CAVs.

This chapter presents a pragmatic data-driven strategy that leverages autoencoders for attack diagnosis. Our methodology's viability is demonstrated through its application to real-time situations using Qcars, an innovative autonomous vehicle platform developed by Quanser Inc. Qcars boasts an array of sensors, encompassing LIDAR, 360-degree vision, depth sensors, IMU, encoders, and user-expandable IO, thereby enabling researchers at the Laboratory for Autonomous Resilient System (LARS) at the University of Windsor to acquire comprehensive datasets. Powered by the efficient NVIDIA Jetson TX2 supercomputer, Qcars provides the impetus for groundbreaking research endeavors [15].

Through meticulous experimentation with QCars, we substantiate the dependability of our AI-centric approach. Real-time application scenarios are harnessed to further endorse the efficacy of our strategy, offering tangible proof of its accomplishments.

1.1 Chapter Organization

The subsequent sections of this Chapter unfold as follows:

In Sects. 2 and 3, the dynamic model of our self-driving car is introduced, shedding light on the intricacies of its communication framework within the vehicle network. This exploration extends further to elucidate the harmonious collaboration that culminates in achieving a consensus. Section 4 focuses on the cyberattacks and offers insight into their diverse manifestations. This discussion lays the foundation for comprehending the nuances of the challenges at hand. Section 5 proposes a methodology aimed at addressing the ADI strategy. Finally, in Sect. 6, we use MATLAB Simulink to illustrate our strategies in action. This section serves as a pivotal bridge between theory and empirical examination establishing the fact that the ML strategies can be effective tools for ADI in CAVs.

2 QCar Model

This section commences with an overview of the dynamic model of QCars, aiming to capture the essence of autonomous vehicles through carefully designed equations. The discussion starts by detailing the vehicle's dynamic model and the corresponding

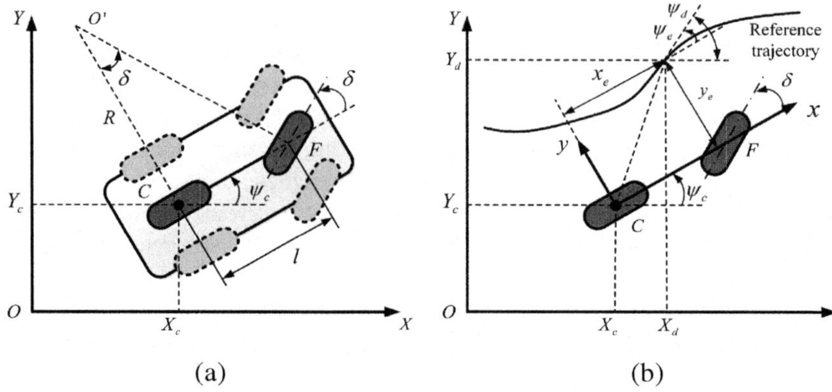

Fig. 1 QCar frame [16]

equations, formulated to emulate the behavior of self-driving cars. Subsequently, the focus shifts towards the motion equations tailored specifically to QCars. Illustrated in Fig. 1a, a notable design choice manifests in the consolidation of front and rear axle wheels into singular entities, strategically positioned at the midpoint of each respective axle. Within this configuration, the front wheel takes on the responsibility of the steering mechanism, while the rear wheel remains statically oriented.

The analysis remains grounded within the domain of planar motions, acknowledging the two-dimensional nature of the vehicle's movement. Consequently, a set of three coordinates within the OXY global coordinate frame, denoted as P_c, efficiently encapsulates the vehicle's position, enabling a precise representation of its trajectory within the plane.

$$P_c = [X_c, Y_c, \psi_c]^T \quad (1)$$

In the given context, the motion of the vehicle is described by a posture represented by coordinates (Xc, Yc) in the global coordinate frame. These coordinates correspond to the global position of the control point C, located at the rear axle's midpoint. Additionally, ψ_c defines the vehicle's orientation, where positive rotation is counterclockwise from the X-axis.

To analyze the low-speed motion of the vehicle, an assumption is made that both wheels roll without slipping [17]. The vehicle's front and rear wheels are represented by perpendicular lines FO' and CO', respectively. The point where these lines intersect, O', is known as the instantaneous rolling center. The radius of the vehicle R is given by:

$$R = l/tan\delta \quad (2)$$

The yaw rate of the vehicle $(\dot{\psi})$ can be expressed in terms of the vehicle's wheelbase (l) and the steering angle of the front wheel (δ), assuming a low vehicle speed.

$$\dot{\psi} = V/R = V/l.tan\delta \quad (3)$$

The kinematic model of the vehicle's motion is derived [18], considering the forward speed at point C, denoted as V.

$$\begin{cases} \dot{X}_c = V \cos \psi \\ \dot{Y}_c = V \sin \psi \\ \dot{\psi} = V/l.tan\delta \end{cases} \tag{4}$$

The Eq. 4 represents the QCar model, which will serve as the basis for the subsequent analysis. In this study, an AI-based methodology will be employed to investigate cyber-security of a fleet of QCar models.

3 Network Dynamics

Consensus control plays a significant role in a multi-agent system aiming to achieve coordination and agreement among multiple agents by ensuring they converge to a common decision or state. Consensus control helps synchronize individual agents' actions and behaviors, enabling them to work together towards a common goal. It allows agents to exchange information, communicate, and make collective decisions in a decentralized manner. Therefore, it is essential for various applications, such as distributed robotics, sensor networks, autonomous vehicles, and swarm intelligence, where multiple agents need to collaborate and reach a consensus on important tasks or objectives [19].

In this study, a network of CAVs is considered, each characterized by two degrees of freedom. The relative position between the vehicles is represented by the vector $\boldsymbol{\xi} = [x,\ y]^T$, while the velocities are described by the vector $\boldsymbol{\eta} = [v_x,\ v_y]^T$. The method outlined in Ren et al. [20] is followed to analyze the dynamics of the multi-agent systems.

Graph theory is applied to multi-agent systems to represent the mathematically dynamic interactions within a network, where agents are the graph's nodes, and an edge between two nodes represents their ability to communicate [21]. The interaction between agents through exchanging information is modeled by graph theory. A directed communication network topology is represented using an adjacency matrix that captures the relationships and information flow among the vehicles. Indeed, the adjacency matrix is a mathematical representation of relationships between nodes or entities. If an element of the adjacency matrix is equal to one, it indicates a connection between the corresponding row and column numbers. Conversely, a zero value in the matrix indicates no connection between the respective entities. In this case, the nodes represent vehicles within the system. By examining the last row and column of the adjacency matrix, we can determine how the followers are connected to the leader.

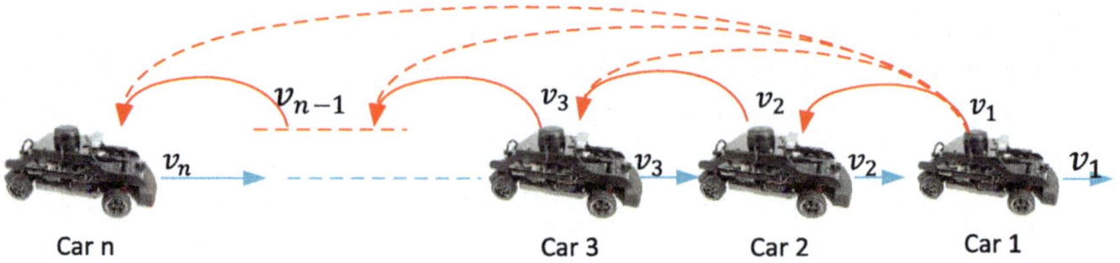

Fig. 2 Communication topology for a platoon of vehicles

$$A = \begin{bmatrix} 0 & 1 & 0 & 1 \\ 0 & 0 & 1 & 1 \\ 0 & 0 & 0 & 1 \\ 0 & 0 & 0 & 0 \end{bmatrix} \tag{5}$$

In the given matrix, the last column specifically represents the communication pattern between the leader and the followers. It indicates that the leader sends data to all the followers in a one-way communication fashion.

Similarly, the last row indicates that the leader does not receive any data from the followers. Instead, the followers only send data to the vehicle ahead of them. This implies that the leader-follower approach enables each vehicle to receive data from both the preceding vehicle and the leader itself.

The plant dynamics for each agent are defined as follows:

$$\begin{bmatrix} \dot{\xi} \\ \dot{\eta} \end{bmatrix} = \begin{bmatrix} 0 & I_{2x2} \\ 0 & 0 \end{bmatrix} \begin{bmatrix} \xi \\ \eta \end{bmatrix} + \begin{bmatrix} 0_{2x2} \\ diag(\frac{1}{m}) \end{bmatrix} \mathbf{u(t)} \tag{6}$$

The direct information exchange topology can be visualized in Fig. 2, illustrating the network's structure and how information is shared among the vehicles.

Consensus Protocols in Networked Multi-agent Systems

The consensus problem is a fundamental coordination challenge in Multi-Agent Systems (MAS), as discussed in the introduction. The core concept of consensus is to steer each agent's information variable toward its neighboring agents' information variable to achieve agreement on specific quantities of interest by utilizing their local information for each agent. In a group of n agents, each agent is represented by an information state, denoted as x_i, and a control input, denoted as u_i. The information states follow single-integrator dynamics can be described by the following equation:

$$\dot{x}_i = u_i, \qquad i = 1, ..., n \tag{7}$$

The control gain given by:

$$u_i = k_i(x_{i_1}, x_{i_2}, ..., x_{i_m}) \tag{8}$$

Each node in the system has a control input k_i, which depends on a subset of nodes. The goal is to address the consensus problem through the discovery of a distributed control protocol. This protocol aims to guide the states of all nodes, denoted as x_i, towards convergence to a common value x_j as time progresses towards infinity ($t \rightarrow \infty$). This convergence condition holds for every pair of nodes i and j ($\forall i, j$).

This shared value is referred as the consensus value. The fundamental consensus algorithm is expressed as follows:

$$u_i = -\sum_{j=1}^{n} a_{ij}(x_i - x_j) \tag{9}$$

where a_{ij} is the elements of the adjacency matrix associated with the communication graph. Consensus is said to be achieved if $\mid x_i(t) - x_j(t) \mid \rightarrow 0$ as $t \rightarrow \infty$. The typical second-order consensus algorithm, which takes into account both position and velocity, is described as follows:

$$u_i = -\sum_{j=1}^{n} a_{ij}[(x_i - x_j) + \gamma(v_i - v_j)] \tag{10}$$

where γ is a control gain, x_i and v_i denote, respectively, the position and velocity of the ith vehicle, the consensus is achieved or reached by the team of vehicles if, for all $x_i(t)$ and $v_i(t)$ and all i, j = 1,...,n, $\mid x_i(t) - x_j(t) \mid \rightarrow 0$ and $\mid v_i(t) - v_j(t) \mid \rightarrow 0$, as $t \rightarrow \infty$.

The following higher-order consensus algorithm, which could include k derivatives of the states, is proposed in [22]:

$$u_i = -\sum_{j=1}^{n} a_{ij}k_{ij}\left[\sum_{k=0}^{i-1} \gamma_k(x_i^k - x_j^k)\right] \qquad i = 1, ..., n \tag{11}$$

The equation is governed by control gains, where $k_{ij} > 0$ and $\gamma_k > 0$. The underlying motivation for Eq. 11 is to steer the information variable and its higher-order derivatives of each vehicle toward the corresponding states of its neighboring vehicles. Consensus algorithms for first-order dynamics to a reference model r are presented in Eq. 12. Assume a team of n vehicles with an additional agent labeled $n + 1$, where agent $n + 1$ is the team leader and vehicles $1, \ldots,$ n are the followers [11].

$$u_i = -\sum_{j=1}^{n} a_{ij}k_{ij}(x_i - \delta_{x_i} - x_j) - a_{i(n+1)}\gamma_k(x_i - \delta_{x_i} - x^r) \tag{12}$$

The given equation represents a system with $n + 1$ states, denoted as $x_i \in \mathbb{R}^m$. The adjacency matrix $A_{n+1} \in \mathbb{R}^{(n+1)\times(n+1)}$ determines the connections between the agents, with a_{ij} representing the (i, j) entry of the matrix. The control gains $k_{ij} > 0$

and $\gamma_k > 0$ play a role in the control mechanism. Furthermore, δ_{x_i} denotes the desired separations between the agents along the x-axis in the state space. In this example, a leader possesses the consensus reference state x^r and disseminates information to three followers.

4 Anomaly/Cyber-Attacks Description

Connected and Autonomous Vehicles (CAVs) integrate a variety of cutting-edge sensing technologies, encompassing radar, GPS, LIDAR, cameras, and more. While these innovations enhance their capabilities, they concurrently render CAVs susceptible to potential attacks. Such vulnerabilities demand heightened attention, primarily attributed to the profound implications they hold for human safety and well-being. Here, anomalies denote deviations from established norms, indicating atypical occurrences that deviate from the expected behavior of a system or network. These deviations serve as indicators of irregular events or patterns that stand out within the operational context. A spectrum of anomalies exists, with particular significance vested in three critical categories: Denial of Service (DoS), Deception, and Replay, as elucidated by Kordestani et al. [6]. These anomalies, when observed, merit special scrutiny due to their potential to compromise the integrity and functionality of CAVs.

1. DoS attacks involve a situation in which an attacker trying to disable an internet-connected device, disrupting legitimate user access to network resources. By overwhelming the target system with a flood of packets or draining its resources, such attacks can render the system unstable or crash it, posing significant risks to the physical systems involved [23].
2. Deception attacks involve the attacker manipulating data packets, either by delivering false packets or injecting false data into legitimate ones, compromising the integrity of control and measurement data. These attacks aim to undermine the reliability of the conveyed data, impacting the system's control performance, and false data injection attacks are a specific type of deception attack [24].
3. eplay attacks involve the intentional repetition or delay of genuine data transmissions, allowing attackers to replay previously collected data from compromised sensors or actuators. These attacks do not require knowledge of the system's design or control and can be represented as time-varying delays with uncertain parameters [25].

Our focus in this study lies on DoS attacks, which will be evident in the subsequent simulation and experimental sections.

5 ADI Methodology

As eluded to before, where a comprehensive understanding of system dynamics is lacking or when the system behavior is complex and nonlinear, data-driven approaches may offer greater efficacy [26]. To effectively detect anomalous events, our proposed solution utilizes an unsupervised learning technique based on a deep

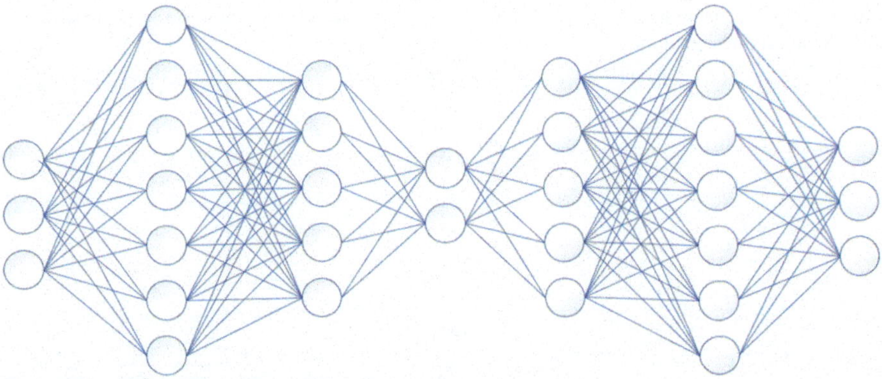

Fig. 3 The structure of proposed autoencoder

autoencoder neural network. This method offers several benefits, making it a valuable tool in anomaly detection. One notable advantage of this approach is its independence from labeled data for training. Indeed, it focuses on learning the underlying patterns of normal behaviors and effectively identifying any abnormal behavior that deviates from these established patterns [27].

This study focuses on a set of states encompassing velocities and relative distances which are the pivotal components in our analysis. These state variables are channelled into our deep autoencoder, visualized in Fig. 3. The neural network encoder design adheres to a specific architecture, featuring layers of 15, 10, and 3 neurons respectively. This design principle persists uniformly throughout the decoder section, imbuing the process with cohesion and structure. The establishment of this architecture was a product of iterative refinement, informed by trial and error, and aimed at achieving optimal reconstruction errors within the confines outlined in the introduction.

Functionally, the Encoder constitutes the input layer, adeptly encoding the input data. On the converse, the Decoder, serving as the output layer, undertakes the task of reconstructing the encoded data. The performance measure for gauging reconstruction accuracy is the comparison between the reconstructed data and the original input. The training phase equips the model with an understanding of normal data patterns, enabling the setting of a specific threshold for reconstruction error. This pivotal threshold subsequently discerns between normal and anomalous data, a principle substantiated by [7].

The deliberate selection of this straightforward yet potent network architecture is underpinned by its rapid detection capabilities, rendering it apt for real-time applications. As the model assimilates the nuances of typical data patterns, the Decoder adeptly reconstitutes the encoded data with nominal loss (reconstruction error). However, when confronted with anomalous data that veers from the established norms, the reconstruction loss escalates, serving as an indicator of anomaly.

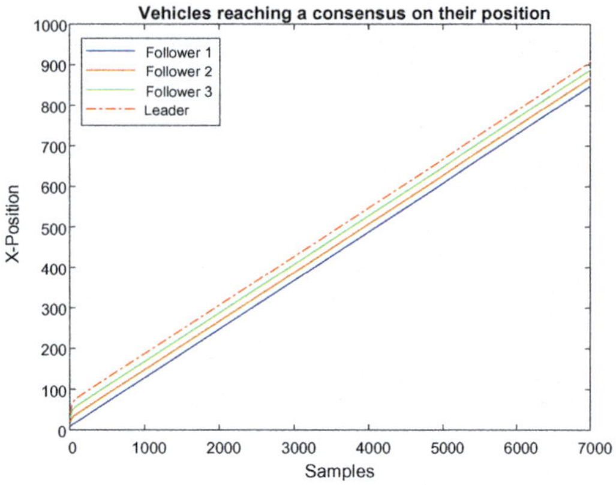

Fig. 4 Vehicles reaching an agreement on their position

Furthermore, in Sect. 6, we use the QCar model as described in Eq. 4 to perform simulations. This QCar-based simulation framework becomes a bedrock for generating a training dataset, which allows us to meticulously design attack scenarios. This strategic approach engenders a comprehensive evaluation of our model's efficacy during the Simulation phase.

6 Simulation Results

6.1 Attack Impact in a Network of Vehicles

The coordination of a fleet of vehicles, employing a leader-follower consensus protocol encapsulated in Eq. 12, forms the basis of our analysis. Our objective here is to unveil the vulnerabilities stemming from potential attacks within the vehicle network. This process is facilitated by the Simulink Consensus Manager block, a pivotal entity that interfaces with the states of the four vehicles in tandem with the desired states. Driven by the tenets of Eq. 12, this block generates essential control laws, which serve as the basis for achieving consensus across the network.

This coordination embraces each vehicle as an active agent within the system. The intricate loop commences with the vehicle's discrete controller, acting as the conduits for channeling real-time feedback to the Consensus Manager block. This iterative dialogue paves the way for a continuous recalibration of states, a dynamic interplay that mirrors the inherent coordination within the vehicular cohort.

A visual representation, as encapsulated in Fig. 4, serves to illustrate the symphony of coordinated motion. This illustration encapsulates the mechanics of aligning positions and sustaining inter-vehicle spacing as they collectively trace a predefined path.

The impact of communication attacks on the network of vehicles is presented in the ensuing discussion. The introduction of an attack into the system causes it to integrate into the output of the consensus control law. Within the contextual framework depicted in Fig. 5, we consider two distinct attack scenarios: a low-magnitude attack characterized by magnitude of 0.05, targeting vehicle #2, and a high-magnitude attack designated as 0.5.

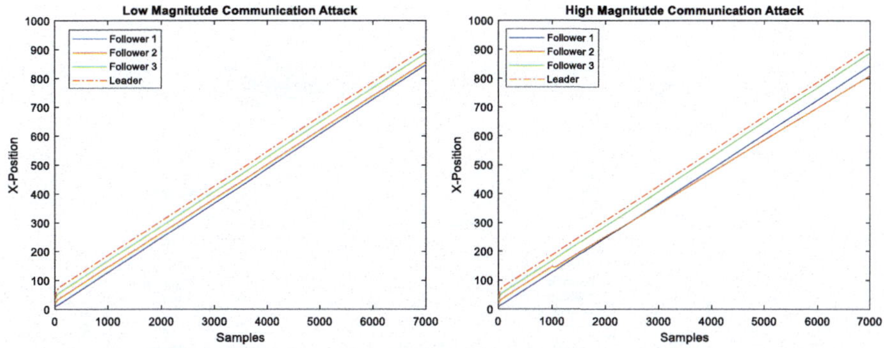

Fig. 5 Impact of communication attack: low-high magnitude attack in a fleet of vehicle

Referring to the network topology outlined in Eq. 5, it is noted that vehicle #2 acts as an information source directed towards vehicle #1. The implications of this arrangement become evident in the scenario of a potential attack. Specifically, the low-magnitude attack imposed on vehicle #2 possesses the capacity to induce deviations in its own trajectory, potentially influencing and misleading vehicle #1. Given that vehicle #1 relies on information derived from both the leader and vehicle #2, this attack has the potential to introduce errors into the coordination. Despite its low magnitude, the cumulative effects can precipitate collision risks if not addressed promptly. In light of this, the bottom line lies in timely detection, an imperative safeguard against the possibility of a collision. The recognition of anomalies becomes a pivotal intervention, serving as a bulwark against potential collisions and preserving the integrity of the vehicular network.

6.2 Attack Detection in a Network of Vehicles

The QCar model represented by Eq. 4 is simulated. The architectural backbone of the platoon is manifested via connections established in a leader-follower configuration, as illustrated in Fig. 2. The strategy underpinned by feedback control mechanisms, synergizes with the detection capability of deep learning, is represented in Fig. 6. Here, the interplay between QCars, regulatory feedback, and the incisive deep learning methodology unfolds, harmoniously unraveling the detection of communication attacks.

Next, the simulation phase yields an output, which, upon juxtaposition with the input, unveils the residual-the hallmark of deviation. The disparity, or residual, arising from this comparison constitutes a prime candidate for the attack detection process. This residual is thoughtfully channeled into the trained autoencoder, an integral component in the cascade of steps dedicated to identifying attacks. In instances

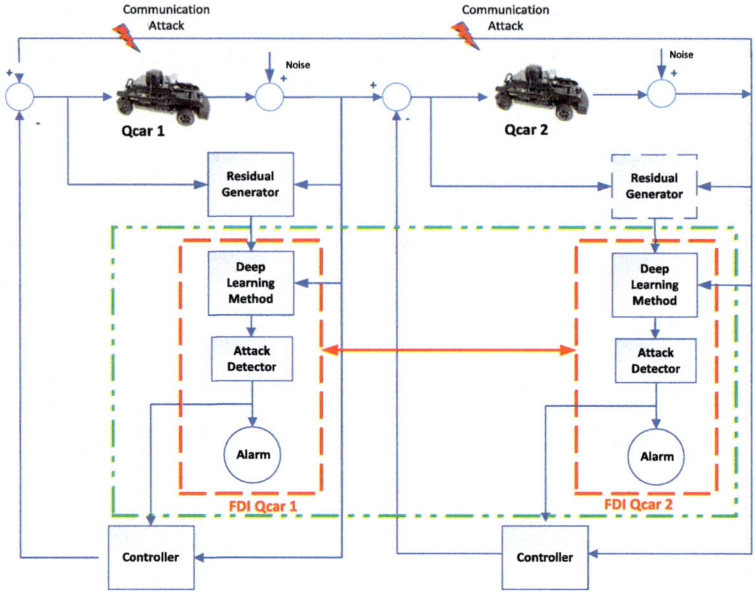

Fig. 6 Communication attack detection framework: deep learning methodology

where the need for mitigation arises, appropriate commands can be dispatched to the controller, effectively influencing and regulating the vehicle's operational trajectory.

A key facet of executing the deep learning methodology described in Sect. 5 entails the definition of five inputs. These inputs span the spectrum of speeds and relative distances characterizing each vehicle's dynamic. The training of the autoencoder constitutes an important step, which draws from the rich set of data harvested during the simulation phase involving the three vehicles.

Figure 7 illustrates the velocity profiles of the leader and two followers, along with the Relative Distance (RD) between the vehicles. Here the focus is on a DoS attack, where network flooding disrupts data transmission and reception, as defined in Sect. 4. Figure 7 shows how DoS attacks affects the data. For vehicle #1, the DoS attack is applied for two intervals around 500 and 2200 which is considered zero, and on vehicle #2, around 1800 and 3100.

Having trained the deep learning network using normal data in an offline setting enables the provided solution to achieve accurate online detection among all relevant agents. In Simulink, the system was trained using normal data and then tested against offline data containing attacks After collecting the data in the simulation step, the proposed autoencoder-based approach for anomaly detection is applied.

As discussed in Sect. 5, the efficacy of our proposed approach hinges upon the utilization of the reconstruction error derived from the autoencoder as a foundation for attack detection. This principle is portrayed in Figs. 8 and 9, offering a tangible exposition of the mechanics.

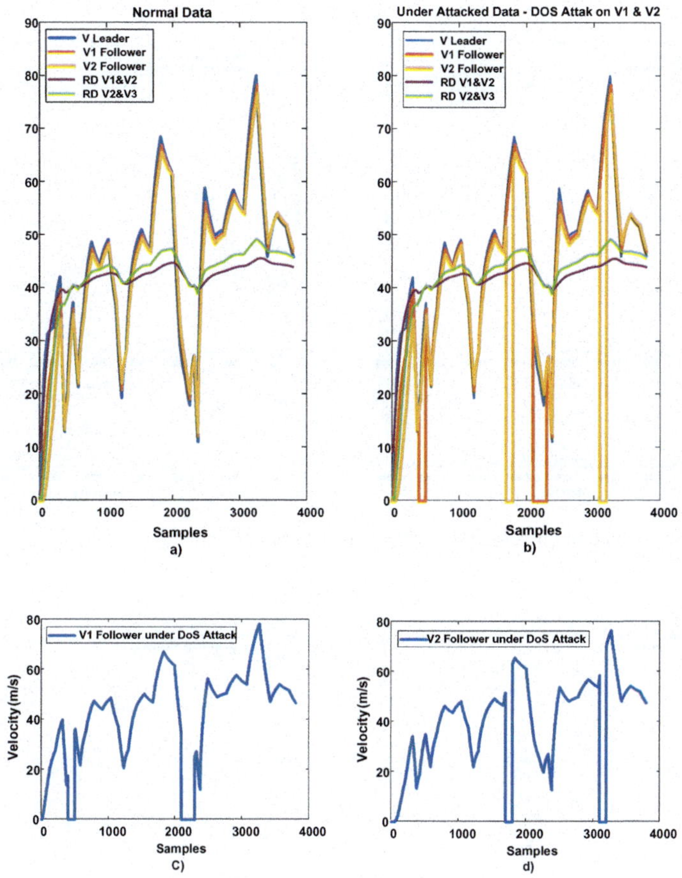

Fig. 7 DoS attacks in the network of vehicles, **a** collected normal dataset, **b** collected dataset under DOS attack, **c** DOS attack on velocity for follower #1, **d** DOS attack on velocity for follower #2

In the absence of anomalies, the fluctuations in error remain notably restrained. For Vehicle 1, the error shift maintains a threshold well below 0.4, while for Vehicle 2, it operates comfortably beneath 1, thereby reflecting the stable behavior under normal circumstances. However, when an attack infiltrates our network, the error metric surges past the defined threshold, unfurling an effective mechanism to discern and subsequently detect abnormal activities within the vehicular network. This visual contrast underscores the method's capacity to sensitively capture and signify deviations, thus reinforcing its role in preserving the integrity of the vehicular ecosystem.

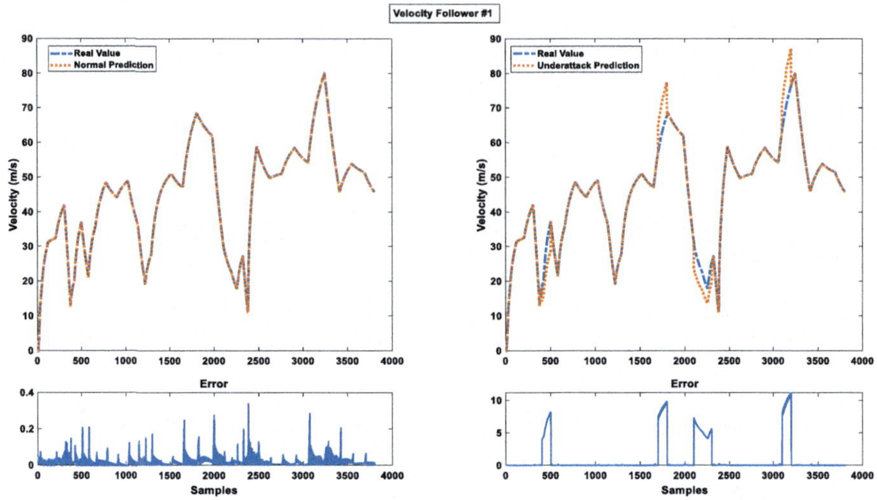

Fig. 8 Attack detection on vehicle #1 within the network of vehicles

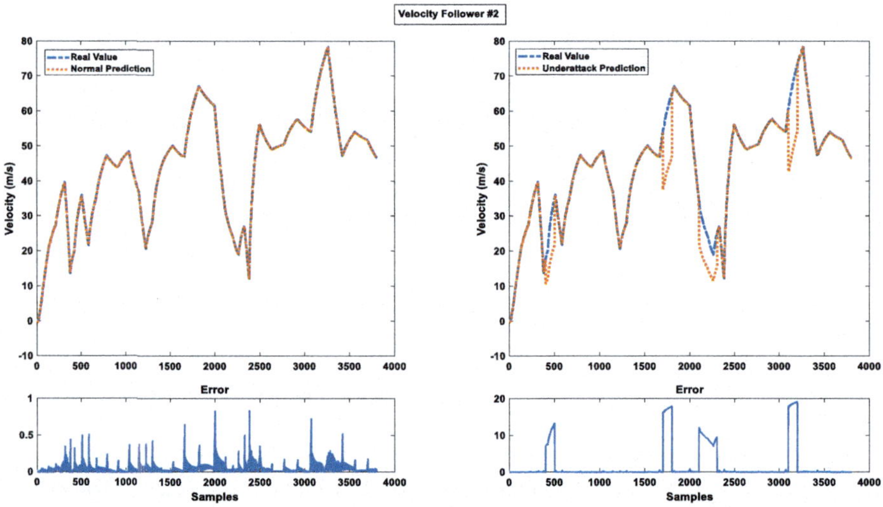

Fig. 9 Attack detection on vehicle #2 within the network of vehicles

7 Experimental Results

The incorporation of Quanser's self-driving vehicles into the Laboratory for Autonomous Resilient System (LARS) at the University of Windsor marks an initial experimental attempt at this line of research. The QCar, a key component of this research, harnesses a dynamic interplay of drive and steering servo motors to navigate its trajectory, with an encoder serving as the feedback conduit. The QCar

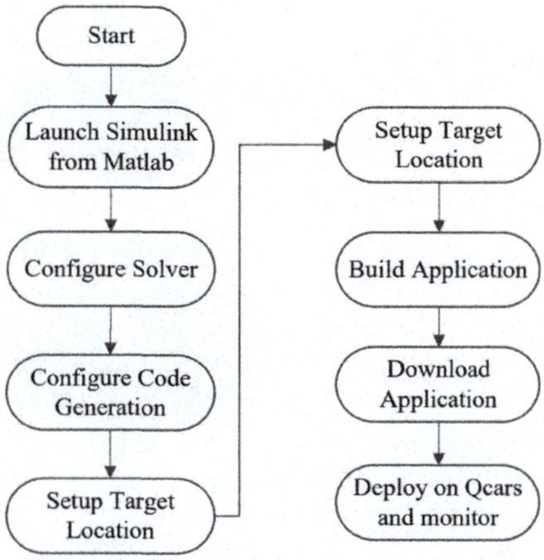

Fig. 10 Process diagram: deploying Simulink Code on QCars

motion is augmented by real-time posture insight, supplied by an onboard 9-axis Inertial Measurement Unit (IMU), as illustrated in Fig. 11. The control action comes through the Simulink blocks, transmuting into executable C-code and embedded within a Linux-based architecture.

The heartbeat of this vehicular ensemble is the NVIDIA Jetson TX2 processor featuring a Quad-Core ARM Cortex-A57 microcontroller. The onboard processor receives commands from the ground station PC and communicates with it through wireless Wi-Fi. Data collected from the IMU and motor encoder is transmitted to the ground station. Additionally, the processor generates pulse-width modulation signals, thereby driving the servo motor's actions.

During the process of validation, the QCar plays a crucial role. It provides us with real-world data that reflects actual situations, allowing us to test and confirm the effectiveness of our data-driven approach. In essence, the QCar serves as a testing ground where our ideas are put to the practical test, showing us whether our approach works successfully in real-life situations. This helps us understand how well our theory aligns with practical outcomes and whether our approach is indeed successful.

Figure 10 presents a comprehensive process showcasing the steps involved, starting from running the Simulink code and successful deployment onto the QCars platform shown in Fig. 11. Figures 12 and 13 show the result of implementation and simulation steps of our proposed approach in velocity attack detection in our self-driving car.

At the Laboratory for Autonomous Resilient System (LARS), the process of gathering data and training our deep autoencoder took place offline.

In Fig. 12, we showcase the successful identification of velocity attacks in our self-driving car. The velocity data collected through wheel encoders, along with the simulated desired speed within the closed-loop control system, is fed to the QCar. Subsequently, the measured speed obtained from the encoder is looped back to the

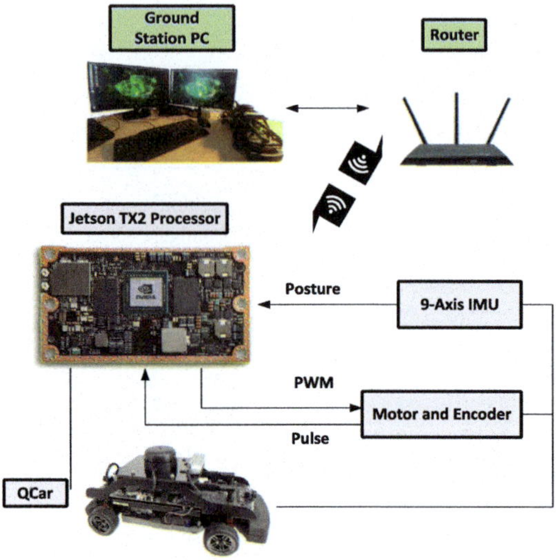

Fig. 11 Schematic of the QCar experiment setup [18]

Fig. 12 Detection of velocity attacks in the QCar

system for comparison against the desired speed. In Fig. 13, we visualize the behavior of the autoencoder error. The blue line corresponds to normal data instances, where the error generated remains fairly consistent. However, when our self-driving car is subjected to an attack, there is a sharp spike in the error, illustrated by the red line. To detect such attacks, we establish a threshold. This threshold is determined based on the maximum error observed in the data that was used to train the autoencoder under normal conditions. When the generated error surpasses this threshold, it serves as an indicator of a potential attack on our data. This mechanism enables us to swiftly pinpoint instances of abnormal behavior in our self-driving car.

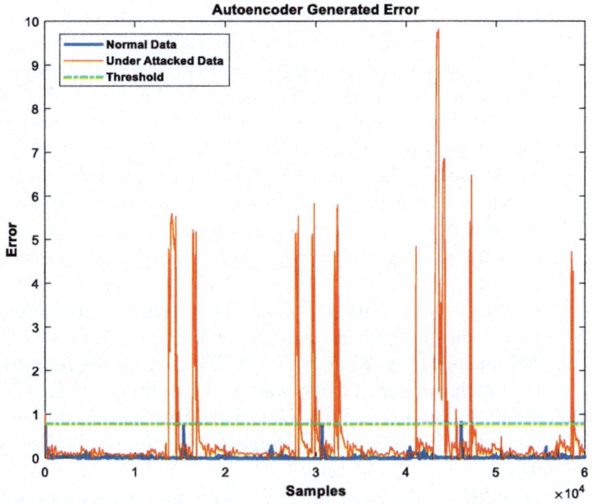

Fig. 13 Velocity attack detection threshold in self-driving car

8 Conclusions

As vehicles become more complex with electronic systems and wireless technologies, security challenges have intensified. The need for robust security in autonomous vehicles, especially those connected to networks, has become paramount. To address this, AI-driven solutions have gained traction due to their adaptability and system-agnostic capabilities.

In conclusion, this chapter proposes a real-time autoencoder-based solution to enhance Autonomous Driving Integrity and Resilience (ADIR) in Connected Autonomous Vehicles (CAV). The successful application within a practical environment demonstrates the viability of our approach. By leveraging AI-driven adaptability, our method offers a robust security enhancement for self-driving vehicles in interconnected landscapes. As autonomous vehicles become integral to modern mobility, our approach contributes to a more secure and reliable automotive ecosystem.

References

1. Ren, K., Wang, Q., Wang, C., Qin, Z., & Lin, X. (2020). The security of autonomous driving: Threats, defenses, and future directions. *Proceedings of the IEEE, 108*(2), 357–372.
2. Ju, Z., Zhang, H., Li, X., Chen, X., Han, J., & Yang, M. (2022). A survey on attack detection and resilience for connected and automated vehicles: From vehicle dynamics and control perspective. *IEEE Transactions on Intelligent Vehicles, 7*(4), 815–837.
3. Qayyum, A., Usama, M., Qadir, J., & Al-Fuqaha, A. (2020). Securing connected and autonomous vehicles: Challenges posed by adversarial machine learning and the way forward. *IEEE Communications Surveys and Tutorials, 22*(2), 998–1026.

4. Qayyum, A., Usama, M., Qadir, J., & Al-Fuqaha, A. (2020). Securing connected and autonomous vehicles: Challenges posed by adversarial machine learning and the way forward. *IEEE Communications Surveys and Tutorials, 22*(2), 998–1026.
5. Bouchouia, M. L., Labiod, H., Jelassi, O., Monteuuis, J.-P., Jaballah, W. B., Petit, J., & Zhang, Z. (2023). A survey on misbehavior detection for connected and autonomous vehicles. *Vehicular Communications, 41*, 100586.
6. Kordestani, M., & Saif, M. (2021). Observer-based attack detection and mitigation for cyber-physical systems: A review. *IEEE Systems, Man, and Cybernetics Magazine, 7*(2), 35–60.
7. Nguyen, D. C., Nguyen, K. D., & Chacko, S. (2022). A real-time explainable anomaly detection system for connected vehicles. In *International conference on internet of things, big data and security.*
8. Hoang, T.-N., & Kim, D. (2022). Detecting in-vehicle intrusion via semi-supervised learning-based convolutional adversarial autoencoders. *Vehicular Communications, 38*, 100520.
9. Alqahtani, H., & Kumar, G. (2022). A deep learning-based intrusion detection system for in-vehicle networks. *Computers and Electrical Engineering, 104*, 108447.
10. Feng, Z., Hu, G., & Wen, G. (2016). Distributed consensus tracking for multi-agent systems under two types of attacks. *International Journal of Robust and Nonlinear Control, 26*, 896–918.
11. Diana, F., & Moncayo, H. (2020). Immunity-based consensus algorithm for increased resilience in multiagent missions. *AIAA Scitech 2021 Forum.*
12. Liu, Y., Zhai, C., & Gao, H. (2017). Autonomous multi-vehicle consensus for high-order systems. In *2017 36th Chinese control conference (CCC)* (pp. 9548–9553).
13. Vlachos, E., & Lalos, A. S. (2022). Admm-based cooperative control for platooning of connected and autonomous vehicles. In *ICC 2022 - IEEE international conference on communications* (pp. 4242–4247).
14. Li, Y., Tang, C., Peeta, S., & Wang, Y. (2019). Nonlinear consensus-based connected vehicle platoon control incorporating car-following interactions and heterogeneous time delays. *IEEE Transactions on Intelligent Transportation Systems, 20*(6), 2209–2219.
15. Educate, Q. C. I. (2020). Self-driving car research studio - quick start guide. https://www.quanser.com
16. *Self-driving car research studio - 3 DOF bicycle model.* Quanser Company Innovate Educate (2020).
17. Rajamani, R. (2011). *Vehicle dynamics and control.* Springer Science & Business Media.
18. Hu, J., Zhang, Y., & Rakheja, S. (2022). Adaptive trajectory tracking for car-like vehicles with input constraints. *IEEE Transactions on Industrial Electronics, 69*(3), 2801–2810.
19. Zhang, Y., Wu, Z.-G., & Shi, P. (2023). Resilient event-/self-triggering leader-following consensus control of multiagent systems against dos attacks. *IEEE Transactions on Industrial Informatics, 19*(4), 5925–5934.
20. Ren, W., & Beard, R. W. (2008). *Distributed consensus in multi-vehicle cooperative control.* Springer.
21. Mesbahi, M., & Egerstedt, M. (2010). *Graph theoretic methods in multiagent networks*. Princeton: Princeton University Press.
22. Ren, W., Moore, K. L., & Chen, Y. Q. (2007). High-order and model reference consensus algorithms in cooperative control of multivehicle systems. *Journal of Dynamic Systems Measurement and Control-transactions of The Asme, 129*, 678–688.
23. Abdollahi Biron, Z., Dey, S., & Pisu, P. (2018). Real-time detection and estimation of denial of service attack in connected vehicle systems. *IEEE Transactions on Intelligent Transportation Systems, 19*(12), 3893–3902.
24. Ding, D., Wang, Z., Han, Q.-L., & Wei, G. (2018). Security control for discrete-time stochastic nonlinear systems subject to deception attacks. *IEEE Transactions on Systems, Man, and Cybernetics: Systems, 48*(5), 779–789.
25. Sun, X., Yu, F. R., & Zhang, P. (2022). A survey on cyber-security of connected and autonomous vehicles (cavs). *IEEE Transactions on Intelligent Transportation Systems, 23*(7), 6240–6259.

26. Zhan, J., Ma, Z., & Zhang, L. (2023). Data-driven modeling and distributed predictive control of mixed vehicle platoons. *IEEE Transactions on Intelligent Vehicles*, *8*(1).
27. Ashraf, J., Bakhshi, A. D., Moustafa, N., Khurshid, H., Javed, A., & Beheshti, A. (2021). Novel deep learning-enabled lstm autoencoder architecture for discovering anomalous events from intelligent transportation systems. *IEEE Transactions on Intelligent Transportation Systems, 22*(7), 4507–4518.

MASTER-XR: Mixed Reality Ecosystem for Teaching Robotics in Manufacturing

Michael Barz, Panagiotis Karagiannis, Johan Kildal, Andoni Rivera Pinto, Judit Ruiz de Munain, Jesús Rosel, Maria Madarieta, Konstantina Salagianni, Panagiotis Aivaliotis, Sotiris Makris, and Daniel Sonntag

Abstract Many industries are transitioning to Industry 4.0 production models by adopting robots in their manufacturing processes. In parallel, Extended Reality (XR) technologies have reached sufficient maturity to enter the industrial applications domain, with early success cases often related to training workers, remote assistance, access to contextual information, and interaction with digital twins. In the future, robots will be increasingly enhanced with XR applications, which requires that industrial workers understand both technologies and use and control hybrid solutions confidently. Specific education and training programs will be essential to this transition, especially for vocational school students and professionals in upskilling. They must learn how to program robots and establish a safe and productive human-robot collaboration. The new EU-funded project MASTER will improve the XR ecosystem for teaching and training robotics in manufacturing by providing an open XR platform that integrates key functionalities like creating safe robotic environments, programming flexible robotic applications, and integrating advanced interaction mechanisms based on eye tracking. It will also provide high-quality training materials for robotics. We report on the project plan, our objectives, and milestones.

M. Barz (✉) · D. Sonntag
Interactive Machine Learning, German Research Center for Artificial Intelligence (DFKI), 66123 Saarbrücken, Germany
e-mail: michael.barz@dfki.de

D. Sonntag
e-mail: daniel.sonntag@dfki.de

Applied Artificial Intelligence, Oldenburg University, 26111 Oldenburg, Germany

P. Karagiannis · S. Makris
Laboratory for Manufacturing Systems and Automation, Department of Mechanical Engineering and Aeronautics, University of Patras, Patras, Greece
e-mail: karagiannis@lms.mech.upatras.gr

S. Makris
e-mail: makris@lms.mech.upatras.gr

J. Kildal · A. Rivera Pinto
Fundaci6n Tekniker, Gipuzkoa, Spain
e-mail: johan.kildal@tekniker.es

M.-R. Alam and M. Fathi (eds.), *Integrated Systems: Data Driven Engineering*,
https://doi.org/10.1007/978-3-031-53652-6_10

Keywords Industry 4.0 · Extended Reality (XR) · Robotics · Worker training · Manufacturing · Human-robot collaboration · Eye tracking

1 Introduction

Many industries adopt robots and Extended Reality (XR) technologies to realize Industry 4.0 production models. This requires industrial workers to understand both technologies, particularly in flexible and collaborative manufacturing settings that require frequent programming of robots and safe shared workspaces. For instance, high awareness of robot actions and intentions can help mitigate stress and potential dangers. Consequently, vocational school students and professionals must learn how to program robots and collaborate with them safely. XR technologies can support the learning process [49] and may become a key technology of future manufacturing processes. According to data published in 2021, XR will create 1.2–2.4 million new jobs in the EU by 2025, and European XR markets are expected to grow between €35 billion and €65 billion by 2025 [50]. Specific education and training programs will be essential for successfully transitioning to XR-based robotic workspaces like the example illustrated in Fig. 1. The presented EU project, MASTER, aims to provide an open XR platform for worker training in robotics and XR-enhanced human-robot collaboration. The platform will facilitate the creation of XR-based training scenarios and materials and provides key functionalities for creating safe robotic environments, programming flexible robotic applications, and integrating advanced interaction techniques based on eye tracking. In this report, we outline the objectives of our project, describe the planned XR platform and its major features, provide insight into didactic considerations for educating workers, and provide an outlook

A. Rivera Pinto
e-mail: andoni.rivera@tekniker.es

J. R. de Munain · J. Rosel
Mondragon Lingua-Alecop S. Coop, Arrasate, Spain
e-mail: juditr@alecop.es

J. Rosel
e-mail: jrosel@alecop.es

M. Madarieta
Virtualware 2007 S.A., Usausuaga 7, 48970, Basauri Bizkaia, Spain
e-mail: mmadarieta@virtualwareco.com

K. Salagianni · P. Aivaliotis
Teaching Factory Competence Center, Patras, Greece
e-mail: salagianni@teachingfactory-cc.eu

P. Aivaliotis
e-mail: aivaliotis@teachingfactory-cc.eu

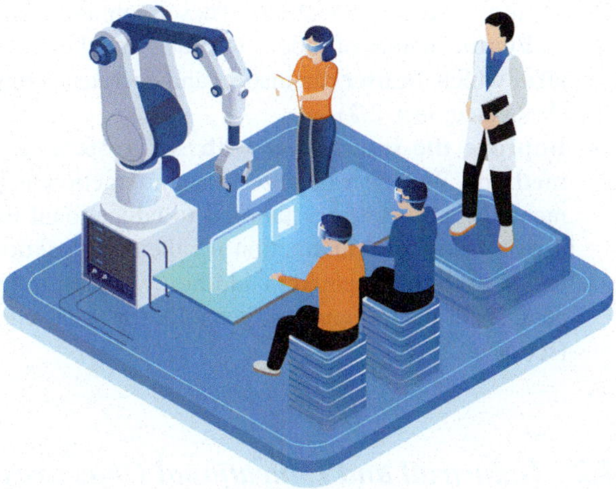

Fig. 1 Exemplary shared workspace in robotic-assisted manufacturing based on XR

on how we plan to evaluate the educational platform in realistic settings through two open calls.

1.1 Technological Objectives

The main objective of the MASTER project is to boost the XR ecosystem for teaching and training of robotics in manufacturing. To achieve this, an open XR platform will be provided, which shall enable creators to deliver rich training content on robotics and integrate key functionalities for creating safe robotic environments, programming flexible robotic applications, and integrating advanced interaction mechanisms. Four technical objectives have been identified, which act as the main pillars of the project:

1. **Provide an open platform for the creation and management of XR content in robotics by non-experts.** We focus on the creation of an XR platform that allows non-expert programmers to create XR learning content and teaching materials on robotics. The targeted stakeholders are robot experts who know about robotic processes and related components. The platform shall enable the transition of this knowledge into XR content, exercises, and learning sessions (see Sect. 2).
2. **Create virtual environment to program and interact with robots in safety configurations.** We address safety aspects in manufacturing, focusing on means to increase the safety and safety awareness of operators in human-robot collaboration, also concerning ergonomics (see Sect. 2.1).
3. **Bridge the gap between the use of virtual and real robots for training and real operation.** We aim to integrate a programming toolbox that allows a user to program a robot easily. A special focus will be on Programming-by-Demonstration (PbD), as well as machine vision assisted techniques, such

as Look-and-Move (L&M). These techniques exploit the benefits of the dual real/virtual nature of robots, the parts that they manipulate, and the environment, all of which the user can inspect and operate in conjunction through mixed reality (MR) (see Sect. 2.2).
4. **Improve the interaction with XR systems using multimodal interaction methods.** We integrate gaze-based interaction techniques in 3D virtual and augmented reality settings tailored to educational use cases. We aim to integrate active gaze-based interaction for the manipulation of, e.g., robots, their digital twins, or digital menus. In addition, we plan to integrate methods for real-time analysis of the human gaze behavior for assessing the learning progress of users (see Sect. 2.3).

1.2 Industrial and Educational Objectives

Besides our technical objectives, we identified one industrial and one educational objective that will guide our development and project activities:

1. **Industrial Objective: Boost companies, especially SMEs and start-ups, to deploy XR technology.** SMEs and start-ups rely on external financial and technical support to reach the market since only 26.6% are able to sustain themselves with their own funds. Thus, the MASTER project focuses on providing financial, technical, and business support for companies in two open calls that will be launched during the project to develop and test new XR technologies for human-robot collaboration (see Sect. 4).
2. **Educational Objective: Create robotics educational material based on state-of-the-art XR technologies.** Educational material is one of the most important resources in any learning process, both in educational and professional fields. The MASTER project aims to deliver state-of-the-art material created using the XR platform. We aim to develop teaching material for XR in multiple formats and adapted to learners in relevant target sectors. The XR contents will address the three core technologies presented in Sects. 2.1, 2.2, and 2.3, and topics of third-party industry partners (see Sect. 3). The material will be validated by experts in robotics and education and by third parties selected in the second open call (see Sect. 4).

2 XR Platform

A key outcome of our project will be an XR platform with three main modules that refer to our technological objectives. The implementation requires consolidating, maturing, and unifying existing technologies from all partners into a single framework. We will extend the VIROO® platform [1], a virtual reality (VR) platform with a focus on industrial applications, for that purpose. It includes tools and services for easy creation, management, and deployment of immersive content for single-user and remote collaborative multi-user settings. For instance, VIROO® *Studio* enables low code VR content creation based on the Unity 3D engine [2], also empowering users without programming experience to create VR content. In addition, it maintains simplicity in certification processes required for industrial applications. VR contents are hosted in a cloud to enable quick and easy deployment and are managed using the *Portal* service. The three modules to be developed in the MASTER project cover the key technologies of the involved research partners. The modules will focus on creating safe robotic environments, programming flexible robotic applications, and integrating gaze-based interaction techniques using eye tracking sensors integrated into modern head-worn XR devices.

2.1 Safe Robotic Environments

In the MASTER project, as described in the technical objectives, one focus is on safety aspects in manufacturing. In general, safety systems, like cameras, curtains, safety mats, etc. are installed in the production lines, to protect the human operators from hazards such as getting hit by a robot or getting trapped by an unexpected robot movement. Nevertheless, such safety sensors do not include MR technologies, but these technologies can be used to provide real-time information on the status of the resources and their operations, to help humans be proactive in their actions. Thus, integrating MR in industrial settings enables the creation of intuitive and effective interaction between robots and humans to increase the operator's safety awareness and support easy robot programming. This can be achieved through the creation of flexible frameworks, able to simulate real-life assembly operations through immersive VR [15]. Combining VR and wearable robotics with exoskeleton technology can be a promising solution that could be used in multiple industrial areas and tasks, leveraging their benefits for the operators [23]. Beyond traditional VR systems that use wearable or portable devices, a 3-wall CAVE [3] consisting of three big projectors and four infrared cameras can be used to track multiple users virtually performing assembly tasks, aiming to visualize key performance indicators (KPIs) in order to support production engineers in decision-making [37]. Using augmented reality (AR) can enable real-time interaction between operators and the workplace. For instance, insights of experienced operators can be integrated into common execution flows through audio commands or manual robot guidance [20], also using smartwatch applications [51]. In addition, the provision of safety aspects, such as notifications or warnings, is a crucial feature for enabling safe human-robot collaboration [29]. AR can be used for visualizing the robot envelope of a static robot inside the operator's field of view using a depth sensor [21].

Although a lot of research has been done on how to exploit AR and VR in human-robot collaboration environments, there is a lack of integrated solutions that facilitate a seamless interaction of robots with human operators in the manufacturing environment. The lack of interfaces to support human operators in cases of unexpected events related to the robot is an important limitation in the existing literature that will be addressed in our project.

2.2 Programming Flexible Robotic Applications

The demand for a high degree of personalization in production is a feature of the Industry 4.0 production paradigm. Satisfying these demands requires that industry operators become agile in generating robot programs without writing code. Various advanced techniques have been proposed in the literature, referred to as **robot teaching** [54]. In the robot teaching paradigm, rather than explicitly writing code, the operator interacts with the robot through natural (i.e., human and unlearned) communication channels to transfer knowledge to the robot. While doing so, the code, corresponding to the skills taught to the robot, is generated automatically behind the scenes. **Kinaesthetic programming-by-demonstration (KPbD)** is a key set of techniques developed to achieve this robot teaching goal. We bring to the MASTER project expertise in developing and studying such techniques, including contributions to the body of relevant prior work. In particular, we have contributed by developing multimodal and multi-channel interactions that merge, in the same collaborative application scenario, a real robot and its holographic digital twin (DT) [45, 46].

In the MASTER project, we will advance the codeless programming of robot actions and routines through natural multimodal interaction with its holographic twin via manual guidance, voice, and gestures. Interactions with the robot to teach new skills will be assisted by a **virtual agent** designed and implemented in the project. Foundational operations for robot routine programming will be first supported as building blocks for the programming of collaborative production processes. These basic operations will support the programming of trajectories for a robot to approach its end effector to a region of interest. Then, support for the precise positioning of the end effector of a robot on selected points of interest is discovered in the region of interest that the robot previously approached. Precise positioning will be achieved with the help of machine vision algorithms that the operator will invoke through interaction with the virtual agent.

To teach approach trajectories, the KPbD technique will be used through interaction with the hologram of the robot. This will involve teaching robot trajectories, actions, and decision-making criteria, by moving the holographic digital twin robot by hand, either in a VR scene with respect to a virtual version of the production environment or in an MR environment within a real production context, which may include the physical version of the robot in addition to its holographic DT. In addition to our references cited above, KPbD in XR is present in the literature as visual MR implementations of the KPbD technique. Using a HoloLens device, pick-and-place action programming is described in [38, 42]. To improve the perception of physicality (absent in the handling of a hologram), the authors provided in a previous project tactile feedback using ultrasound actuators, resulting in a visuotactile interaction that improved the user experience (UX) of handling a hologram [45]. In VR, a description of a controller-free implementation of similar pick-and-place interactions exists [32]. To implement precise positioning capabilities for the end effector, machine vision-assisted techniques will be used, such as a **look and move (L&M) technique**. With an onboard camera calibrated with its end effector, the robot visually inspects the region of interest from close-up positions. The user will engage with the virtual agent to perform detailed visual analysis using machine vision applications that can identify features of interest in the image captured by the robot. When programming approach and precise-positioning trajectories, XR technologies will allow the user to preview simulated visualizations of the executions of the trajectories. This will help the user decide about the proposed trajectories' correctness before adding them to the program. This can all be done through interaction with the holographic version of the robot. The operator will use these tools by interacting with the virtual agent through an interface in the XR device, which will also support voice interaction based on semantic technologies, for which we also bring in expertise (e.g., [33]). This agent will monitor the progress of the approach-and-position sequence and provide guidance and step-by-step instructions to help the user interpret intermediate results, decide on the next steps, and complete the programming task.

This module of the MASTER project will be developed for and validated with two highly relevant real-world production scenarios: a machine tending scenario and a line following scenario. The **machine tending scenario** will involve the robot picking unsorted parts and feeding them one by one in an orderly way to a specific point of the production process. For that, the robot must first perform an approach to the region of interest in which the parts to be picked are located. Then, using machine vision analysis, the robot must identify features of the parts that lead it to identify a grasping point. After that, the robot will grasp a part (completing the L&M routine) and feed it to the process. With this module, the operator will use KPbD and L&M functionalities and, by interacting with the interface of the virtual agent, generate a behavior tree as the executable program for the robot to perform the machine tending task. The **line following scenario** will reflect tasks in which the robot must follow lines with high precision, such as when applying welding along a line or when fixing a crack in the material. For that, using the KPbD functionality of the module, the robot will first approach the region containing the line's starting point that will be followed. Then, the robot will find the starting point of the line that it will follow. This will be done with machine vision image analysis features of the L&M functionality of the module, invoked through interaction with the virtual agent. This interaction will be used to define the criteria for the robot to follow the line while acting on it (welding, applying a chemical treatment, etc.). Criteria to determine that the end of the line has been reached will be determined in a similar way. The whole sequence of actions and all decision criteria will be added to the executable behavior tree.

2.3 Interaction Based on Eye Tracking

In the MASTER project, we implement gaze-based interaction methods for XR and investigate their impact on educational XR interfaces for training workers in robotics. Our goals include reducing extraneous cognitive load [34, p. 39] induced by the XR-based education system to enable better learning and to integrate a reliable method for, e.g., observing a user's learning progress. We bring in our expertise in cyber-physical manufacturing systems [47], human-in-the-loop systems [7], and, particularly, active and passive gaze-based interfaces. In active gaze-based interfaces, users explicitly move their eyes to interact with a digital system, for instance, for object selection or manipulation. Passive gaze-based interfaces implicitly observe users' eye movements to adapt the interaction to their current state, such as cognitive load. In addition, we aim to develop novel interactive machine learning (IML) methods that facilitate the continuous improvement of user models or interaction methods in the MASTER project.

Eye trackers have evolved from pure diagnostic sensors to input devices for real-time interactive systems [16]. This was partially driven by advances in eye tracking hardware over the last decade concerning the devices' affordability, performance, and form factor. However, the characteristics of the human visual system, including the eyes, are essential to this development. Directing the gaze to a specific spot is very efficient, and the gaze direction implicitly reveals where a human allocates their limited attentional resources [22, p. 26]. This visual attention can be attracted by generally appealing, bottom-up factors and goal- or task-oriented, top-down factors [13]. Modern eye trackers can capture and stream a user's gaze directions in real-time, allowing developers to use the human gaze as an input to interactive systems. The gaze signal can be captured in stationary settings using remote eye trackers and in mobile settings using head-mounted eye trackers. Remote eye trackers have a fixed position in the world, usually attached to a screen, and allow little to no movement by a user. Head-mounted eye trackers are worn by the user like glasses and allow more freedom of movement. Eye trackers in MR glasses are head-mounted eye trackers. Still, interaction settings can be stationary if users do not move. From stationary to mobile interaction settings, gaze can be employed as an active input modality or as a passive sensor-based cue, which spans the design space for gaze-informed multimodal interaction introduced by [43, pp. 382–384]. Gaze-informed multimodal interaction directly relates to the theory of multimodal-multisensor interfaces, i.e., interfaces that "combine one or more user input modalities with sensor information" [39, p. 3].

Active Gaze-Based Interfaces When the human gaze signal is used as an active input mode, i.e., in active gaze-based interfaces, a user's intentional eye movements directly impact the interaction. Examples include gaze-based selection [48] and text input [19]. We plan to integrate an active gaze-based interaction method that supports the educational use case. This can be achieved by transferring an existing interaction method to an MR setting or by optimizing an existing MR interaction technique based on eye tracking for the educational use case at hand. We aim to investigate the impact of using eye tracking versus no eye tracking in educational settings. We assume that using eye gaze for interacting with MR interfaces can reduce the required effort for the interaction and leaves more capacities for learning. The related cognitive load theory states that reducing extraneous cognitive load can lead to better learning outcomes [34, 53]. Recent examples of active gaze-based interfaces in MR include the selection of virtual objects [31], the manipulation of virtual objects, i.e., scaling, rotating, and translating them [41], the interaction with virtual menus [44], and gaze-based text entry systems [30].

Passive Gaze-Based Interfaces When the human gaze signal is used as a passive input mode, i.e., in passive gaze-based interfaces, naturally occurring eye movement behavior is monitored and can indirectly impact the interaction. Typically, the gaze signal is used to infer information about the user, such as the ongoing activity [28] or the induced cognitive load of an activity [24], to tailor the interaction to the user and their current state. This includes two stages, the detection of certain states and a suitable adaptation to them. Good results have been observed in estimating participants' expertise based on their eye movements [14]. Other researchers investigated whether the change in knowledge can be observed from eye movements [11] and whether user confusion can be detected [27]. In prior work, we investigated whether eye movements and fixated visual stimuli can be used to predict the target of a visual search [9] or the relevance of a text snippet read by a user [6]. Also, we implemented and evaluated methods for detecting what objects have been fixated based on head-mounted eye trackers [8]. First attempts have been made to adapt user interfaces based on a user's preferences determined through eye tracking. One example concerned the presentation of details for flat advertisements based on the perceived relevance per user [18]. Such monitoring methods can complement multimodal learning analytics tools, which aim at analyzing students' learning behavior and progress [12]. We aim to identify a user characteristic that can be reliably tracked by observing a user's eye movements in the context of teaching robotics via XR headsets. However, our experiences show modeling user states is connected to certain risks. Typically, it is hard to identify generic patterns that generalize across different users, either because we cannot collect sufficient data through user studies or because eye movement behavior is connected to individual experiences and varies significantly between users. Our goal in the MASTER project is to measure a user characteristic relevant to the educational use case with high validity and reliability. We can imagine showcasing our final method in an adaptive educational user interface that adapts to the state of this characteristic or in a dashboard for teachers that visualizes the characteristic for multiple learners.

Interactive Machine Learning (IML) In interactive machine learning systems, the user engages in a two-way interaction with a machine learning system. The goal is to improve the model based on user feedback [17]. Such human-in-the-loop machine learning systems are very relevant when transferring non-perfect machine learning systems to real use cases in the industry. Typically, the datasets for training are not large enough to reach a sufficient level of accuracy in predictions, or the task is very sensitive to wrong classifications and requires human quality assurance. We aim to apply the principle of interactive machine learning to machine learning models in the context of gaze-based interaction. For instance, passive gaze-based interfaces rely on models that take human eye movements as input to predict a certain user characteristic or to understand which elements of a virtual or real scene are viewed by a user in which order. Such models typically suffer from a low number of training samples because collecting and annotating data is expensive and time-consuming. We envision a system for interactive data exploration or interactive modeling of user characteristics based on recorded or real-time eye movement data. One concrete example is the semi-automatic annotation of mobile eye tracking data based on computer vision models [8, 25, 40, 52]. Such models must be adjusted for every new context. IML can enable education experts with no prior machine learning experience to apply machine learning technology [5, 26].

3 Didactic Approach

Educational technology is "the study and ethical practice of facilitating learning and improving performance by creating, using, and managing appropriate technological processes and resources" [10, p. 1]. The use of XR technology as an effective educational technology in diverse educational and training settings has been the subject of extensive research for many years. Alnagrat et al. [4] presented a related survey that highlights the potential benefits of XR for training and education, such as greater user engagement, lower (operational) costs, better knowledge acquisition and retention, training as often as needed, and training in safe environments. Recent advances and increased affordability of XR technologies have led to a proliferation of related educational resources. The major challenges remain in corporate training, technical and professional higher education, and, generally, where complex technical skills or high-stake learning goals are required. An example is teaching robotics in manufacturing, assisted by XR. Our vision in the MASTER project is that industrial workers, vocational and higher education students of robotics, robotics teachers and trainers, and other stakeholders in real-world education and training environments will be immersed in high-quality teaching and learning experiences based on XR technology.

To this end, the XR platform will extend the existing VIROO® platform with additional functionalities arising from the technological areas of the research partners. These include the creation of safe robotic environments, the programming of flexible robotic applications, and the integration of gaze-based interaction techniques. These technologies must be merged into an effective educational technology with the learning content from specific educational use cases. This shall promote the learning outcomes formulated in the targeted robotics domains in our open call projects. The educational system will be configured functionally in compliance with the principles of effective teaching for complex learning in technical contexts, as these have been operationalized in instructional models like the ones introduced in [35, 36].

The didactic approach in the XR platform is to have problem classes and associated problem sequences related to robotics study targets, such as collaborative robot programming. This content and supporting information shall enable progressive and self-directed learning of students. It will be presented with pertinent immersive and realistic characteristics and required interactions. The system will be equipped with functionalities to manage the teaching and learning processes and facilitate the work of teachers and administrators, including capabilities for content management, student access management, and tracking students' activity and learning progression. In addition, the XR platform will also allow teachers, trainers, or instructional designers to plan new activity scenarios, teaching and learning sequences, and XR content. The XR platform and educational content will be conceived considering audiences' specific expectations, requirements, and feedback. They will be formatively evaluated during the second open call, as described next.

4 Evaluation via Open Calls

The MASTER project aims to develop and deliver a high-quality platform and training content to effectively support the project objectives. This platform will be supported by two open calls. The first one targets technical SMEs to integrate their XR developments into the platform, enhancing its capabilities. The second one shall attract educational institutes that can prepare training material using the above tools and test the proposed functionalities. A validation methodology will be proposed and iteratively improved during the project to achieve this. Through this experimental validation process, the project will focus on evaluating user acceptance, usability, and trainees' perceptions of safety. It shall provide the necessary feedback throughout the design and development phases of the XR platform and training material, ensuring continuous improvement and refinement. The first open call addresses technical challenges, supporting and funding satellite projects focusing on the technologies being developed. This way, a knowledge-sharing channel will be created while raising awareness among various companies and organizations regarding the project's advancements and potential. The technologies developed by the core partners of the MASTER project and those provided by the open call participants will be evaluated via experiments designed and implemented during the project's lifecycle. The second open call addresses educational challenges. The developed technologies and the created training content will be evaluated through empirical user studies. This approach allows an inclusive evaluation of the platform and the training content, considering the various perspectives and needs of our broad target audience. Findings from these studies will be used to improve the final platform and didactic materials through an iterative design process.

5 Conclusion

In this paper, we presented the EU-funded project MASTER. We summarized the aim of the project, which is to advance the application of innovative XR technologies in robotics education for manufacturing companies. Six objectives were defined, namely, four technical, one educational, and one industrial. The technical objectives include the development of an open XR platform that is extended by key functionalities for creating safe robotic environments, programming flexible robotic applications, and integrating advanced interaction techniques based on eye tracking. We also described the planned technological developments in detail. The educational objective addresses the creation of related learning material. In this context, we detailed our didactic considerations and the challenges connected to educating workers. The industrial objective addresses the support and the transfer of our technologies to SMEs and start-up companies. Further, we described the evaluation methodology. We planned two open calls: the first enables technology transfer and evaluation in realistic settings; the second focuses on the assessment of the educational material that will be created by open call beneficiaries. News and recent developments concerning the MASTER project will be published on our project website https://www.master-xr.eu.

Acknowledgements This work is co-funded by the European Union (EU) under grant number 101093079 (project MASTER, https://www.master-xr.eu). Thanks to Abdulrahman Mohamed Selim and Nathalie Barz for the very useful feedback on the initial draft and to the European Science Communication Institute (ESCI) gGmbH for the permission to use the image.

References

1. Enterprise VR made easy with VIROO® (2023). https://www.virtualwareco.com/viroo/
2. Unity Real-Time Development Platform | 3D, 2D, VR & AR Engine (2023). https://unity.com
3. Virtual Reality CAVE Solutions (2023). https://steantycip.com/vr-cave/
4. Alnagrat, A., Che Ismail, R., Syed Idrus, S. Z., & Abdulhafith Alfaqi, R. M. (2022). A review of extended reality (XR) technologies in the future of human education: Current trend and future opportunity. *Journal of Human Centered Technology*, *1*(2), 81–96 (2022). https://doi.org/10.11113/humentech.v1n2.27
5. Barz, M., Bhatti, O. S., Alam, H. M. T., Nguyen, D. M. H., & Sonntag, D. (2023). Interactive fixation-to-AOI mapping for mobile eye tracking data based on few-shot image classification. In *Companion proceedings of the 28th international conference on intelligent user interfaces. IUI '23 Companion* (pp. 175–178). New York, NY, USA: Association for Computing Machinery. https://doi.org/10.1145/3581754.3584179, event-place: Sydney, NSW, Australia
6. Barz, M., Bhatti, O. S., & Sonntag, D. (2022). Implicit estimation of paragraph relevance from eye movements. *Frontiers in Computer Science, 3*. https://doi.org/10.3389/fcomp.2021.808507, https://www.frontiersin.org/article/10.3389/fcomp.2021.808507
7. Barz, M., Poller, P., Schneider, M., Zillner, S., & Sonntag, D. (2017). Human-in-the-loop control processes in gas turbine maintenance. In V. Mařík, W. Wahlster, T. Strasser, & P. Kadera (Eds.), *Industrial applications of holonic and multi-agent systems* (pp. 255–268). Cham: Springer International Publishing.

8. Barz, M., & Sonntag, D. (2021). Automatic visual attention detection for mobile eye tracking using pre-trained computer vision models and human gaze. *Sensors*, *21*(12), 4143 (2021). https://www.mdpi.com/1424-8220/21/12/4143, number: 12 Publisher: Multidisciplinary Digital Publishing Institute
9. Barz, M., Stauden, S., & Sonntag, D. (2020). Visual search target inference in natural interaction settings with machine learning. In: A. Bulling, A. Huckauf, E. Jain, R. Radach, & D. Weiskopf, (Eds.), *ACM symposium on eye tracking research and applications. ETRA '20 full papers* (pp. 1–8). Association for Computing Machinery (2020). https://doi.org/10.1145/3379155.3391314
10. Betrus, A. K., Branch, R. M., Doughty, P., Molenda, M., Pearson, R., Persichitte, K. A., Rezabek, L. L., Robinson, R. S., Stoddard, C., & Januszewski, A. (2008). Definition. In A. Januszewski, & M. Molenda, (Eds.), *Educational technology: A definition with commentary* (2nd edn., p. 384). Routledge. https://www.routledge.com/Educational-Technology-A-Definition-with-Commentary/Januszewski-Molenda/p/book/9780805858617
11. Bhattacharya, N., & Gwizdka, J. (2018). Relating eye-tracking measures with changes in knowledge on search tasks. In *Proceedings of the 2018 ACM symposium on eye tracking research and applications. ETRA '18*. New York, NY, USA: Association for Computing Machinery. https://doi.org/10.1145/3204493.3204579, event-place: Warsaw, Poland
12. Blikstein, P. (2013). Multimodal learning analytics. In *Proceedings of the third international conference on learning analytics and knowledge. LAK '13* (pp. 102–106). New York, NY, USA: Association for Computing Machinery. https://doi.org/10.1145/2460296.2460316, event-place: Leuven, Belgium
13. Borji, A., & Itti, L. (2013). State-of-the-art in visual attention modeling. *IEEE Transactions on Pattern Analysis and Machine Intelligence, 35*(1), 185–207. https://doi.org/10.1109/TPAMI.2012.89
14. Castner, N. J. (2020). Gaze and visual scanpath features for data-driven expertise recognition in medical image inspection. Dissertation, Universität Tübingen. https://doi.org/10.15496/publikation-51677, https://publikationen.uni-tuebingen.de/xmlui/handle/10900/110301, accepted: 2020-12-07T07:53:09Z ISBN: 9781742181882
15. Dimitropoulos, N., Togias, T., Michalos, G., & Makris, S. (2020). Framework enabling the design of virtual environments used for simulation of assembly operations. *Procedia Manufacturing*, *51*, 571–576 (2020). https://doi.org/10.1016/j.promfg.2020.10.080, https://www.sciencedirect.com/science/article/pii/S2351978920319375
16. Duchowski, A. T. (2018). Gaze-based interaction: A 30 year retrospective. *Computers and Graphics, 73*, 59–69 (2018). https://doi.org/10.1016/j.cag.2018.04.002, https://www.sciencedirect.com/science/article/pii/S0097849318300487
17. Dudley, J. J., & Kristensson, P. O. (2018). A review of user interface design for interactive machine learning. *ACM Transactions on Interactive Intelligent Systems, 8*(2). https://doi.org/10.1145/3185517, place: New York, NY, USA Publisher: Association for Computing Machinery
18. Feit, A. M., Vordemann, L., Park, S., Berube, C., & Hilliges, O. (2020). Detecting relevance during decision-making from eye movements for UI adaptation. In *ACM symposium on eye tracking research and applications. ETRA '20 full papers*. New York, NY, USA: Association for Computing Machinery. https://doi.org/10.1145/3379155.3391321, event-place: Stuttgart, Germany
19. Feng, W., Zou, J., Kurauchi, A., Morimoto, C. H., & Betke, M. (2021). HGaze typing: head-gesture assisted gaze typing. In *ACM symposium on eye tracking research and applications. ETRA '21 full papers*. New York, NY, USA: Association for Computing Machinery. https://doi.org/10.1145/3448017.3457379, event-place: Virtual Event, Germany
20. Gkournelos, C., Karagiannis, P., Kousi, N., Michalos, G., Koukas, S., & Makris, S. (2018). Application of wearable devices for supporting operators in human-robot cooperative assembly tasks. *Procedia CIRP, 76*, 177–182. https://doi.org/10.1016/j.procir.2018.01.019, https://www.sciencedirect.com/science/article/pii/S2212827118300325
21. Hietanen, A., Pieters, R., Lanz, M., Latokartano, J., & Kämäräinen, J. K. (2020). AR-based interaction for human-robot collaborative manufacturing. *Robotics and Computer-*

Integrated Manufacturing, 63, 101891. https://doi.org/10.1016/j.rcim.2019.101891, https://www.sciencedirect.com/science/article/pii/S0736584519307355

22. Holmqvist, K., & Andersson, R. (2011). *Eye tracking: A comprehensive guide to methods, paradigms and measures*. Lund, Sweden: Lund Eye-Tracking Research Institute.
23. Karvouniari, A., Michalos, G., Dimitropoulos, N., & Makris, S. (2018). An approach for exoskeleton integration in manufacturing lines using virtual reality techniques. *Procedia CIRP, 78*, 103–108. https://doi.org/10.1016/j.procir.2018.08.315, https://www.sciencedirect.com/science/article/pii/S2212827118312289
24. Klingner, J. M.: Measuring cognitive load during visual tasks by combining pupillometry and eye tracking. Ph.D., Stanford University, Department of Computer Science. http://purl.stanford.edu/mv271zd7591
25. Kumari, N., Ruf, V., Mukhametov, S., Schmidt, A., Kuhn, J., & Küchemann, S. (2021). Mobile eye-tracking data analysis using object detection via YOLO v4. *Sensors, 21*(22). https://doi.org/10.3390/s21227668, https://www.mdpi.com/1424-8220/21/22/7668
26. Kurzhals, K. (2021). Image-based projection labeling for mobile eye tracking. In ACM symposium on eye tracking research and applications. New York, NY, USA: Association for Computing Machinery. https://doi.org/10.1145/3448017.3457382
27. Lallé, S., Toker, D., & Conati, C. (2021). Gaze-driven adaptive interventions for magazine-style narrative visualizations. *IEEE Transactions on Visualization and Computer Graphics, 27*(6), 2941–2952. https://doi.org/10.1109/TVCG.2019.2958540
28. Li, Y., Liu, M., & Rehg, J. M. (2018). In the eye of beholder: Joint learning of gaze and actions in first person video. In V. Ferrari, M. Hebert, C. Sminchisescu, & Y. Weiss (Eds.), *Computer vision - ECCV 2018* (pp. 639–655). Cham: Springer International Publishing.
29. Lotsaris, K., Fousekis, N., Koukas, S., Aivaliotis, S., Kousi, N., Michalos, G., & Makris, S. (2021). Augmented reality (AR) based framework for supporting human workers in flexible manufacturing. *Procedia CIRP, 96*, 301–306 (2021). https://doi.org/10.1016/j.procir.2021.01.091, https://www.sciencedirect.com/science/article/pii/S2212827121001190
30. Lystbæk, M. N., Pfeuffer, K., Grønbæk, J. E. S., & Gellersen, H. (2022). Exploring gaze for assisting freehand selection-based text entry in AR. *Proceedings of the ACM on Human-Computer Interaction, 6*(ETRA). https://doi.org/10.1145/3530882, place: New York, NY, USA Publisher: Association for Computing Machinery
31. Lystbæk, M. N., Rosenberg, P., Pfeuffer, K., Grønbæk, J. E., & Gellersen, H. (2022). Gaze-hand alignment: Combining eye gaze and mid-air pointing for interacting with menus in augmented reality. *Proceedings of the ACM on Human-Computer Interaction, 6*(ETRA). https://doi.org/10.1145/3530886, place: New York, NY, USA Publisher: Association for Computing Machinery
32. Masurovsky, A., Chojecki, P., Runde, D., Lafci, M., Przewozny, D., & Gaebler, M. (2020). Controller-free hand tracking for grab-and-place tasks in immersive virtual reality: Design elements and their empirical study. *Multimodal Technologies and Interaction, 4*(4) (2020). https://doi.org/10.3390/mti4040091, https://www.mdpi.com/2414-4088/4/4/91
33. Maurtua, I., Fernández, I., Tellaeche, A., Kildal, J., Susperregi, L., Ibarguren, A., & Sierra, B. (2017). Natural multimodal communication for human-robot collaboration. *International Journal of Advanced Robotic Systems, 14*(4), 1729881417716043. https://doi.org/10.1177/1729881417716043, publisher: SAGE Publications
34. Mayer, R. E. (Ed.) (2014). *The Cambridge handbook of multimedia learning*. Cambridge handbooks in psychology (2 edn.). Cambridge: Cambridge University Press. https://doi.org/10.1017/CBO9781139547369
35. van Merrienboer, J. (1997). *Training complex cognitive skills: A four-component instructional design model for technical training*. Englewood Cliffs, NJ: Educational Technology Publications.
36. van Merriënboer, J. J. G., & Kirschner, P. A. (2012). *Ten steps to complex learning: A systematic approach to four-component instructional design* (2nd edn.). New York: Routledge. https://doi.org/10.4324/9780203096864
37. Michalos, G., Karvouniari, A., Dimitropoulos, N., Togias, T., & Makris, S. (2018). Workplace analysis and design using virtual reality techniques. *CIRP Annals, 67*(1), 141–144

(2018). https://doi.org/10.1016/j.cirp.2018.04.120, https://www.sciencedirect.com/science/article/pii/S0007850618301446
38. Ostanin, M., Mikhel, S., Evlampiev, A., Skvortsova, V., Klimchik, A. (2020). Human-robot interaction for robotic manipulator programming in mixed reality. In *2020 IEEE international conference on robotics and automation (ICRA)* (pp. 2805–2811). https://doi.org/10.1109/ICRA40945.2020.9196965
39. Oviatt, S., Schuller, B., Cohen, P. R., Sonntag, D., Potamianos, G., & Krüger, A. (2017). Introduction: Scope, trends, and paradigm shift in the field of computer interfaces. In *The handbook of multimodal-multisensor interfaces* (pp. 1–15). New York, NY, USA: Association for Computing Machinery and Morgan & Claypool. https://doi.org/10.1145/3015783.3015785
40. Panetta, K., Wan, Q., Kaszowska, A., Taylor, H. A., & Agaian, S. (2019). Software architecture for automating cognitive science eye-tracking data analysis and object annotation. *IEEE Transactions on Human-Machine Systems, 49*(3), 268–277. https://doi.org/10.1109/THMS.2019.2892919, https://ieeexplore.ieee.org/document/8633339/
41. Pfeuffer, K., Mayer, B., Mardanbegi, D., & Gellersen, H. (2017). Gaze + pinch interaction in virtual reality. In *Proceedings of the 5th symposium on spatial user interaction. SUI '17* (pp. 99–108). New York, NY, USA: Association for Computing Machinery. https://doi.org/10.1145/3131277.3132180, event-place: Brighton, United Kingdom
42. Quintero, C. P., Li, S., Pan, M. K., Chan, W. P., Machiel Van der Loos, H., & Croft, E. (2018). Robot programming through augmented trajectories in augmented reality. In *2018 IEEE/RSJ international conference on intelligent robots and systems (IROS)* (pp. 1838–1844). https://doi.org/10.1109/IROS.2018.8593700
43. Qvarfordt, P. (2017). Gaze-informed multimodal interaction. In *The handbook of multimodal-multisensor interfaces: Foundations, user modeling, and common modality combinations* (Vol. 1, pp. 365–402). Association for Computing Machinery and Morgan & Claypool (2017). https://doi.org/10.1145/3015783.3015794
44. Reiter, K., Pfeuffer, K., Esteves, A., Mittermeier, T., & Alt, F. (2022). Look and turn: One-handed and expressive menu interaction by gaze and arm turns in VR. In *2022 symposium on eye tracking research and applications. ETRA '22*. New York, NY, USA: Association for Computing Machinery. https://doi.org/10.1145/3517031.3529233, event-place: Seattle, WA, USA
45. Rivera-Pinto, A., Kildal, J., & Lazkano, E. (2020). Multimodal mixed reality impact on a hand guiding task with a holographic cobot. *Multimodal Technologies and Interaction, 4*(4) (2020). https://doi.org/10.3390/mti4040078
46. Rivera-Pinto, A., Kildal, J., & Lazkano, E. (2023). Toward programming a collaborative robot by interacting with its digital twin in a mixed reality environment. *International Journal of Human-Computer Interaction, 1–13*. https://doi.org/10.1080/10447318.2023.2221599, publisher: Taylor & Francis
47. Sonntag, D., Zillner, S., van der Smagt, P., & Lörincz, A. (2017). Overview of the CPS for smart factories project: Deep learning, knowledge acquisition, anomaly detection and intelligent user interfaces. In S. Jeschke, C. Brecher, H. Song, & D. B. Rawat, (Eds.), *Industrial internet of things: Cybermanufacturing systems* (pp. 487–504). Cham: Springer International Publishing. https://doi.org/10.1007/978-3-319-42559-7_19
48. Stellmach, S., & Dachselt, R. (2012). Look and touch: Gaze-supported target acquisition. In *Proceedings of the SIGCHI conference on human factors in computing systems* (pp. 2981–2990). New York, NY, USA: Association for Computing Machinery. https://doi.org/10.1145/2207676.2208709
49. Thees, M., Altmeyer, K., Kapp, S., Rexigel, E., Beil, F., Klein, P., Malone, S., Brünken, R., & Kuhn, J. (2022). Augmented reality for presenting real-time data during students' laboratory work: comparing a head-mounted display with a separate display. *Frontiers in Psychology, 13* (2022), https://www.frontiersin.org/articles/10.3389/fpsyg.2022.804742
50. Vigkos, A., Pauer, A., Bevacqua, D., Turturro, L., & Kulesza, M. (2021). XR and its potential for Europe. Technical report, ECORYS, Brussels. https://xreuropepotential.com/assets/pdf/ecorys-xr-2021-report.pdf

51. Villani, V., Sabattini, L., Battilani, N., & Fantuzzi, C. (2016). Smartwatch-enhanced interaction with an advanced troubleshooting system for industrial machines. *IFAC-PapersOnLine, 49*(19), 277–282. https://doi.org/10.1016/j.ifacol.2016.10.547, https://www.sciencedirect.com/science/article/pii/S2405896316321371
52. Wolf, J., Hess, S., Bachmann, D., Lohmeyer, Q., & Meboldt, M. (2018). Automating areas of interest analysis in mobile eye tracking experiments based on machine learning. *Journal of Eye Movement Research, 11*(6). https://doi.org/10.16910/jemr.11.6.6, https://bop.unibe.ch/JEMR/article/view/4415, section: Articles
53. Zhou, J., Yu, K., Chen, F., Wang, Y., & Arshad, S. Z. (2018). Multimodal behavioral and physiological signals as indicators of cognitive load. In *The handbook of multimodal-multisensor interfaces: signal processing, architectures, and detection of emotion and cognition* (Vol. 2, pp. 287–329). Association for Computing Machinery and Morgan & Claypool. https://doi.org/10.1145/3107990.3108002
54. Zhou, Z., Xiong, R., Wang, Y., & Zhang, J. (2020). Advanced robot programming: A review. *Current Robotics Reports, 1*(4), 251–258. https://doi.org/10.1007/s43154-020-00023-4

Semantically Enriched Interfaces for Product Configuration

Stefan Berlik, Philipp Hartmann, and Jannick Stranghöner

Abstract This paper highlights the challenges of product customization in the context of mass customization, particularly the combinatorial explosion of product variants that arises when trying to offer lot size one production. A knowledge graph can be used to store the required product knowledge including all potential products that can be manufactured. This paper proposes a method of reducing the complexity of the resulting configuration problem using automated reasoning. The approach decreases the number of individuals and by extension the number of product variants involved in the problem by materializing constraint results and grouping equivalent specifications. As opposed to many existing approaches, the contribution fully solves the underlying constraint satisfaction problem inside the knowledge graph without the need of an external solution. The approach still benefits from facilitated data interchange and the simple implementation of various downstream tasks.

1 Introduction and Motivation

Mass customization as a new production concept combines the individuality of one-off production with the efficiency of mass production. Customer-specific requirements can be met with customized products without having to sacrifice the high precision and low costs of mass production. The advantages of this approach are clear: new customers can be won and higher margins achieved. However, variant diversity and individualization also have their price and go hand in hand with high complexity, requiring a modularized product portfolio and redesigned business and

S. Berlik (✉) · P. Hartmann · J. Stranghöner
Hochschule Bielefeld, University of Applied Sciences and Arts, Bielefeld, Germany
e-mail: stefan.berlik@hsbi.de

P. Hartmann
e-mail: philipp.hartmann@hsbi.de

J. Stranghöner
e-mail: jannick.stranghoener@hsbi.de

M.-R. Alam and M. Fathi (eds.), *Integrated Systems: Data Driven Engineering*,
https://doi.org/10.1007/978-3-031-53652-6_11

production processes. The effective and efficient realization of the mass customization approach therefore requires accompanying measures.

From a conceptual point of view, the focus of interest is initially on the configuration process with its core component, the configurator. Based on a product knowledge base, it makes it possible to interactively design the desired product. Ideally, no specialized technical knowledge is required for this. This means that sales staff and even customers can quickly find 'their' product without the help of product engineers. The configurator first ensures that valid specifications are created—but can also generate other documents required during the production process, thus reducing the error rate and lead time.

The implementation process is both complex and time-consuming, requiring various steps such as analyzing and modeling the company's product range as a knowledge base, selecting appropriate configuration software, programming software components, and continuously improving the configuration system. The use of artificial intelligence methods could prove to be a critical factor for success, for example to partly generate the product knowledge base automatically. To achieve this, it is necessary to first define the syntax and semantics of the underlying knowledge graph in a suitable manner.

Having a well defined knowledge graph, where each part is annotated with a set of properties, one can reason about the compatibility of parts. Compatibility is determined by considering the interactions between these properties, which may include factors such as physical dimensions, mechanical properties, electrical properties, and protocol properties. To simplify this process, interfaces can be used to encapsulate these factors, and compatibility between two interfaces can be established through a set of rules that describe how their properties interact.

However, even with these measures, determining compatibility can be a challenging task as it involves considering a large number of factors, leading to a combinatorial explosion of possible product variants. As such, we propose augmenting the interface concept to help streamline this process.

2 Related Work

Our review of the related work begins with an overview of the fundamental concepts and theories related to the topic, followed by a discussion of previous studies and research works that have been conducted in the area of interest. We will provide a critical analysis of the key findings, methodologies, and contributions of the existing research, highlighting the gaps in the current knowledge that our research aims to fill.

Configuration systems allow valid product variants to be specified in accordance with customer requirements. The configuration knowledge about the products is usually recorded manually by the product manager and the product developers in meetings in the form of a Product Variant Master. Transformed into a set of rules (constraints), this results in the basis of an expert system. For the configuration of less complex products, systems based on maximum BOMs (bill of materials) have also proven their worth. [1, 2]

In the context of configuration systems, an *interface* is a boundary or a point of interaction between different components or modules of a system. It defines a set of rules, protocols, and conventions that ensure physical fit and govern the exchange of information and commands between these components. Interfaces play a crucial role in configuration systems as they enable the seamless integration of different hardware and software components, allowing them to work together in a coordinated and efficient manner [1, 3].

Knowledge graphs (KG) are directed, annotated multigraphs that, as knowledge bases, are the key component for a whole range of knowledge-driven applications, such as question answering [4], decision support systems [5], recommender systems [6], just to mention a few. Among the possible technologies that can be used to model the knowledge base of a configuration system, KGs stand out as a particularly suitable method because they link entities and concepts into relationships and represent them as graphs with defined syntax and semantics [7]. Processing knowledge in the form of KGs is one of the key skills for computers to capture the meaning of entities and the relationship between entities, much like humans do [8]. Knowledge graph construction has been an attractive research topic for decades, leading to many general KGs such as DBPedia [9] or Google Knowledge Vault [10], but also numerous domain-specific knowledge bases as for example WordNet [11].

An *ontology* as an explicit, formal specification is used to conceptualize a delimited domain of discourse. For this purpose, the concepts and properties of the domain and their relations are precisely described via axioms and constraints in languages such as OWL/RDF [12]. This offers the advantage for knowledge graphs to implement and thus connect to other common ontologies like GoodRelations [13]. This results in many advantages, especially in the context of configuration. One of the key benefit is sharing and advertising of semantically defined configurations [13]. Thus companys could configure every product any company can offer. Additionally, this process can be automated since ontologies allow semantic interpretation for algorithms.

Ontologies to describe the topology, structure, functions and constraints of customizable products are plentiful. There have been several attempts towards a general ontology of configuration. Altuna et al. [14] introduce several key concepts, such as components, ports, resources and constraints. Similarly, Yang et al. [15] subdivided their ontology in three sub-ontologies, separating the product's topology and functions from the constraints set and the customer requirements. These ontologies utilize traditional modeling techniques such as UML, which cannot fully be mapped to Semantic Web Languages such as OWL. For this reason, the following ontologies are formulated exclusively in OWL.

Domain agnostic upper ontologies such as the Manufacturing Service Description Ontology (MSDL) and even GoodRelations exist, but they lack expressiveness to fully capture the intricacies of configurable products. Meaningful alignment to these upper ontologies is useful nonetheless to facilitate reuse and sharing of product knowledge.

One ontology that proves to be particularly useful is the Building Product Ontology (BPO) [16], for several reasons. Though initially developed for the buildings domain, it is generic enough to be applied to other products while providing useful terms for describing part-whole relationships and inter-connections between components. It also introduces a secondary instantiation mechanism (besides the standard instance-of relationship in OWL) via the component and entity class, where the latter is a realization or specific instance of the former. This can be applied to the configuration problem in a generic fashion, where a product family can be described as an assembly of components, whereas a specific product variant is an assembly of entities, each realizing a different component. This way, both the abstract product family itself as well as its instances are first order citizens of the graph and may have further information attached to them. To give an example, a configurable computer may be composed of the components CPU, Motherboard and RAM, among other components. A specific variant of this computer could be composed of the entities CPU A, Motherboard B and RAM C, each being an instance of the respective component. BPO allows us to describe both the topology, interconnections and attributes of the generic PC as well as the concrete variant in the same knowledge base. Due to these reasons, alignment of concepts with BPO will be established subclassing the component and entity class, and reusing as many object properties as feasible.

However, the ontology provides no way of expressing constraint knowledge. Some contributions model constraints via rules, either using SWRL rules (Semantic Web Rule language) [15] or more often SHACL shapes (Shapes Constraint Language) [17], to validate a given configuration or return valid value ranges. While the idea is certainly attractive in that constraint knowledge in the form of standardized rule languages can be exchanged easily between applications, reasoning at runtime is computationally expensive and difficult to scale in the context of a web configurator. Other applications solve the computationally hard part (the constraint satisfaction problem, CSP) in an offline phase and, by explicitly materializing the compatibility, mutual in- or exclusivity of entities, query that information at runtime. One approach that stood out to us was the *Configuration as Linked Data* ontology (COLD), which addressed the configuration of cars at Renault. The ontology considers partially defined products (PDPs), i. e., configurations where one or more component is not specified, as first order citizens of the graph as well. This naturally leads to the implementation of features such as auto-completion and conflict resolution. A large drawback of this approach is that the number of possible PDPs is huge (even larger than the number of possible products, which is 10^{20} for the Renault configurator), which is why the authors use partial and potentially inaccurate indexing.

In conclusion, relevant approaches to configuration systems that utilize Semantic Web Technologies from various perspectives, such as software engineering, symbolic AI, and business, have been reviewed. In this process the challenges of existing approaches were highlighted, such as the difficulty of expressing constraint knowledge through Semantic Web Languages without explicit runtime reasoning. We generally agree with the authors of COLD that PDPs deserve their status of first order objects, as that significantly simplifies downstream applications, and reuse their concepts while attempting to alleviate the computational complexity of the resulting PDP graph.

3 Product Configuration with Discrete Choices

When dealing with customizable products on an industrial scale, the number of potentially viable product variants may often exceed 10^{20}, making it infeasible to list every potential variant a customer could order. Instead, product families are characterized by a set of variables that represent the relevant descriptive attributes: A configurable computer may be fully characterized by the choice of CPU, Motherboard and GPU, among other variables. In a completely defined product, each such variable is assigned one and only one value. For the purposes of this paper, only discrete variables that have a finite list of possible values are of concern.

The large number of potential product variants, i.e., the cartesian product of variable ranges, is reduced by a set of constraints, usually by orders of magnitude. Solving for the valid variable ranges is therefore a CSP, which is, similar to related downstream tasks, computationally hard (SAT being NP-complete).

A configurator application is the primary way of presenting such complex, interdependent product ranges to customers. It is a decision tool that guides users to desirable and valid product configurations, assisting them in incrementally refining their choices from a set of remaining, valid options. Each step of the configuration process represents a valid partially defined product, in the sense that it can be completed into an existing product without altering any of the current selections made by the user.

A product configuration engine whose vocabulary is specified using Semantic Web languages provides many additional benefits. It allows seamless integration of product information, such as the product topology and its dependencies, sales and manufacturing information and customer preferences. This is not restricted to internal knowledge about the product; a car manufacturer could integrate publicly available product data from its suppliers and reason with it, without the need to define specific interfaces or build elaborate pipelines. Additionally the Semantic Web tech stack offers powerful rule languages, that may be used to reason about the validity of configurations, supply alternatives or complete partially defined configurations.

3.1 Requirements Towards a Configuration Engine

Functional requirements towards configurators that are commonly discussed in literature, and the proposed solution should comply with these requirements:

- **Completeness of inference** should be guaranteed. This means that all consequences of user choices are inferred, without leading into dead ends or requiring backtracking. [18]
- **Full information** concerning the consequences of a choice should be communicated to the user. This includes implied options as well as pricing and delivery information.
- **Freedom of choice**. The user should have the ability to decide which component to select at any given point, to make negative choices and to choose no option at all. [18]
- **Permissiveness and conflict resolution**. The user is allowed to change its mind about previous selection, even when that choice is inconsistent. It is the configurator's task to restore consistency and find the minimal set of former choices that have to be revised. [13]
- **No prerequisite knowledge**. Any choice should have an associated help to inform the user about each option. [18]
- **Automatic completion** from any point in the configuration process, optionally concerning metrics like lowest price.
- **Low response time** for any requests/queries in any point in the configuration process. Pargamin [18] recommends less than two seconds in the worst case.

4 Proposed Concept

This chapter outlines our primary contribution, which is a set of generic rules to automatically infer a graph of PDPs from a set constraints. This graph serves as the underlying data structure of a configuration system and simplifies several downstream tasks. On an ontological level, we align our concepts with BPO and introduce a semantic extension to the interface concept. This enables us to treat multiple distinct products as one unit during compatibility evaluations, thereby reducing complexity and runtime. The terms introduced do not refer to OWL classes and properties, but are to be understood on a purely conceptual level, to be refined and published in further contributions.

4.1 Configuration as Graph Traversal

A key concept adopted from the COLD ontology are partially defined products. Starting from an empty root configuration, a graph of PDPs is constructed by incrementally adding specifications to the nodes and restricting outgoing connections to valid PDPs only. The authors point out three advantages of this:

1. The basic configuration process is reduced to mere graph traversal. Many features defined in Fig. 3.1 are then straightforward to implement, e.g., auto-completion, which corresponds to a shortest-path problem (potentially weighted on price).
2. Each node in the graph is uniquely identifiable and can be completed to a fully defined product with the same specifications. A crawling agent may index these PDPs, which are optionally aligned with GoodRelations, to improve visibility and interactivity on the web.
3. A web configurator is trivially implemented via REST, because each URI corresponds to a unique point in the configuration process, which reduces the state information to a minimum.

We do not however reuse or align with the COLD ontology. The aforementioned compatibility, in- and exclusivity relationships and by extension the possible PDPs have to be added manually on the basis of a solved CSP. Exchanging rules, e.g., why and on what basis two entities are incompatible, is not straightforward; the constraint knowledge is effectively separate from the rest of the product knowledge that resides in the graph. Our contribution attempts to combine these by automatically inferring the entire PDP Graph from the root configuration without the need of an external system. For this to work the CSP can still be calculated offline, but via reasoning inside the KG instead of highly optimized bitwise operations. The following sections outlines how to solve this for industrial scale problems while reducing the complexity of the PDP graph in the process.

4.2 *Materializing Constraints as Entity Sets*

Earlier it was stated that reasoning about valid entity ranges at runtime is undesirable. How then are constraints incorporated into the knowledge base? Note that even without the full PDP graph, the rules themselve are not actually needed to determine valid entity ranges given the already selected components. Instead only the resulting restriction on the range is needed, which can be materialized by grouping mutually inclusive entities from different components.

This is best illustrated using an example. Consider the PC configuration example in Fig. 1. One rule may state that, for any valid PC variant, the socket attribute of the CPU must match the socket attribute of the Motherboard. To materialize the range restriction, a group for every possible socket value is created. Next, both CPU and Motherboard entities are attached to the group that match their socket value. The term *inclusivity groups* is used to describe these groups, as an entity in component class A is inclusive with all of its counterparts from component class B, and vice versa. Semantically, these groups correspond to a single constraint, as such two components may be connected by more than one set of inclusivity groups if there are more dependencies between their attributes. Querying the valid entity range for a component A, given a set of chosen entities is trivial. It requires querying all inclusivity groups that are connected to the already selected entities, and computing

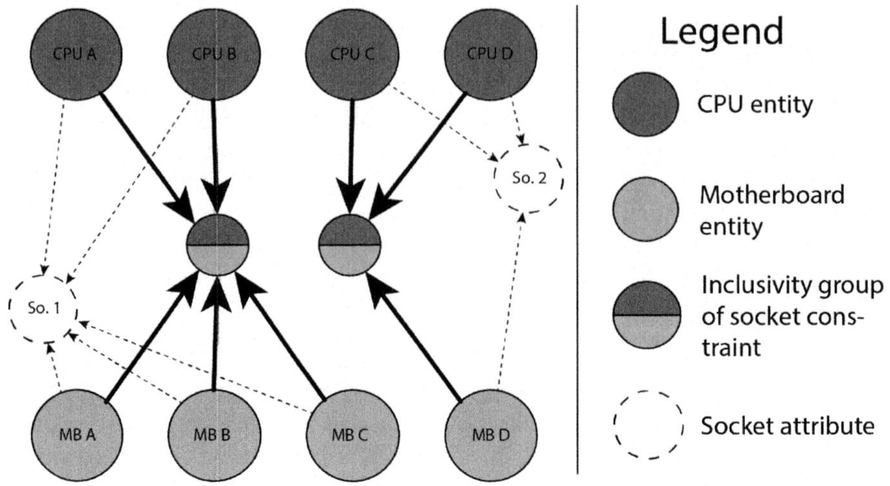

Fig. 1 The figure illustrates entities for CPU and Motherboard components, each possessing a socket attribute representing a specific socket value. The result of a simple constraint, requiring matching sockets, is materialized by inclusivity groups, one for each value of the socket property

the intersection of entities that are connected to the inclusivity groups and are an instance of component A.

Note that the assignment of entities to inclusivity groups is the only thing the product expert has to do. The required rules are semantically similar to standard SWRL rules that would validate a given product variant. The remaining framework, i.e., generating the PDP graph, requires only a small number of generic rules, which are outlined below.

4.3 *Rewriting the Problem*

It is not feasible to directly construct the PDP graph from inclusivity groups. Nonetheless, it is possible to perform reasoning on the compatibility of entities w.r.t. the rule set by examining the inclusivity groups they are linked to. If the set of all inclusivity groups that entities of the same component belong to are the same, the entities are equivalent in the sense that all attributes that are relevant for the configuration are identical. They may thus be treated as one entity in the configuration process. The term used to describe this abstracted entity is *equivalence group*.

The inclusivity groups can also be used to reason about the compatibility of two equivalence groups, i.e., whether or not they would produce empty ranges for any component when they are selected together.

Figure 2 illustrates the relationship between inclusivity and equivalence groups. Note that the number of equivalence groups is always at least equal to the number of

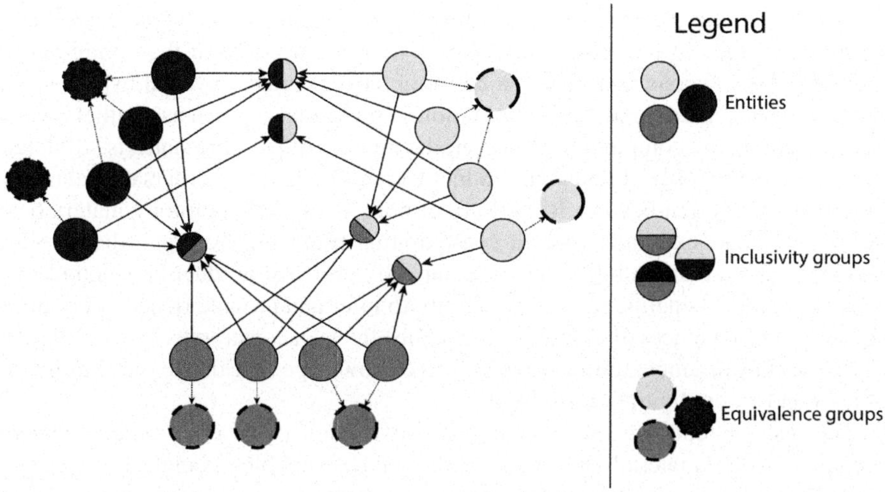

Fig. 2 The concept presented in Fig. 1 is extended to a third component here. Three constraints are imposed, resulting in three distinct sets of inclusivity groups, each referenced by only two components as a consequence of two-component constraints. Entities belonging to the same component and inclusivity groups are linked to the same equivalence group

entities for each component. For components that are largely independent of other components, such as the color of a car, most entities are treated as one singular entity. The size of the PDP graph is significantly decreased by redefining each node to specify equivalence groups as opposed to individual entities, as outlined in the original paper. Furthermore, the entire graph can be reasoned into existence by recursively creating new PDPs using compatible equivalence groups and pruning dead ends.

Inferring equivalence groups and PDP nodes requires dynamic creation of new individuals as well as computing intersections inside the rule body, which no rule language in the Semantic Web stack supports. Nevertheless, certain reasoning engines capable of processing RDF data provide these functionalities, such as RDFox, which was utilized for an initial prototype.

5 Conclusion and Further Work

Mass customization can be a significant advantage for companies as it allows for the production of lot size one with mass production pricing, resulting in increased market share and the acquisition of new customers. However, mass customization requires diverse and complex prerequisites, including a production line that can fulfill lot size one production and a customer-friendly configuration system. The latter requires a complex and enriched knowledge base about all potential products that can be manufactured, which can be realized as a knowledge graph. In this regard, the *Con-*

figuration as Linked Data (COLD) ontology implements a graph based approach, where each node in the graph corresponds to a unique state in the configuration process. This way, configuration can be formulated as graph traversal, which offers many benefits regarding the implementation of related tasks, such as conflict resolution and automatic completion. However, the number of possible states is still huge, surpassing the number of potential product variants by multiple orders of magnitude (which itself is often beyond 10^{15}). Furthermore, the COLD ontology materializes the results of an externally solved configuration problem. We argue that explicit reasoning about compatibility via a semantically enriched product knowledge base offers additional benefits, such as the ability to infer and provide richer information about the consequences of user choices and the effects of constraints. Product experts also benefit from a more uniform and intuitive knowledge base due to the elimination of a secondary configuration solution.

The task of reducing the combinatorial explosion of product variants can be accomplished by materializing range restrictions of variables as inclusivity groups, each corresponding to a constraint between two or more components. For example, in a computer configuration, the constraint that CPU and Motherboard entities must have the same socket can be represented by creating one inclusivity group for each socket value and linking the corresponding entities. These entities can be further combined into equivalence groups based on their shared inclusivity groups, resulting in a smaller set of individuals to consider in the configuration graph.

To fully assess the potential of the proposed approach, further investigations are needed, particularly with respect to its scalability to industrial scale applications, which pose unique challenges due to the exponential increase in the number of product variants with larger data sets. It should be noted that the efficiency of the approach may depend on the specific domain, as the benefits of combinatorial reduction using equivalence groups may vary depending on the context. However, it is noteworthy that the proposed approach is always superior to the trivial approach that constructs the PDP graph directly from the individual entities. Moreover, a concrete ontology model is needed, which should consider mapping as many types of constraints as possible and integrating them accordingly in the configuration with equivalence groups. Alignment with other essential ontologies should also be considered.

References

1. Hvam, L., Mortensen, N. H., & Riis, J. (2008). *Product customization*. No. 978-3-540-71449-1 in Springer Books. Springer. https://doi.org/10.1007/978-3-540-71449-1
2. Felfernig, A., Hotz, L., Bagley, C., & Tiihonen, J. (2014). *Knowledge-based configuration: From research to business cases* (1st edn.). San Francisco, CA, USA: Morgan Kaufmann Publishers Inc. https://doi.org/10.1016/C2011-0-69705-4
3. Blaha, M., & Rumbaugh, J. (1991). *Object-oriented modeling and design with UML*, 2nd edn. Upper Saddle River, NJ: Pearson Education. Rev. ed of: Object-oriented modeling and design/James Rumbaugh. 1991.

4. Hao, Y., Zhang, Y., Liu, K., He, S., Liu, Z., Wu, H., & Zhao, J. (2017). in *Proceedings of the 55th annual meeting of the association for computational linguistics (Volume 1: Long papers)* (pp. 221–231). Vancouver, Canada: Association for Computational Linguistics. https://doi.org/10.18653/v1/P17-1021
5. Dikaleh, S., Pape, D., Mistry, D., Felix, C., & Sheikh, O. (2018). *Proceedings of the 28th annual international conference on computer science and software engineering, CASCON '18* (pp. 344–346) USA: IBM Corp. https://doi.org/10.5555/3291291.3291332
6. Zhang, F., Yuan, N. J., Lian, D., Xie, X., & Ma, W. Y. (2016). in *Proceedings of the 22nd ACM SIGKDD international conference on knowledge discovery and data mining, KDD '16* (pp. 353–362). New York, NY, USA: Association for Computing Machinery. https://doi.org/10.1145/2939672.2939673
7. Hogan, A., Blomqvist, E., Cochez, M., D'amato, C., Melo, G. D., Gutierrez, C., Kirrane, S., Gayo, J. E. L., Navigli, R., Neumaier, S., Ngomo, A. C. N., Polleres, A., Rashid, S. M., Rula, A., Schmelzeisen, L., Sequeda, J., Staab, S., & Zimmermann, A. (2022). *ACM Computing Surveys, 54*(4), 1. https://doi.org/10.1145/3447772
8. Paulheim, H. (2017). *Semantic web, 8*(3), 489–508. https://doi.org/10.3233/SW-160218
9. Auer, S., Bizer, C., Kobilarov, G., Lehmann, J., Cyganiak, R., & Ives, Z. (2007). K. Aberer, K.S. Choi, N. Noy, D. Allemang, K.I. Lee, L. Nixon, J. Golbeck, P. Mika, D. Maynard, R. Mizoguchi, G. Schreiber, & P. Cudré-Mauroux (Eds.), *The semantic web* (Vol. 6, pp. 722–735). Berlin: Springer. https://doi.org/10.1007/978-3-540-76298-0
10. Dong, X. L., Gabrilovich, E., Heitz, G., Horn, W., Lao, N., Murphy, K., Strohmann, T., Sun, S., & Zhang, W. (2014). *The 20th ACM SIGKDD international conference on knowledge discovery and data mining, KDD '14, New York, NY, USA - August 24 - 27, 2014* (pp. 601–610). Evgeniy Gabrilovich Wilko Horn Ni Lao Kevin Murphy Thomas Strohmann Shaohua Sun Wei Zhang Geremy Heitz
11. Fellbaum, C., (Ed.), *WordNet: An electronic lexical database*. Language, speech, and communication. Cambridge, MA: MIT Press.
12. Schneider, T., & Šimkus, M. (2020). KI - Künstliche Intelligenz *34*. https://doi.org/10.1007/s13218-020-00686-3
13. Chevalier, E., & Servant, F. P. (2012). In E. Simperl, P. Cimiano, A. Polleres, O. Corcho, & V. Presutti (Eds.), *The semantic web: Research and applications* (pp. 603–617). Berlin: Springer.
14. Altuna, A., Cabrerizo, A., Laresgoiti, I., Nieves, P., & Sastre, D. (2004). (pp. 69–80). https://doi.org/10.1007/978-3-540-30196-76
15. Yang, D., Miao, R., Wu, H., & Zhou, Y. (2009). *Expert systems with applications, 36*, 4399. https://doi.org/10.1016/j.eswa.2008.05.026
16. Wagner, A. (2020). Linked product data: Describing multi-functional parametric building products using semantic web technologies. Ph.D. thesis. https://doi.org/10.13140/RG.2.2.17992.88327
17. Bischof, S., Schenner, G., Steyskal, S.,& Comploi-Taupe, R. (2018). Integrating semantic web technologies and asp for product configuration.
18. Pargamin, B. (2002). *Configuration workshop at ECAI-2002*. France: Lyon.

Knowledge Technology in Material Engineering

Recent Progress on Effect of Different Parameters on Thin Films for Enhanced Superconducting Properties

Navid Nasajpour-Esfahani, Hamid Garmestani, Ngoc L. Tran, Yuzhou Chen, Elnaz Jamshidi, As'ad Alizadeh, and Reza Nabiun

Abstract This paper presents an overview of thin film superconductors materials, their properties, and potential issues. Thin film superconductors are materials that exhibit superconductivity at temperatures (70–90 K) much higher than those of conventional superconductors (1–10 K). They are typically composed of thin layers of superconducting materials, such as high-temperature superconductors, deposited on a substrate. The properties of thin film superconductors depend on the materials used, the thickness of the film, and the substrate. Potential applications of thin film superconductors include high-speed computing, energy storage, and medical imaging. This paper provides an overview on the properties of thin film superconductors materials and the essential parameters include films thickness, films growth, substrates, chemical reactions, dislocations, deposition techniques, film materials, different types of defects, boundaries, second phases, multilayers, thermal expansion coefficient, and annealing. Each of these parameters affects the physical and chemical properties of thin film superconductors. Therefore, this paper provides a fundamental knowledge about thin film superconductors and their important parameters that affect their properties for researchers to overcome the issues and limitation of superconductors in the future.

N. Nasajpour-Esfahani (✉) · H. Garmestani · N. L. Tran · Y. Chen · E. Jamshidi
Department of Materials Science and Engineering, Georgia Institute of Technology, 771 Ferst Dr. NW, Atlanta, GA 30332, USA
e-mail: Navid_nasajpour@yahoo.com; nesfahani3@gatech.edu

A. Alizadeh
Department of Civil Engineering, College of Engineering, Cihan University-Erbil, Erbil, Iraq

R. Nabiun
School of Civil Engineering, Iran University of Science and Technology, Tehran, Iran

M.-R. Alam and M. Fathi (eds.), *Integrated Systems: Data Driven Engineering*,
https://doi.org/10.1007/978-3-031-53652-6_12

1 Introduction

YBCO has been the traditional high-temperature superconducting material, while Nb_3Sn is the lower-temperature superconducting material with an extensive industrial application for superconducting magnets. One of the essential high-temperature YBCO thin film issues is the in-field pinning enhancement. Superconducting properties can also be enhanced by sequential sputtering, followed by annealing. For example, Nb_3Sn thin films were created by that method at 850–1200 °C. When the Nb_3Sn films were annealed at 950 °C for 3 h, the highest critical temperature was 17.86 K. After annealing above 1000 °C for 3 h; the superconducting performance was degraded due to the evaporation of Sn, which resulted in the creation of Nb-rich film [1]. The superconducting properties of NbN thin films were analyzed using three different substrates: Si, Al_2O_3, and MgO. It was shown that Al_2O_3 and MgO substrates could improve the superconductivity of NbN film with the superconducting transition temperature of 15.2 and 13.1 K, while Si substrate offered a lower critical temperature of 8.3 K [2]. A thin layer of magnetoresistive material $La_{0.67}Ca_{0.33}MnO_3$ (LCMO), deposited on $SrTiO_3$ substrate to evaluate the superconducting properties of YBCO. It was reported that LCMO not only highly affects the superconducting but also the normal state of the YBCO [3]. Thin film superconductors have the potential to be used in many applications, including energy devices, electronic applications, and transport services. Several factors affect the superconductivity in thin film materials. The key feature of superconductors is that the resistance is zero at critical temperatures, which provides zero energy loss. This paper overviews the primary characteristics and parameters that should be considered for producing film superconductors.

2 Effect of Thin Film Thickness

Naturally oxidized Si (100) was used as a substrate, and pure Nb (t_{Nb}) and hybrid Nb (t_{Nb})/Cu (5 nm)/Co (40 nm) films with t_{Nb} of 20 and 100 nm were grown on this substrate. As a result, the superconducting critical temperatures were 5 and 7 K for $t = 20$ and 100 nm, respectively. These temperatures were reduced by increasing a magnetostatic effects of the Co layer [4]. Sputtering of nanocrystalline tungsten thin films with an N_2:Ar mixture at room temperature enhanced the superconductivity of the films. For tungsten films thicker than 17 nm, the superconducting critical temperature ($T_c = 4.7$ K) was thickness independent while T_c was decreased below this thickness, and at a 4 nm thickness, the T_c was 3.1 K [5]. Using Molecular-beam epitaxy (MBE), $TaSe_2$, a metallic 2D material, was crystallized into 3R-$Ta_{1+x}Se_2$, a pure 3R phase, which could not be found as a pure material phase in nature. Compared with all the polymorphs of $TaSe_2$, the 3R-$Ta_{1+x}Se_2$ film had the highest superconducting T_c of 3 K. It was shown that T_c decreased with thickness reduction [6]. Further, the superconducting properties of MgB_2 (single crystalline) thin films with different thicknesses were analyzed under the effects of oxygen (O) ion irradiation. As a result, an interesting exponential relation was found between T_c and thickness because of conventional proximity. In most irradiated films, the field performance critical current density was enhanced [7]. $Fe_{1.05}Te_{0.5}Se_{0.5}$ superconducting films with two different thicknesses were fabricated on $LaAlO_3$ substrates using the PLD method. The values of activation energy and upper critical field for the thicker film (185 nm) were higher than a film with a lower (74 nm) thickness. Also, the magnetotransport data analysis suggested that due to the defects and higher content of Fe_2O_3, a flux pinning behavior was improved in films with a higher thickness [8]. Besides, heat transfer for thin films with high T_c could be superior by one magnitude to bulk samples in both normal and superconducting states [9].

3 Mechanism of Film Growth

In superconducting systems, two different types of interfaces occur which are natural microstructural features like interphase boundaries, grain boundaries, and twin boundaries, or artificially fabricated boundaries such as heterostructures, composites' interfaces, and substrate-superconductor interfaces. Researchers have used different techniques to evaluate the interfaces and to understand the mechanism of oxide growth on thin film superconductors [10]. These methods include Rutherford backscattering spectroscopy (RBS) [11], reflection high energy electron diffraction (RHEED) [12], transmission electron microscopy (TEM) [13], scanning tunneling microscopy (STM) [14], and X-ray diffraction (XRD) [15]. Also, it is essential to know some information about the background of the interfaces, which could start with the definition of epitaxy. Epitaxy can be defined as an orientation in which the growth of crystals occurs on a crystalline substrate [10]. It was shown that a small misfit between the substrate and crystal deposition could be necessary to create an epitaxy [16]. It was also demonstrated that an epitaxy could occur for large misfits [17]. The growth of epidictic thin film has essential effects on the final film's microstructure, which leads us to know more about different forms of epitactic growth. Three various epitactic growth forms are shown in Fig. 1.

Each of these forms has specific features that need to be addressed. Layer-by-layer growth, known as Frank-van der Merwe, and occurs when a strong interaction and zero misfits exist between substrate and film atoms. However, the misfit is non-zero when clusters can be seen on a bare substrate known as Volmer-Weber, and on some uniform layers known as Stranski–Krastanow. Further, in all cases, pseudomorphic growth is an initial deposition that strained to accommodate the substrate elastically. It is essential to mention that this growth is related to the strain, which ensures small coherency at the substrate–film interface [10]. There are two main factors in selecting a substrate. The first one is that the substrate should have a high critical current and high transition temperature. Second, it should be a good fit for technological applications [10]. However, in many systems, these two factors were not both satisfied.

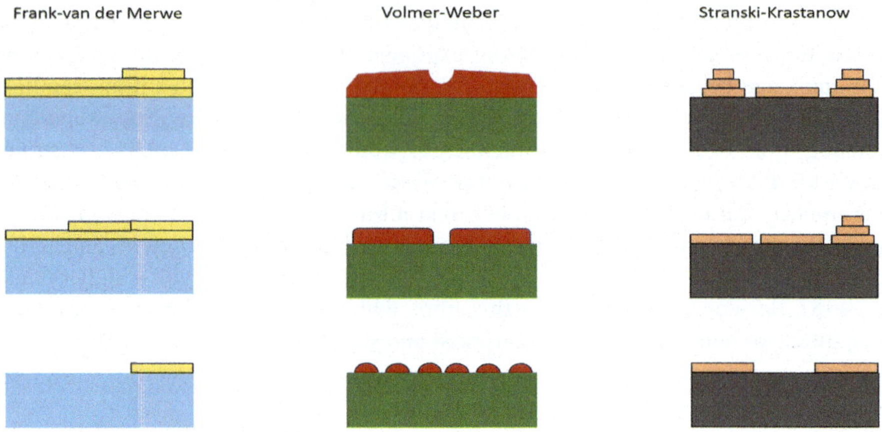

Fig. 1 Different forms of epitaxial growth

For example, a very high level of crystalline perfection was observed in the deposition of $YBa_2Cu_3O_{7-\delta}$ thin film on $SrTiO_3$ substrate by using RBS, while a very high dielectric constant (>300) at 20 °C was observed for $SrTiO_3$ which makes that useless for many applications such as microwave devices [18]. In addition to these two factors, surface steps are critical and play an essential role in different situations. For example, they could affect the surface diffusion process by providing active sites for catalysis [19]. A schematic of the stepped surface is seen Fig. 2.

Many studies, which have been performed on misfits between substrate and film growth, are based on utilizing various materials. For example, the growth of $YBa_2Cu_3O_{7-\delta}$ on $SrTiO_3$ and $LaAlO_3$ showed a lattice misfit of 2.3 and −1.1% [10]. The other parameter to evaluate the growth of films is moiré fringe patterns which are formed because of overlapping two crystals, and it is interference among diffracted beams. There are two different moiré fringes patterns called parallel and rotation moiré patterns [10]. It is essential to mention that a slight rotation in one

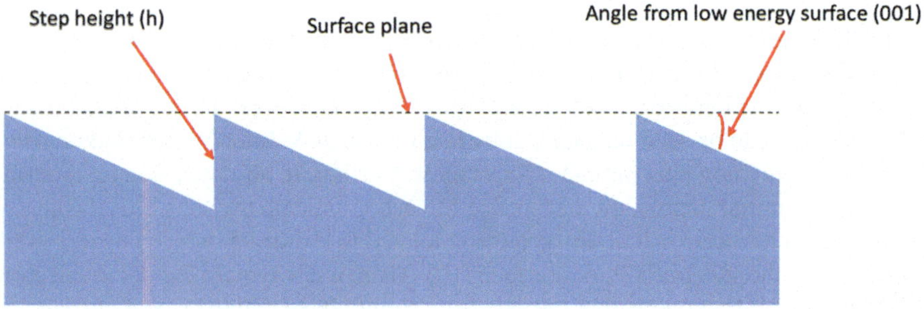

Fig. 2 Schematic of a stepped surface

grating could magnify the rotation of the moiré patterns. Moiré patterns could identify microstructural defects in epitactic deposits, which primarily means that during the growth of film, generated threading dislocations are revealed in moiré patterns [10]. The excellent properties of thin films compared to the bulk samples are ascribed to a perfect flux pinning and preferred crystallographic orientation, which leads to solid behavior. Further, the most critical factors affecting nucleation and growth are substrate orientation, substrate choice, chemical considerations, lattice mismatch, pressure including total and partial pressure, surface preparation, deposition rate and temperature, film thickness, and cooling protocol [10].

4 Substrates

The substrate is one of the most critical parameters affecting the conductive properties of thin films. Researchers have used different materials as the substrate, but each has shown different behavior and orientation during nucleation. For example, it was observed that annealing of MgO (100) substrates at 1200 °C caused to increase surface roughness, no matter whether they were cleaved or polished initially [20]. Some studies also showed that film could grow much faster in [010] and [100] in comparison to [001] [21]. A possible answer to that could be because the materials could be added to the edge easier in contrast to the addition of materials to the top of the completed unit cell. Therefore, it will grow faster once nucleation of the α-axis grain occurred on the c-axis growing surface, and as a result, the α-axis materials could cover the surface [10]. In addition, researchers have also observed that the type of substrate affects the α-axis grain direction. In the $NdGaO_3$ substrate, 90% of α-axis grains were only aligned in one direction of the substrate's plane, but for $SrTiO_3$, the grains were aligned in two different directions with 90° rotation [22].

5 Chemical Reactions

One of the most parameters to determine the morphology and rate of growth of a solid–fluid interface is supersaturation. Supersaturation is much lower in vapor–solid interfaces in comparison with liquid–solid interfaces. The growth of crystals at low effective supersaturations characterizes dislocation-mediated or spiral growth [10]. In Fig. 3, the relation between supersaturation and growth rate is schematically shown. At low supersaturation conditions, the possibility of birth and spread mechanism is low, which means that it is very unlikely to have nucleation-mediated growth. Increasing the supersaturation conditions makes it more likely to see two-dimensional nucleation on the completed surface. Therefore, at critical supersaturation $\Delta\mu^*$, a transition from dislocation growth to nucleation growth areas occurs. Finally, unexpected growth could occur if supersaturation is enhanced. During the last regime, the dendritic type of growth happens due to uncontrolled nucleation growth. This

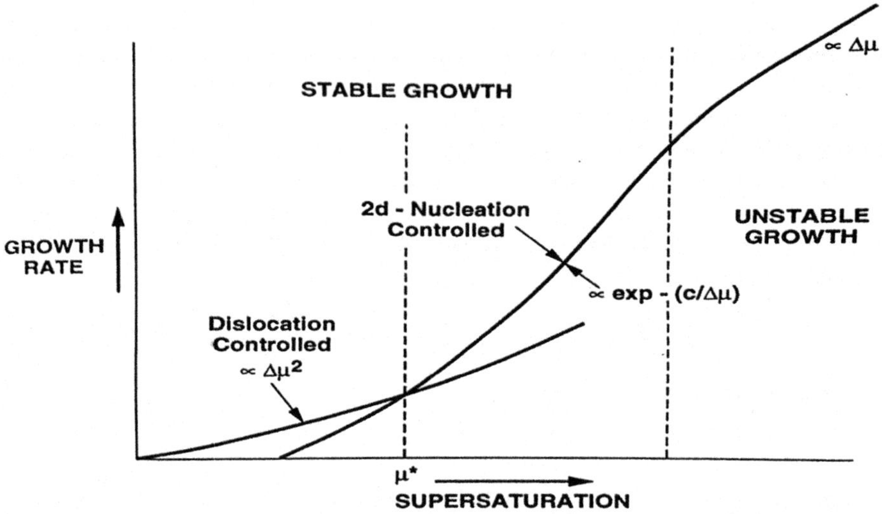

Fig. 3 Dependency of supersaturation with the growth rate [10]

dendritic type reaches to vapor phase rapidly, usually giving an undesirable and rough surface [10]. Further, three main reactions have been determined for YBCO, which should be avoided due to the leaving of non-superconducting products. The first type of reaction includes those which cause the substrate's atoms substitution into the superconductor lattice. The second type is the passivation layer, which results from secondary phase formation and prevents further interaction. Finally, those reactions that the second phase is formed without passivation layers [10]. For example, decreasing susceptibility was observed due to substituting Ti and Sr into the YBCO films on $SrTiO_3$ substrate [23]. Also, Si substrate has been seen as highly reactive when YBCO films are deposited on it. Therefore, it is essential to separate Si substrate and YBCO films with a proper buffer layer [10]. However, in some substrates, such as Y_2O_3, and CeO_2, the chemical reactions and interdiffusion are absent and do not affect the YBCO films at the corresponding deposition temperature [24]. Besides, it is well understood that there is a correlation between the oxygen content and valency of the copper with crystal structure and transition temperature [25].

6 Lattice Mismatch and Dislocations

Researchers have observed that due to the intersection of dislocation with the surface's growth planes, a helicoid surface is produced [10]. The growth of 500 nm thick $YBa_2Cu_3O_7$ film on MgO substrates contained 0.1–1% of 45° misaligned materials. It was also observed that the amount of rotated MgO materials quickly decreased by increasing the thickness [26]. Similarly, $YBa_2Cu_3O_7$ films were deposited on

$SrTiO_3$ using the hollow cathode sputtering method. The obtained images from STM were almost interchangeable with the deposition of the same film on MgO substrates [27, 28]. It has also been reported that the surface dislocations' density and growth of film are proportional, and the thickness has much better correlations than the deposition rate [27]. The slight lattice mismatch between $Sn_{1-x}In_xTe$ (~6.33–6.26 Å) and BaF_2 (6.196 Å) warrants a thorough investigation of the in-plane lattice parameter [29]. In contrast to the 5a surface reconstruction observed for growth on a (001) LSAO substrate (Fig. 4c), the films under tensile strain exhibit 4a surface reconstructions for the growths on (100) STO and (001) LSGO substrates (Fig. 4a, b), respectively. These results indicate that the lattice mismatch between the LSCO film and the substrate influences the atoms' arrangement at the surface: the larger the in-plane lattice constant of the substrate, the smaller the size of the coherent atomic arrangement [30].

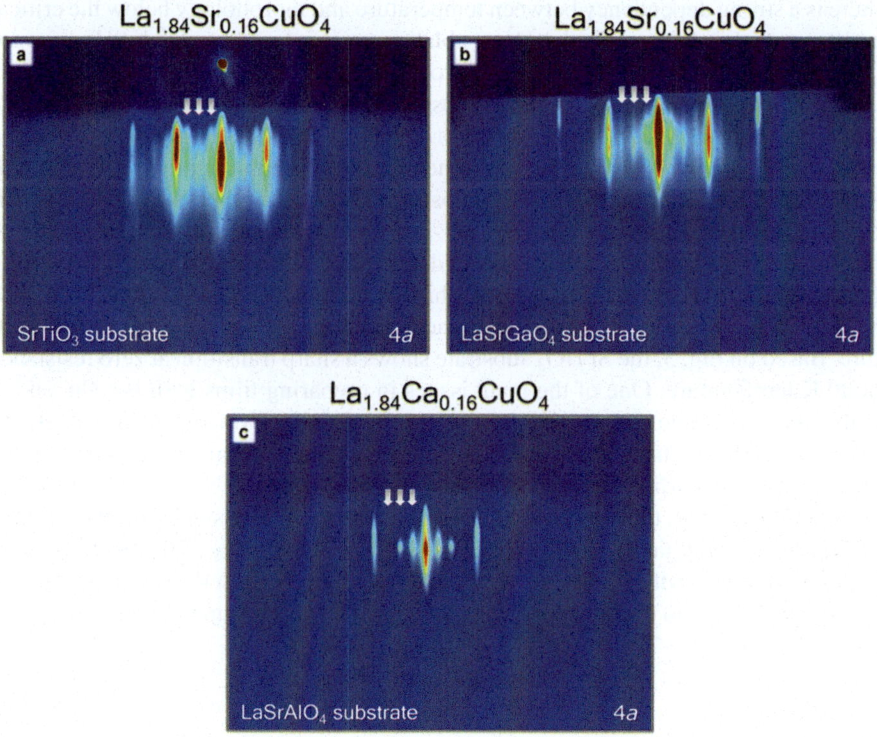

Fig. 4 Reflection high energy electron diffraction images for growth of $La_{1.84}Ca_{0.16}CuO_4$ on **a** STO substrate, **b** LSGO substrate, **c** LSAO substrate [30]

7 Film Deposition by Different Techniques

The deposition of films depends on different parameters, including deposition rate and temperature. $YBa_2Cu_3O_7$ films on $LaAlO_3$ and $SrTiO_3$ were deposited by the laser-ablated method, and the spiral growth was seen at the lower optimum growth temperature (680 °C vs. 730 °C) [31]. Another study observed that the microstructure growth of $YBa_2Cu_3O_7$ films on $NaGaO_3$ substrates was strongly correlated with deposition conditions [32]. Researchers have also found that the reduction of deposition temperature, especially on perovskite substrates, is caused to the perpendicular growth of film in the α-axis. A great example was observed for the deposition of $YBa_2Cu_3O_7$ on $SrTiO_3$ substrate. At 640 °C, the $YBa_2Cu_3O_7$ had pure α-axis orientation, but the mixed α-axis and c-axis perpendicular orientations were achieved between 640 and 740 °C. The reason for this alteration in the orientation has yet to be determined, but this is very common in thicker films of $YBa_2Cu_3O_7$ [33]. Further, there is a strong dependency between temperature and morphology below the critical temperature. In addition to pulsed laser ablation, researchers found pulsed laser evaporation (PLE) as a powerful method to produce high-quality thin films. TlBaCaCuO thin films were deposited on different substrates, such as Y-ZrO_2, Poly Al_2O_3, MgO, and $SrTiO_3$. It has been found that the formation of phases strongly depends on the heat treatment conditions, which could include reaction time duration and temperature. It has also been deduced that the substrates' material and post-annealing conditions affect the films' microstructure [34]. YBaCuO superconducting thin films with high J_c and T_c were produced by e-beam deposition technique by using BaF_2 source material instead of metallic barium on three different substrates (MgO, ZrO_2, and $SrTiO_3$). Four probe methods measured the dependency of resistance to the temperature. Based on Fig. 5, the $SrTiO_3$ substrate shows a sharp transition at zero resistance at 90 K temperature. One of the main issues in preparing films with Ba, Cu, and Y is that they are prone to degrade T_c. Also, during a period, loss of superconductivity might occur due to the conversion of barium-to-barium oxide and the reaction of this component with carbon dioxide and water [35].

In another study, the $Nd_{1-x}Sr_xNiO_3$ thin films were deposited with the reactive molecular beam epitaxy (MBE) technique [36]. Spinel $LiTi_2O_4$ (001) thin films were also grown epitaxially on $MgAl_2O_4$ (001) substrates by pulsed laser deposition in an ultrahigh vacuum chamber (to promote the growth of the superconducting phase) [37].

8 Materials Aspects

It is essential to know that the thickness of the film affects the size of surface grains in a way that by increasing the thickness of the film, the grain size is increased. This means that in the first layer, some grains grow faster than others, and these overgrown grains will grow much slower. The number of overgrowing grains in the

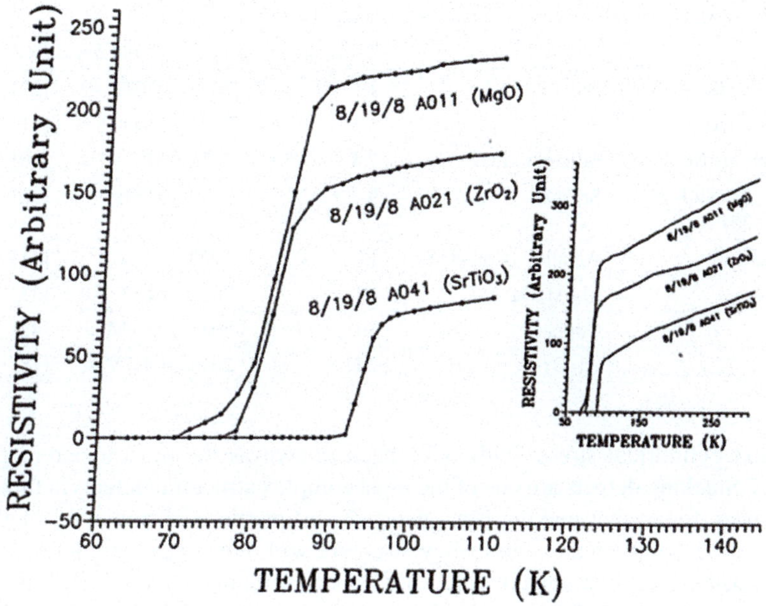

Fig. 5 The correlation between temperature and resistance for YBaCuO films on MgO, ZrO_2, and $SrTiO_3$ substrates [35]

first layer is minimized if there are only a few numbers of screw dislocations at the early stages [21]. It has been observed that the growth of films on MgO substrates usually have small fractions of material that are misorientated around the c-axis. This affects the electrical properties such as current–voltage behavior, surface resistance, and critical current profoundly, even if a small fraction of the material exists. The amount of misoriented material fraction depends on substrate surface preparation and deposition conditions [36, 38]. In another study, TEM was used to study the microstructure of thin films. It was seen that there were connected islands with one, two, or three-unit cells all over the surface. By adding more material to these nuclei, they spread out all over the surface to gain 10 nm thickness instead of becoming taller nuclei [39]. Table 1 shows a summary of various materials with their corresponding T_c and deposition techniques.

9 Different Types of Defects

Different types of defects have been determined by researchers. Growth dislocations during the sputtering process, granularity because of uncontrolled nucleation, stepped surface, which leads to variation, and terraces in the thickness of films are some of the most important defects [21]. In addition, screw dislocations with highly distorted

Table 1 A review of thin-film superconductors materials with the lowest T_c

Films	Substrates	T_c (zero) (K)	Year	Technique	References
$Sn_{1-x}In_xTe$	BaF_2	1.5–3.5	2022	MBE	[29]
$Nd_{0.8}Sr_{0.2}NiO_2$	$SrTiO_3$	3.5	2021	PLD	[31]
Amorphous MoSi	Silicon oxide	7.4	2021	MS	[40]
$Nd_{1-x}Sr_xNiO_2$ (x = 0.12–0.25)	$SrTiO_3$	9–15	2020	MBE	[36]
TlBaCaCuO	Ag alloy (Consil 995)	12	1990	RF-MS	[41]
$LiTi_2O_4$	$MgAl_2O_4$	13	2022	PLD	[37]
NbN	Copper	16.1	2021	DC-MS	[42]
MgB_2	Al_2O_3	24	2019	HPCVD	[18]

core and cylindrical stress fields have been known as the most apparent growth defects. Stacking defects are one of the most complex structural defects in thin films. It has been found that making films using the evaporation or sputtering method at low pressure could create a less wavy structure with fewer stacking defects. These defects usually are conjugated with other crystallographic defects [21]. It is known that c-axis films made by laser ablation have more stacking faults compared to the α-axis films [43].

10 Boundaries and Grain-Boundary Junctions'

The grain boundaries' structure greatly affects the superconducting structure, and its relationship with low critical currents is important to be understood. It was found that decreasing the critical currents is significant for grain boundary angles greater than 5°–10° compared to intragranular currents [44]. Generally, the variation of grain sizes depends on different parameters such as substrate, film thickness, growth mechanism, and deposition temperature. X-ray rocking curve could measure the grains' average mosaic speared by determining the tilt's angle out of the plane (ω) and rotation angles in the plane (ϕ). Antiphase boundaries (APBs) have been known as boundaries that are normal to the interfaces, and they shift across the c/3 or c/6 boundaries between (001) planes at c-axis films. These boundaries tend to heal themselves at stacking fault intersections, which is caused to compensate the APB shift [21]. α-axis films mostly contain 90° domain boundaries compared to c-axis films [33]. 90-degree boundaries could be divided into two categories; the first is base plane faced boundaries with a parallel interface to one of the grain's (001) planes; the second type are boundaries between two (001) planes at 45° [21].

11 Second Phases

A second phase could affect the superconducting properties either negatively or positively. Second phases could act like pinning sites; therefore, if small second phase precipitates distribute evenly, the critical current properties will be increased. In addition, a second phase could be nucleation sites for other orientations or precipitants. It is important to know that deposition processes are known as non-equilibrium processes, and therefore it is possible that phase diagrams could not predict the formation of second phase. Second phases in bulk material have seen a tendency to be large and incoherent particles, but this is quite different for in situ coated films. In in-situ films, the second phases are usually seen as highly oriented, small, and semi-coherent or coherent particles [21]. Identifying Y_2O_3 precipitates (100 A) in YBCO films has been investigated by several researchers. These films were deposited at a specific temperature and pressure, and during a non-equilibrium process, precipitates were formed. Precipitates were small and distributed, which could act as the pinning agent. However, the role of these residues and other microstructural defects is still ambiguous [45, 46]. Further, it was reported that the $Nd_{0.8}Sr_{0.2}NiO_3$ films were difficult to grow due to the formation of a secondary phase [31].

12 Multilayers and Their Properties

The fabrication of heterostructures with superconducting compounds is essential for superconducting Josephson technology. Therefore, the fabrication of superconductor-insulator-superconductor (SIS) or S-normal metal-S (SNS) junctions is important. It is necessary to know that the insulating or normal metal barrier must be continuous, and the metal contact should have low resistance at the top layer [21]. The first fabrication of multilayers of YBCO superconductors was made in c-axis orientation using $SrTiO_3$ or PrBCO by laser deposition and sputtering [47, 48]. Different techniques include X-ray diffraction, ion channeling spectroscopy [47], Auger profiling, secondary ion mass spectroscopy, and TEM [49]. The main focus in both α-axis and c-axis oriented heterostructures is to investigate the vertical transport properties in Josephson junctions' fabrication [21]. It should be noted that $Nd_{1-x}Sr_xNiO_2$ is in the infinite-layer phase; as such, a small anisotropy value is predictable [33].

13 Thermal Expansion Coefficient

It is important to know that in dealing with high-temperature superconductors must consider a proper match between the thermal expansion coefficients of materials to prevent severe cracking [50]. The variation of the thermal expansion coefficient during cooling could introduce different strains in the film. If this variation is slight, these strains will be accommodated as lattice strains or twins, but the more significant variation of thermal expansion coefficient leads to crack formation and propagation [50]. The strain can be relieved by stacking faults and dislocations at the deposition temperature in (110) $YBa_2Cu_3O_{7-\delta}$ and $PrBa_2Cu_3O_{7-x}$ films on (110) $SrTiO_3$ substrates [51].

14 Annealing

Researchers determined the effect of annealing temperature for films produced by laser ablation processing. Two BSCCO films on ZrO_2 substrates were used. For one set, the deposition was done at room temperature while another film deposition was occurred at 300 °C. For the first set, annealing at 810 °C for 10 min could produce films with the optimum quality. It was also observed that the resistance decreased by increasing the temperature. For the second set, the film was annealed at 790 °C for 10 min. The effect of annealing temperature was more significant than the previous set-in order to produce optimum quality films [52].

It was shown that increasing the annealing time for the deposition of BiPbSr-CaCuO increased the non-zero resistance temperature [53]. In another study, the effect of annealing on YBCO film was determined. Three different processes were used (Fig. 6). For process B, the grain size was around 1–2 mm, and the structure was well-oriented. The initial step (step 1) in process B was essential to form a microstructure. Also, process B was more successful than other processes for higher Cu concentration, and the surface roughness was less than 180 nm [54]. In another study, the tube was evacuated and sealed before annealing at 340 °C for 100 min. There was no direct contact between the samples and CaH_2 during the treatment. Although some unknown factors seemed to be influencing the properties and the reproducibility of superconductivity in thin films, all specimens exhibited superconductivity after a gentle post-annealing under an H_2 atmosphere produced by CaH_2 [33]. It is widely believed that the high-vacuum annealing process can remove the extra vertex oxygen from the tip of a Cu_2O layer in an NCCO crystal structure, thus preventing the extra vertex oxygen from becoming the solid scattering center and affecting the formation of a Cooper electron pair [38].

15 Conclusion

Thin film superconductors have the potential to revolutionize many industries, from energy storage to computing. They offer a wide range of advantages, including higher critical temperatures, improved thermal stability, and increased flexibility. With further research and development, thin film superconductors could become

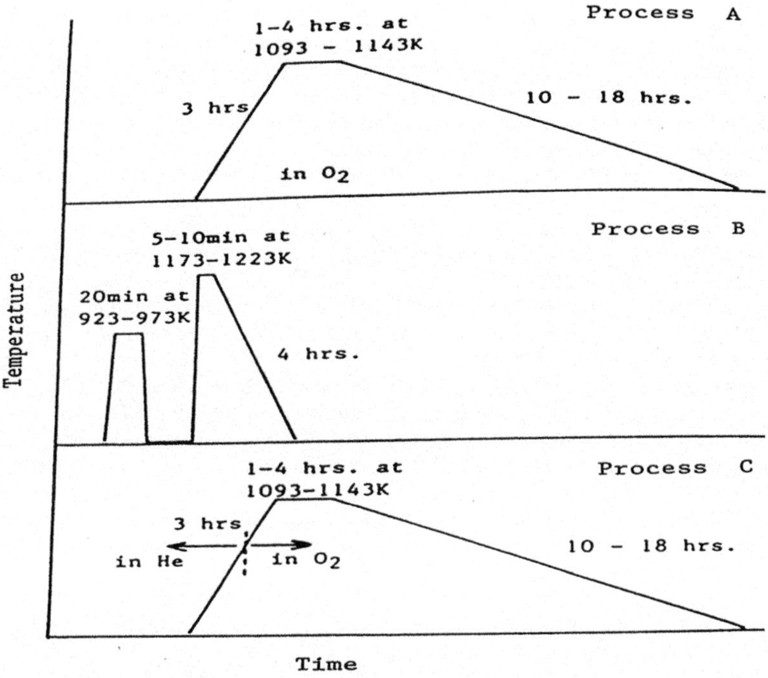

Fig. 6 Different annealing processes on YBCO films [54]

a significant player in the future of technology. In this paper, the main parameters that could affect the resistance of thin film superconductors were discussed. Also, different thin film materials were reviewed. Finally, there are many important features for materials selection such as electric properties, materials availability, critical temperature, materials cost, materials chemical reactions, and many other parameters that should be considered for further research.

References

1. Sayeed, M. N., Pudasaini, U., Reece, C. E., Eremeev, G., & Elsayed-Ali, H. E. (2019). Structural and superconducting properties of Nb_3Sn films grown by multilayer sequential magnetron sputtering. *Journal of Alloys and Compounds, 800*, 272–278. https://doi.org/10.1016/j.jallcom.2019.06.017
2. Volkov, S., et al. (2019). Superconducting properties of very high quality NbN thin films grown by pulsed laser deposition. *Journal of Electrical Engineering, 70*(7), 89–94. https://doi.org/10.2478/jee-2019-0047
3. Hjelmeland, T. B., Bazioti, C., Gunnæs, A. E., Volkov, Y., & Mikheenko, P. (2019). Effect of colossal magnetoresistance material $La_{0.67}Ca_{0.33}MnO_3$ on superconducting properties of $YBa_2Cu_3O_{7-\delta}$ thin films. *Journal of Magnetism and Magnetic Materials, 487*. https://doi.org/10.1016/j.jmmm.2019.165335

4. Paschoa, A., Gonzalez, J. L., & Passamani, E. C. (2020). Nb and Nb/Co films: Testing a new approach to determine the Co stray field and its influence on superconducting properties. *Current Applied Physics, 20*(7), 888–894. https://doi.org/10.1016/j.cap.2020.04.011
5. Hofer, J. A., & Haberkorn, N. (2019). Superconductivity in nanocrystalline tungsten thin films growth by sputtering in a nitrogen-argon mixture. *Thin Solid Films, 685*(April), 117–122. https://doi.org/10.1016/j.tsf.2019.06.014
6. Tanaka, Y., et al. (2020). Superconducting 3R-$Ta_{1+x}Se_2$ with giant in-plane upper critical fields. *Nano Letters, 20*(3), 1725–1730. https://doi.org/10.1021/acs.nanolett.9b04906
7. Pham, D., Jung, S. G., Tran, D. H., Park, T., & Kang, W. N. (2019). Effects of oxygen ion implantation on single-crystalline MgB_2 thin films. *Journal of Applied Physics, 125*(2). https://doi.org/10.1063/1.5061772
8. Kumar, R., Mitra, A., & Varma, G. D. (2019). Study of vortex glass-liquid transition in superconducting Fe(Te, Se) thin films on $LaAlO_3$ substrates. *Journal of Applied Physics, 125*(19). https://doi.org/10.1063/1.5093284
9. Castillo-López, S. G., Villarreal, C., Esquivel-Sirvent, R., & Pirruccio, G. (2022). Enhancing near-field radiative heat transfer by means of superconducting thin films. *International Journal of Heat and Mass Transfer, 182*. https://doi.org/10.1016/j.ijheatmasstransfer.2021.121922
10. Rudmann, D. A., & Shinde, S. L. (2012). *Interfaces in high-Tc superconducting systems*. Springer.
11. Geerk, J., Linker, G., & Meyer, O. (1995). *Thin films of high-temperature superconductors: Application-oriented studies of growth and properties*. Springer. https://doi.org/10.1007/978-94-015-8459-3_6
12. Terashima, T., et al. (1990). Reflection high-energy electron diffraction oscillations during epitaxial growth of high-temperature superconducting oxides. *Physical Review Letters, 65*(21), 2684–2687. https://doi.org/10.1103/PhysRevLett.65.2684
13. Norton, M. G., & Carter, C. B. (1991). The nucleation and heteroepitactic growth of $YBa_2Cu_3O_{7-\delta}$ thin films on MgO. *Journal of Crystal Growth, 110*(3), 641–651. https://doi.org/10.1016/0022-0248(91)90303-M
14. Mannhart, J., Schlom, D. G., Gerber, C., Anselmetti, D., & Bednorz, J. G. (1991). Screw dislocations in high-Tc films. *Nature, 354*, 737–740.
15. Zheng, J. Q., Wang, X. K., Shih, M. C., Lee, S. J., Williams, S., So, J., Dutta, P., Ketterson, J. B., & Chang, R. P. H. (1990). In-situ synchrotron study of growth of Y-Ba-Cu-O thin films on $SrTiO_3$.
16. Royer, L. (1928). Recherches experimentales sur l'epitaxie ou orientation mutuelle de cristaux d'especes differentes. *Bulletin Soc Fr Mineral Crystallogr, 7*, 7–159.
17. Schulz, L. G. (1951). Growth of alkali halide crystals from the vapor phase and from solution onto substrates of mica. *Acta Crystallographica, 4*(6), 483–486. https://doi.org/10.1107/s0365110x5100163x
18. Meyer, O., Geerk, J., Li, Q., Linker, G., & Xi, X. X. (1990). Epitaxial growth analysis of YBaCuO thin films by ion backscattering and channeling spectrometry. *Nuclear Instruments and Methods in Physics Research Section B, 45*(1–4), 483–487. https://doi.org/10.1016/0168-583X(90)90881-T
19. Tasker, P. W., & Duffy, D. M. (1984). The structure and properties of the stepped surfaces of MgO and NiO.
20. Raistrick, I. D., Hawley, M., Beery, J. G., Garzon, F. H., & Houlton, R. J. (1991). Microstructure and growth mechanism of thin sputtered films of $YBa_2Cu_3O_7$ on MgO substrates. *Applied Physics Letters, 59*(24), 3177–3179. https://doi.org/10.1063/1.105776
21. Norton, M. G., & Carter, C. B. (1991). The nucleation and heteroepitactic growth of $YBa_2Cu_3O_{7-\delta}$ thin films on MgO.
22. Young, K. H., & Sun, J. Z. (1991). Synthesis of in-plane aligned a-axis $YBa_2Cu_3O_{7-\delta}$ thin films. *Applied Physics Letters, 59*(19), 2448–2450. https://doi.org/10.1063/1.105991
23. Gupta, D., Lacey, J. A., & Laibowitz, R. B. (1991). Migration of Sr at the $YBa_2Cu_3O_{7-\delta}$ epitaxial film and (100) $SrTiO_3$ substrate interface. *Defect and Diffusion Forum, 75*, 79–88. https://doi.org/10.4028/www.scientific.net/ddf.75.79

24. Oishi, A., et al. (1991). Y_2O_3 single crystalline substrate for epitaxial growth of high TC superconducting thin films. *Applied Physics Letters, 59*(15), 1902–1904. https://doi.org/10.1063/1.106183
25. Farneth, W. E., Bordia, R. K., McCarron, E. M., III., Crawford, M. K., & Flippen, R. B. (1988). Influence of oxygen stoichiometry on the structure and superconducting. *Solid State Communications, 66*(9), 953–959.
26. Moeckly, B. H., Russek, S. E., Lathrop, D. K., Buhrman, R. A., Li, J., & Mayer, J. W. (1990). Growth of $YBa_2Cu_3O_7$ thin films on MgO: The effect of substrate preparation. *Applied Physics Letters, 57*(16), 1687–1689. https://doi.org/10.1063/1.104137
27. "gerber1991".
28. Growth Mechanism of Sputtered Films of YBa,Cu30, Studied by Scanning Tunneling Microscopy.
29. Wang, J., et al. (2022). Observation of coexisting weak localization and superconducting fluctuations in strained $Sn_{1-x}In_xTe$ thin films. *Nano Letters, 22*(2), 792–800. https://doi.org/10.1021/acs.nanolett.1c04370
30. Sci, J. V., Suyolcu, Y. E., Christiani, G., & Gemperline, P. T. (2022). Engineering ordered arrangements of oxygen vacancies at the surface of superconducting La_2CuO_4 thin films Engineering ordered arrangements of oxygen vacancies at the surface of superconducting La_2CuO_4 thin films. 013214(November 2021). https://doi.org/10.1116/6.0001473
31. Norton, D. P., Lowndes, D. H., Zheng, X.-Y., Zhu, S., & Warmack, R. J. (1991). Scanning tunneling microscopy of pulsed-laser-deposited $YBa_2Cu_3O_{7-b}$ epitaxial thin film: Surface microstructure and growth mechanism.
32. Pamplin, B. (1979). Preparation and properties of solid state materials. *Progress in Crystal Growth and Characterization, 2*, 247–248. https://doi.org/10.1016/0146-3535(81)90035-6
33. Eom, C. B., Marshall, A. F., Laderman, S. S., Jacowitz, R. D., Geballe, T. H., & H.- Packard. (1990). Epitaxial and smooth films of a-axis $YBa_2Cu_3O_7$. [Online]. www.sciencemag.org
34. Meyer, D., Woollam, J. A., Liou, S. H., Ianno, N. J., Johs, B., & Thompson, D. (1989). Pulsed laser evaporation of Tl-Ba-Ca-Cu-O films. *Science and Technology of Thin Film Superconductors.*
35. Chang, R. P. H., Ketterson, J. B., Wang, X. K., Lee, S. J., Sheng, K. C., Shen, Y. H., & Song, S. N. (1989). Preparation of thin films YBaCu(F)O with high T_c and J_c. *Science and Technology of Thin Film Superconductors.*
36. Gu, Q., et al. (2020). Single particle tunneling spectrum of superconducting $Nd_{1-x}Sr_xNiO_2$ thin films. *Nat Commun, 11*(1). https://doi.org/10.1038/s41467-020-19908-1
37. Xue, H., et al. (2022). Fourfold symmetric superconductivity in spinel oxide $LiTi_2O_4$(001) thin films. *ACS Nano, 16*(11), 19464–19471. https://doi.org/10.1021/acsnano.2c09338
38. Jia, J., Liu, C., Guo, M., & Lei, L. (2022). Preparation of Nd1.85Ce0.15CuO4 superconducting thin film by pulsed laser deposition. *Journal of Vacuum Science & Technology A, 40*(2), 023407. https://doi.org/10.1116/6.0001540
39. Streiffer, S. K., Lairson, B. M., Eom, C. B., Clemens, B. M., Bravman, J. C., & Geballe, T. H. (1991). Microstructure of ultrathin films of $YBa_2Cu_3O_{7-\delta}$ on MgO.
40. Liu, Z., Luo, B., Zhang, L., Hou, B., & Wang, D. (2021). Vortex dynamics in amorphous MoSi superconducting thin films. *Superconductor Science and Technology, 34*(12), 125014. https://doi.org/10.1088/1361-6668/ac2eb0
41. Arendt, P., et al. (1990). Highly-textured Tl-Ba-Ca-Cu-O polycrystalline superconducting films on Ag substrates. *Science and Technology of Thin Film Superconductors 2*, 47–56, Springer. https://doi.org/10.1007/978-1-4684-1345-8_6
42. Gayathri, V., Amaladass, E. P., Vinod, K., Sathyanarayana, A. T., Geetha Kumary, T., & Mani, A. (2022). Variation in the superconducting properties of Bi1.75Pb0.25Sr2Ca2Cu3O10+δ in proximity with $Pr_{0.5}Ca_{0.5}MnO_3$. *Cryogenics (Guildf), 126*(August), 103542. https://doi.org/10.1016/j.cryogenics.2022.103542
43. Ramesh, R., et al. (1990). Structural perfection of Y-Ba-Cu-O thin films controlled by the growth mechanism. *Applied Physics Letters, 57*(10), 1064–1066. https://doi.org/10.1063/1.104282

44. Dimos, D., Chaudhari, P., & Mannhart, J. (1990). Superconducting transport properties of grain boundaries in $YBa_2Cu_3O_7$ bicrystals.
45. Catana, A., Broom, R. F., Bednorz, J. G., Mannhart, J., & Schlom, D. G. (1992). Identification of epitaxial Y_2O_3 inclusions in sputtered $YBa_2Cu_3O_7$ films: Impact on film growth. *Applied Physics Letters, 60*(8), 1016–1018. https://doi.org/10.1063/1.106507
46. Ahn, B. T., Lee, V. Y., Beyers, R., Gur, T. M., & Huggins, R. A. (1990). Quaternary phase relations near $YBa_2Cu_3O_{6+x}$ at 850 °C in reduced oxygen pressures.
47. Poppe, U., Prieto, P., Schubert, J., Soltner, H., & Urban, K. (1989). Epitaxial multilayers of $YBa_2Cu_3O_7$ and $PrBa_2Cu_3O_7$ as a possible basis for superconducting electronic devices.
48. Lowndes, D. H., Norton, D. P., & Budai, J. D. (1990). Superconductivity in nonsymmetric epitaxial $YBa_2Cu_3O_{7-x}/PrBa_2Cu_3C_{7-x}$ Superlattices: The superconducting behavior of Cu-O bilayers.
49. Venkatesan, T., et al. (1990). Epitaxial $Y_1Ba_2Cu_3O_{7-y}/Y_{1-x}Pr_xBa_2Cu_3O_{7-y}$ heterostructures. *Applied Physics Letters, 56*(4), 391–393. https://doi.org/10.1063/1.103291
50. Fork, D. K., et al. (1990). High critical currents in strained epitaxial $YBa_2Cu_3O_{7-\delta}$ on Si. *Applied Physics Letters, 57*(11), 1161–1163. https://doi.org/10.1063/1.104225
51. Olsson, E., Gupta, A., Thouless, M. D., Segmüller, A., & Clarke, D. R. (1991). Crack formation in epitaxial [110] thin films of $YBa_2Cu_3O_{7-\delta}$ and $PrBa_2Cu_3O_{7-x}$ on [110]$SrTiO_3$ substrates. *Applied Physics Letters, 58*(15), 1682–1684. https://doi.org/10.1063/1.105110
52. Adrian, F. J., Moorjani, K., Kim, B. F., Bohandy, J., Agostinelli, E., Phillips, T. E., & Green, W. J. (1989). The effect of annealing conditions on laser deposited superconducting Bi-Sr-Ca-Cu-O thin films. *Science and Technology of Thin Film Superconductors.*
53. Gu, H., Lin, T. X., Wang, S. Z., Yin, D. L., & Xiong, G. C. (1989). Preparation and characterizations of Bi-Pb-Sr-Ca-Cu-O superconducting thin films. *Science and Technology of Thin Film Superconductors*, 199–203.
54. Huang, Y., Vittoria, C., Ryder, D. F., Jr., Marzik, J., Benfer, R., Spurgeon, W., & Ryu, J. (1989). Effect of annealing conditions on the structural and superconducting properties of Y-Ba-Cu-O films. *Science and Technology of Thin Film Superconductors*, 245–252.

Intelligent Lumber Production (Sawmill 4.0): Opportunities, Challenges, and Pathways to Adoption

Vahid Nasir, Sohrab Rahimi, Ahmad Mohammadpanah, Eric Hansen, and Farrokh Sassani

Abstract In transiting from automated to smart lumber manufacturing, the sawmilling industry should adopt cutting-edge technologies for efficient and cost-effective lumber production. This manuscript reviews the opportunities and challenges the sawmilling industry is facing to embrace a smart, data-driven manufacturing approach. Accordingly, the main steps in lumber manufacturing are introduced, and opportunities to practice smart production in sawmills, kilns, and planer mills are explained. Sawmills benefit from sensing and machine vision technology combined with artificial intelligence (AI) for efficient log breakdown, process health and equipment condition monitoring, and smart edging/trimming. The kiln drying process benefits from moisture pre- and post-sorting, in which nondestructive evaluation (NDE) of lumber moisture content is crucial. Smart lumber grading can be practiced using NDE-based techniques combined with AI. Despite the unprecedented opportunities, the lumber industry struggles to leverage big data analytics value and power. The advancement in the Internet of Things (IoT), cyber-physical systems, and cloud computing should highlight the importance of digital twins in sawmilling. AI and big data analytics will be used in the future, not just for smart quality control, processing monitoring, and optimization but for smart design and planning, material distribution and tracking, customization and flexible manufacturing, and market analysis and positioning.

Keywords Wood manufacturing · Industry 4.0 · Artificial intelligence · Machine vision technology · Data-driven smart manufacturing

V. Nasir · E. Hansen
Department of Wood Science and Engineering, College of Forestry, Oregon State University, Corvallis, USA

V. Nasir · A. Mohammadpanah · F. Sassani (✉)
Department of Mechanical Engineering, University of British Columbia, Vancouver, Canada
e-mail: Sassani@mech.ubc.ca

S. Rahimi
Department of Wood Science, University of British Columbia, Vancouver, Canada

M.-R. Alam and M. Fathi (eds.), *Integrated Systems: Data Driven Engineering*,
https://doi.org/10.1007/978-3-031-53652-6_13

1 Introduction

In transitioning from automated to smart lumber manufacturing, sawmills should adopt cutting-edge technologies and embrace industry 4.0-based production practices. While this transition offers unprecedented opportunities to the wood and lumber industry, it is accompanied by challenges mainly related to capital investment cost, lack of skilled labor familiar with Industry 4.0 components, and lack of needed infrastructure for robust connectivity. Manufacturers must become acquainted with the main components of Industry 4.0 and its driving principles to ensure their smooth transition to advanced manufacturing practices. The main components of Industry 4.0 relevant to wood manufacturing are shown in Fig. 1. Adopting the Internet of Things (IoT) through establishing an interconnected network of devices, sensors, and types of machinery facilitates data sharing, remote control, and seamless information flow. Data storage, sharing, and processing were revolutionized using cloud computing, enabling remote access to online and historical production data. The interconnection between the physical objects and the digital models using cyber-physical systems offers remote control and real-time data exchange and decision-making.

Robotics and automation are core components in reducing manual labor and human error, thus enhancing productivity. Artificial intelligence (AI) can provide autonomous decision-making, process optimization, and intelligent solutions for complex manufacturing challenges. Leveraging the power and value of data can revolutionize production by practicing data-driven smart manufacturing [69]. However, while manufacturers increasingly use sensors and data acquisition devices, they still need help interpreting stored data to improve products and processes [32]. Advancements in machine learning and big data analytics can provide manufacturers with the necessary insight for operation optimization, predictive maintenance, and overall efficiency improvement. Lumber manufacturers must acknowledge the opportunities and challenges of adopting Industry 4.0 and data-driven smart manufacturing practices while pursuing sustainable growth and competitiveness in an evolving marketplace.

This study examines the opportunities and challenges encountered by the sawmilling industry in its pursuit of data-driven manufacturing. The paper delves into the potential for implementing intelligent production techniques in key areas, namely sawmills, kilns, and planer mills, for smart grading. A typical sawmill involves a series of interconnected processes and operations that transform raw logs into finished lumber products, as summarized below [78]:

Raw material → Log yard/pond → Long log deck → Debarking → Bucking → Headrig—primary log breakdown → Resaws—secondary breakdown → Edging → Trimming → Rough green sort/stack → Drying → Planer → Grade and trim → Sort → Stack and pack → Store → Shipping.

The sawmilling Industry would benefit from Industry 4.0 and data-driven smart manufacturing in multiple ways, as shown in Fig. 2.

Smart, data-driven manufacturing can streamline production by reducing labor costs and minimizing human error resulting in higher production rates. Through data analytics, manufacturers can analyze historical production data to identify trends,

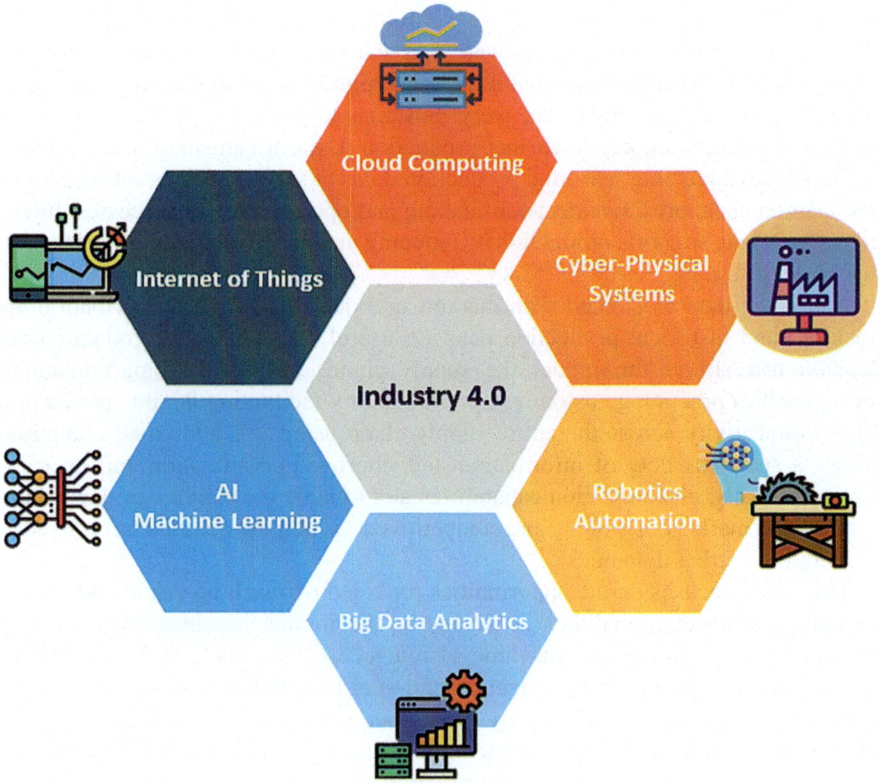

Fig. 1 Some of the main components of Industry 4.0

Fig. 2 The main benefits of and key reasons for lumber manufacturers to embrace Industry 4.0

optimize cutting patterns, and improve yield. Machine vision technology combined with log breakdown modeling software that analyzes factors such as log size and shape, along with equipment capabilities, can optimize cutting decisions, minimize waste, and maximize lumber recovery. Sustainable practices by mitigating environmental impacts are key to cleaner production. Data-driven smart manufacturing benefits from analyzing real-time production data for smart quality control, preventive maintenance, proactive decision-making and optimization of processes. Intelligent monitoring supports production by reducing unwanted downtime and extending machinery lifespan.

Analyzing data associated with markets and demand, material distribution and tracking, and historical production data are useful in customized production and demand traceability throughout the supply chain. Data-driven smart manufacturing enables practicing customized production by facilitating flexible production. Also, connectivity across the entire supply chain using IoT and cloud computing allows a seamless flow of information and improved coordination and decision-making. Finally, customization capabilities and reliable traceability are key factors in improved market positioning and competitiveness, helping manufacturers to adapt to changing market dynamics.

This study focuses on the opportunities provided by machine vision and sensing technology with data analytics in advancing the lumber manufacturing industry. Specifically, it uses AI and machine vision technology, such as X-ray and CT scanning tools, for log characterization and smart breakdown. Also, the opportunities offered by sensing technology to advance sawing monitoring are explained, and the shortcomings of fully embracing the developed techniques by industry are discussed. In addition, state-of-the-art research on smart kiln-drying for minimizing the moisture variation in kiln-dried lumber and improving drying efficiency through adopting data-driven models are discussed. Also, opportunities offered by nondestructive evaluation (NDE) methods for intelligent lumber grading are introduced, and their corresponding challenges are debated. Finally, the study critically discusses the obstacles and limitations the lumber manufacturing industry faces in transitioning to data-driven, smart manufacturing and outlines the direction for future research and development to facilitate a smooth transition.

2 Smart Sawmill

Log sawing can benefit from different technologies for efficient log breakdown, increased recovery and lumber value, reduced downtime, and smart equipment maintenance. This section reviews the opportunities machine vision and sensing technology provides for smart log breakdown and intelligent sawing monitoring.

2.1 Log CT Scanning

Machine vision technology, such as log computed tomography (CT) scanning, can be used to optimize log breakdown and sawing strategies. Log CT scanning offers detailed information on knot characteristics and other internal defects. Laser scanning and imaging systems can capture external log characteristics that provide information about log exterior features such as variation in diameter, taper, and different surface defects. Internal defects detection is useful for log sorting, targeted processing, and increased value. Also, laser and CT scanning could be used for optimizing sawing strategies and finding the best log breakdown pattern. This is done by adjusting the saw positioning resulting in improved lumber recovery and better lumber grade. An automated sawing machine can support sawmills with machine vision technology and sawing pattern simulation software for accurate log positioning and saw alignment. Such automation in production with minimal human error should enhance accuracy and consistency in the production line. Data collected from such a system could be used for continuous process improvement. Historical data could be analyzed to identify the challenges, optimize sawing strategies, and continuously improve lumber recovery and quality.

The effectiveness of log CT scanning for improving sawing efficiency was reported in the literature. Fredriksson et al. [19, 20] stated that log CT scanning can identify unsuitable logs for a certain product at an early production stage. Belley et al. [4] showed that sawing optimization by log CT scanning and knot detection algorithm increases the lumber recovery and yields better lumber grade resulting in higher efficiency and profitability. Rummukainen et al. [65] also discussed that CT scanning before bucking could provide additional value for a sawmill by optimizing log rotation. Log rotation based on CT scan data resulting in increased value was also reported by Berglund et al. [6]; however, they discussed that the value increase depends on the rotation error of the sawing machine and price differences between grades. Fredriksson [18] studied handling the positioning errors during sawing optimization of Scots pine and Norway spruce logs using CT scanning. Fredriksson [17] discussed that sawing position based on log CT scanning could increase value recovery by 13% compared to that based on outer log shape showing potential for higher value recovery and profit. [25] used CT scanning for pith detection in hardwood logs. Gazo et al. [29] used the X-ray CT scanned reconstructed logs for knot characterization and value analysis of hem-fir forests (Fig. 3). Their models estimate the standing tree value based on the tree parameters and knot characteristics. By leveraging log CT scanning and machine vision technologies, sawmills can achieve precise log characterization, optimize log breakdown strategies, and increase lumber recovery and value.

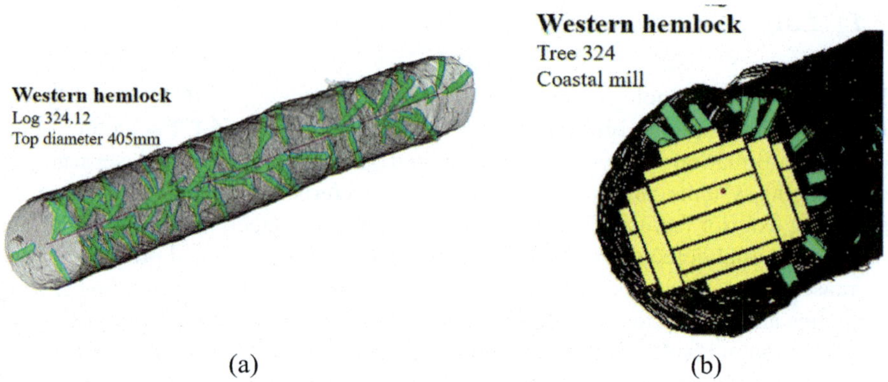

Fig. 3 Log CT scanning can characterize knots (**a**). Examples of cutting patterns from simulation results of western hemlock (**b**) (From [29])

2.2 Sawing Monitoring

Wood sawing is impacted by different factors, including but not limited to wood species parameters (density, moisture content (MC), etc.), cutting tool parameters (cutting speed, saw blade material, number of saw teeth, rake and clearance angles, etc.), and feed factors (feed speed, cutting direction, depth of cut, etc.), as explained by Nasir and Cool [43]. These factors and their interactions can impact the quality of sawn lumber, power consumption during the sawing process, dust emissions, and saw blade conditions. Wood sawing under sawmilling conditions is different from other types of wood machining due to the extreme cutting conditions for lumber manufacturing. Studying wood sawing to resemble sawmilling conditions involves practicing very high feed speed (up to 180 m/min), resulting in higher sawing deviation and waviness in sawn lumber (Fig. 4a) and increased cutting power [42]. The dynamic behavior of the cutting tool can further complicate the sawing process. Mohammadpanah and Hutton [37–40] identified the critical and flutter instability speeds of guided circular saws. Nasir et al. [51] showed the complex role of rotation speed in lumber waviness during lumber circular sawing. They showed that the proximity of the rotation speed to the saw critical or flutter instability speeds would impact the waviness of the sawn lumber. The sawing process can further be complicated by climate conditions resulting in sawing frozen (or partially frozen) logs. While sawing frozen logs requires higher cutting energy [54], less waviness is reported during sawing frozen or green lumber than kiln-dried wood [46].

The role of different cutting parameters and their interactions still need to be better understood. Developing an intelligent framework for process conditions and health monitoring is crucial for sharp and well-maintained cutting tools, high production rates, and optimal lumber quality. Wood sawing monitoring was studied in the literature with the objective of monitoring:

(a) (b)

Fig. 4 **a** Sawn lumber with waviness. **b** Schematic of a guided circular saw monitored with a microphone and contact sensor (such as accelerometer or acoustic emission) placed on the saw guide

- cutting power
- lumber waviness
- dust emission
- tool temperature
- tool wear condition

Cutting power can be monitored using a current sensor, which is inexpensive and easy to install. However, its limited bandwidth range and low-pass filter nature may restrict its applications for minor tool breakage [9]. Direct measurement of lumber waviness requires employing a laser displacement sensor following the sawing process, which may not be a readily available option in sawmills. Laboratory-scale research showed the potential of using sound [48], vibration [44], and acoustic emission [49] sensors for monitoring lumber waviness. Sound signals vary by cutting conditions [35]. The challenge is processing signals polluted by the background noises in a sawmill, which requires sophisticated signal pre-processing (e.g., denoising). Vibration or acoustic emission sensors could be placed on the saw guided (Fig. 4b).

Cutting conditions can significantly impact chip formation and dust emission [43]. The average chip thickness reduction was reported to increase generated airborne dust [45]. Sawdust is a byproduct of lumber manufacturing. Airborne dust is also an occupational health concern and safety hazard. While a dust monitor can be used to assess the overall mass concentration of the airborne dust (Fig. 5a), acquired signals from an acoustic emission sensor placed on the saw guide were shown to successfully predict the generated airborne dust during the circular sawing of Douglas-fir lumber [45].

Tool condition monitoring is an essential preventive maintenance task to ensure the performance of sawing processes by minimizing unscheduled tool changes and downtime. A key indicator of tool condition is tool temperature. Increased tool

Fig. 5 **a** Dust monitor can be used to detect the generated airborne dust during the lumber circular sawing process. **b** Wireless temperature sensors placed on the saw guide to measure the saw eye and rim temperature

temperature could compromise the blade stiffness and may result in burn marks on sawn lumber. A non-contact temperature sensor may not be helpful at sawmills due to the cutting zone's excessive amount of sawdust and coolant. Recently, a contact temperature sensor was designed (Fig. 5b) to monitor the tool temperature at saw-eye and saw-rim [41]. It is worth noting that the temperature gradient (between the saw-eye and saw-rim) is also of uttermost significance, which should be minimized. The temperature sensor could measure the cooling capacity of saw guides and the rate of heating to cooling over time [41]. Nasir et al. [53] studied the effect of cutting parameters on the tool temperature and attempted to predict the temperature using the sensory features extracted from sound, vibration, and acoustic emission signals. They fed the multi-sensorial features into a random forest predictive model for temperature prediction. Finally, a sensor fusion approach could be implemented for tool wear monitoring. Nasir et al. [52] developed a sensor fusion-based model using power, sound, vibration, and acoustic emission signals to classify the tool wear condition (sharp vs. dull). They showed the effectiveness of a tree-based ensemble model (random forest and gradient boosting machine) for tool condition monitoring. In a related study, a similar multi-sensorial model could classify frozen lumber from green lumber during the circular sawing process [50].

Despite the reported effectiveness of the developed monitoring systems under laboratory conditions, their performance in harsh sawmilling conditions is not yet well evaluated. The reported studies mainly focus on developing data-driven models using a small sample size and machine learning models such as group method of data handling (GMDH), multilayer perceptron (MLP) neural network, adaptive neuro-fuzzy inference system (ANFIS), or tree-based ensemble learning models (e.g., random forest, gradient boosting machine). Models developed under controlled laboratory conditions may not correctly reflect the process uncertainty in real-life situations [47]. The gap between the laboratory and industry mainly involves the size of the acquired dataset and its challenges to satisfactorily address the process uncertainty and the type of suitable data-driven model. Accordingly, some classic machine learning models might be replaced with models capable of handling big data (e.g., deep learning models). The laboratory-scale developed models might be used as a baseline in training models from the sawmill data. The literature lacks studies on incremental and transfer learning to address the model's veracity when dealing with process health and condition monitoring in sawmills. Finally, the harsh sawmilling conditions make data acquisition challenging, increasing the risk of sensor failure due to dust and coolant or cutting tool vibration. Topics such as sensor fusion and multi-sensorial features, sensor redundancy, or system fault versus sensor fault must be better studied in sawmills when developing reliable data-driven monitoring models.

3 Smart Kiln-Drying

In lumber production process, wood drying is the most time/energy-consuming step that requires a significant portion of the total energy in sawmills [80]. Wood drying is crucial to improve timber mechanical properties and dimensional stability while facilitating storage, reducing transportation costs, mitigating the risk of biodegradation, and improving the machinability of lumber [10, 66]. Conventional heat-and-vent wood drying is a process of random characteristics owing to numerous boards with varied levels of thermo-physical–mechanical properties simultaneously dried.

Temperature, relative humidity, and air velocity remarkably impact moisture loss rate. As aforesaid, in commercial kiln drying, a population of green boards with different anatomy and characteristics enters a kiln and remains there until reaching the target average MC. Two phenomena transpire when the drying cycle ends: a moisture variation (moisture gradient) between the core and shells in each dried board and a moisture variation between boards in the same kiln drying batch [2]. Lack of moisture uniformity within a board (high moisture gradient in thickness) significantly lowers the quality of the dried wood, worsening drying defects and may cause further issues in remanufacturing [57, 68]. Also, the non-uniformity of final MC between boards remarkably affects the monetary value of the dried product. Boards with final MC far below the target moisture have a lower commercial/market value because over-drying increases the risk of drying defects such as checks, splits, and shape distortion. Moreover, boards with final MC much higher than the average cannot suit structural applications, considered wet or semi-dried at low prices [11]. Figure 6 shows an example of moisture variation in a drying run, representing over-dried, acceptable, and under-dried categories.

Sorting boards before or after kiln-drying improves moisture uniformity and product quality. Researchers have applied pre- and post-sorting strategies to the wood drying on different scales to prevent over- and under-drying, dimensional instability, and low-grade recovery [8, 60]. Concerning pre-sorting, Elustondo et al. [14] sorted hem-fir lumber before drying into three groups by electric capacitance and weight, achieving reduced residence time by ~10% and over-dried boards to practically zero. In another study, Berberović and Milota [5] pre-sorted Western hemlock logs based on basic density, initial MC, heartwood percentage, and annual ring angle. Their

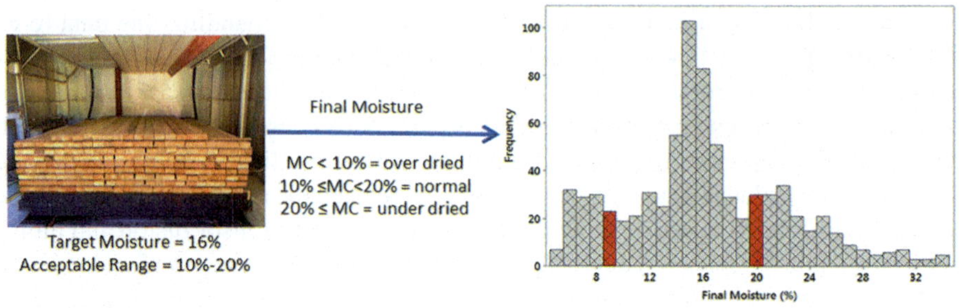

Fig. 6 Kiln-drying may result in over- and under-dried lumber. Figure show western hemlock lumber in a kiln where the target MC is 16%, with an acceptable moisture range of 10–20%. The histogram of the boards' final MC shows the distribution of over- and under-dried boards (data from [55])

findings showed that pre-sorting could reduce drying time and alleviate moisture variability. Regarding post-sorting, Elustondo et al. [13] applied the dry-sort-re-dry (DSRD) method to baby-square Pacific Coast hemlock, reducing drying time by 4.6% and reducing timber degradation by 4.7%. However, sawmills have daunting challenges applying sorting methods as they require time and labor costs, which could outweigh the advantages of sorting strategies. Therefore, finding optimum breakpoints and efficient sorting techniques is paramount, where wood drying modeling and Industry 4.0 could play a pivotal role.

The successful implementation of Industry 4.0 in kiln-drying requires accurate and adaptable measuring tools and equipment to collect and record data integrated with smart analysis systems. Precise moisture measurement in wood, especially above the fiber saturation point, is still challenging. To date, researchers have used several methods for measuring moisture in wood [71], such as the gravimetric method [31], electric property measurement [18], low field nuclear magnetic resonance [36], X-ray computed tomography [76], neutron imaging [64], magnetic resonance imaging [63], and near-infrared spectroscopy. However, there is still a gap between the laboratory scale research and sawmilling condition on employing a fast and reliable non-destructive moisture measurement method that can fully address the challenges associated with mass production scale at sawmills, high MC level in green lumber and measurements on boards with different thickness.

Industry 4.0 technologies enable collecting and analyzing vast amounts of data during kiln-drying. Integrated sensors capture temperature, humidity, airflow, and MC data, providing real-time insights into the drying progress. Advanced analysis and machine learning algorithms can process this data to determine patterns, optimize drying factors, and estimate drying times more precisely. This data-driven approach facilitates proactive decision-making, minimizes product defects, and improves drying efficiency.

Controlling and modeling drying criteria is an arduous task, mainly due to the complexity of the drying process. However, wood drying quality is closely interrelated with the integrity and accuracy of the final moisture. In order to develop efficient drying methods, understanding and improving drying quality factors, especially final MC and its spread, are crucial [79, 81]. Systematic methods for evaluating MC assist wood scientists in optimizing drying schedules. Elustondo and Avramidis [12] compared three approaches for wood drying simulation, i.e., Cronin's method, Monte Carlo method, and continuous probability density. Watanabe et al. [77] used the bootstrap method for the probabilistic evaluation of final MC in kiln-dried boards, and showed that the moisture data did not follow a classical probability distribution, indicating that nonparametric analysis should be employed. Besides, the bootstrap method is perhaps suitable for analyzing data derived from drying processes and the strength data of structural timber due to the model's simplicity, which widens its application range. Rahimi et al. [57] predicted final moisture variation based on initial moisture variation and target moisture using the polynomial fitting method.

Although traditional methods predicted MC to some degree of accuracy, they could not deal with several inputs to create a holistic predictive model. Therefore, more recently, researchers applied machine learning methods to predict and classify MC after drying. Rahimi and Avramidis [56] used different artificial neural network (ANN) methods and found the group method of data handling (GMDH) best to predict final moisture based on initial weight, initial moisture, and basic density. Using a decision tree method, Rahimi et al. [58] classified kin-dried Western hemlock into three categories: over-dried, acceptable, and under-dried. In another study, Rahimi et al. [59] utilized support vector regression (SVR) and ANFIS and reported that the former method is a robust predictive model while the latter one had unsatisfactory results when dealing with multiple inputs. Figure 7 shows an example of inputs and outputs in a moisture-classifying approach. Finally, Rahimi et al. [61] employed machine learning methods to predict and classify the final MC using wood properties and kiln-drying parameters. Results revealed that although the impact of the wood properties is more pronounced in the predictive model, drying parameters such as schedule, conditioning, and post-storage improve the robustness of the predictive approach. Further studies are required to develop wood drying models, making the drying process more compatible with Industry 4.0 technologies.

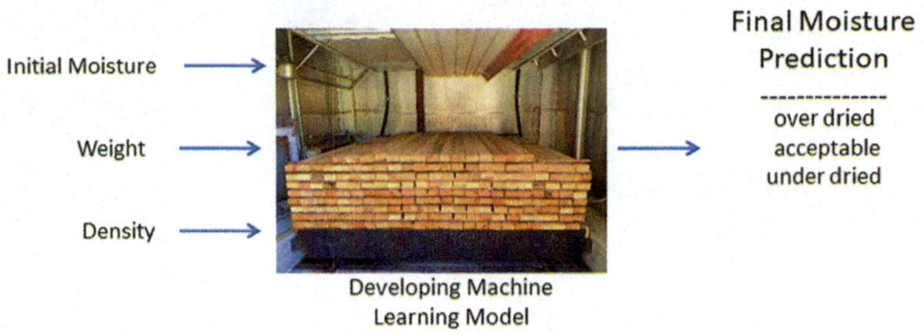

Fig. 7 Lumber properties such as initial MC, density, and weight before the drying process could be used to train a machine learning model to predict the risk of having over- and under-dried lumber after drying [58]. The model could be helpful for lumber pre-sorting. The key to this approach is the fast and reliable measurement of MC in green lumber

In summary, smart kiln-drying in wood manufacturing promises to advance the industry through enhanced efficiency, quality control, and data-driven decision-making. Integrating Industry 4.0 technologies enables automation, real-time monitoring, advanced analytics, and optimization.

4 Smart Grading

Overall, higher-grade lumber represents fewer defects and larger, clear, defect-free areas [70]. Lumber grading is crucial and would benefit from advances in machine vision technology and data science, from visual inspection to an automated grading system. Apart from appearance, the surface defects of sawn lumber, such as knots or cracks, impact its strength, a critical factor for structural timber. Also, the technology can be used for smart trimming and edging to increase lumber value. Trimming or edging could be practiced to eliminate the major defects when they are located near the end or an edge of the board [24]. Data acquisition for grading purposes consists of collecting surface images of sawn lumber, which should then be pre-processed (e.g., de-noising) and extracting informative features from the images. The extracted and selected features can then be fed into an intelligent decision-making model for defect detection and lumber grading. Bharati et al. [7] used a lumber vision system containing a light source and scan camera (Fig. 8a) to detect sawn lumber defects such as knots, splits, wane, pitch, and bark pockets (Fig. 8b), using multivariate image analysis.

Kline et al. [30] employed a laser profile detector, color cameras, and an X-ray scanner for automated lumber grading of red oak boards. Thomas [70] developed an

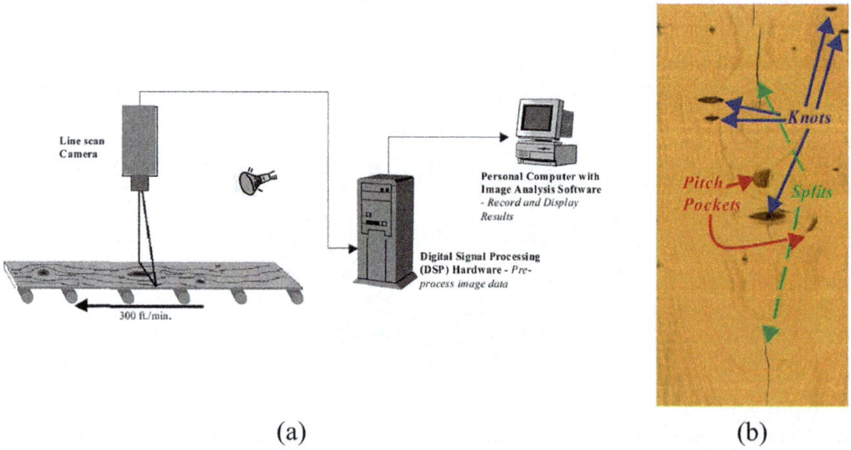

Fig. 8 **a** Schematic of a vision system for lumber grading. **b** Example of different types of defects in lumber (Reprinted with permission from [7], Copyright ©2003, American Chemical Society)

ANN model for real-time hardwood lumber grading. Gazo et al. [24] used a Microtec Goldeneye 300 Multi-Sensor Quality Scanner™ for automated hardwood lumber grading. They reported that the automated grading system accuracy was 92.22% on-grade and 99.50% on-value. They used the scanner at a speed of 980 lineal ft/min, suitable for a sawmill with an annual production rate of 20–25 million board feet of lumber. Deep learning has been used in recent years for sawn lumber grading. Wang et al. [75] reviewed the main deep learning approach for quality control of sawn lumber. Accordingly, the deep algorithms used include target detection based on region proposals such as R-CNN (Region with Convolutional Neural Network), those based on regression such as YOLO (You Only Look Once) or Single Shot Multibox Detector algorithms [75]. Using CNNs for lumber defects detection was also reported in the literature [1, 26–28].

Sensing technology and AI could also be applied to producing machine stress-rated (MSR) lumber used in manufacturing engineered wood products. E-rated lumber could be produced based on the American National Standards Institute ANSI/AITC A190.1 requirement, the American National Standard for Wood Products—Structural Glued Laminated Timber [23]. The fact that E-rated grades are not commodity products and are produced following marketing discussions with customers [23] limits the production of these products. Different technologies could be used for the mechanical grading of lumber. Galligan and McDonald [23] discussed using a stress wave system for measuring the dynamic modulus of elasticity (MOE) of a board, using an X-ray system and density profile for grading, or identifying the lumber grade by the response to piece stiffness by continuously, mechanically flexing the board over a set of rolls as the board passes flatwise through the machine [23]. Li et al. [34] studied density-based lumber grading for cross-laminated timber (CLT) manufacturing. Schajer [67] utilized X-ray scanning to obtain the density profile and estimate the wood strength for lumber strength grading. Wang et al. [74] studied the grading of softwood lumber using the ultrasonic wave method for calculating the dynamic MOE. Rais et al. [62] estimated the dynamic MOE using the longitudinal vibration (resonance) method for grading European beech logs and lumber. Fujimoto et al. [21, 22] used near-infrared (NIR) spectroscopy to predict the wood properties and showed the potential of the method for the online grading of sawn lumber. Ayanleye et al. [3] also used NIR spectroscopy to predict the MOE and strength of Douglas-fir and western hemlock and showed higher prediction accuracy when collecting NIR spectra from lumber with a rough surface (versus smooth surface following surface planing). They also used ANN and ANFIS to develop an intelligent predictive model.

Wave propagation methods have been widely practiced in estimating the mechanical properties of lumber, showing their high potential for lumber grading. Fathi et al. [15] used a guided Lamb wave propagation method for accurate estimation of the elastic and viscoelastic properties of wood. Fathi et al. [16] used the Lamb wave velocity as an input of a machine learning method for predicting the wood MOE and bending strength. Nondestructive testing and machine learning can be combined to provide sorting strategies for specific applications. An example is the research conducted by van Blokland et al. [73], where lumber properties measured after kiln drying was used to predict check formation and bending properties in lumber following 2.5 years of weathering exposure. Blokland et al. [72] studied the opportunities for lumber pre-selection at sawmills to minimize internal check formation in weathered timber. Sorting strategies could go beyond the criteria of mechanical grading and provide intelligent decision-making strategies for specific grading applications. However, the main challenge still concerns market demand for these sorted products. In addition, Galligan and McDonald [23] discussed that grading machines might necessitate reevaluation of existing planer mills and related facilities. Also, the capital machinery, operation, quality control, maintenance, and personnel training costs associated with these machines and technologies should be considered when calculating the cost and benefits of such technologies.

5 Limitations and Prospect

Gaining insights into the barriers that hinder the successful implementation of Industry 4.0 strategies is an initial stride toward enabling companies to embrace and deploy novel technologies. Adopting these technologies is crucial for mills to enhance operational efficiency and maintain relevance in intensifying global competition. However, it is first vital to understand the challenges and then plan accordingly to address the limitations. Kusiak [32] discussed that most companies need help with meaningful data interpretation and operation in isolation. They further discussed that data science is rarely taught to engineering or business students. Kusiak [32] suggested that manufacturers should be advised on the type of data to collect, the type of sensors to use, and their placement in production. Challenges associated with data collection should also be considered. The acquired data can only sometimes be readily processed as it may be polluted by noise. Data collected from multiple sources may not necessarily be synchronized, and sensor failure can result in missing data that further complicates data processing. The developed models should also address the uncertainty in the production line. The sawmill industry should practice concepts such as transfer and incremental learning. Successful implementation of these strategies requires well-trained employees and fruitful collaboration between academia and industry. Finally, the industry must be assured about the cost and benefits and how data-driven smart manufacturing would impact their investment, costs, and revenue.

Recent work shows the industry is ill-prepared to embrace Industry 4.0-based production practices [33]. Overall, awareness of current advanced manufacturing technologies is relatively low. In turn, strategic approaches to embracing Industry 4.0 and smart manufacturing are quite rare, at least within surveyed primary manufacturers in the United States [33]. Regarding expertise, the highest level of proficiency was in manufacturing process monitoring and data analysis, while the lowest was in robotics. The study underscores the significant lag in adopting Industry 4.0 technologies within the primary wood products sector in the United States, falling far short of its full potential. Among the critical obstacles identified as impeding the adoption of new technologies were the need for a more skilled workforce and clarity regarding the financial benefits associated with investments in these advancements. As automation and digitization increasingly become prerequisites for competitiveness, the industry must expedite its transition to Industry 4.0. Failure to enhance awareness in this regard places the US industry at risk of losing its competitive edge [33].

Legg et al. [33] suggest that companies inevitably stagnate if their workforce lacks the necessary expertise to embrace new technologies, and there is a need for more qualified experts who can champion and facilitate the adoption of advanced technological solutions. Given the shortage of skilled labor in the market, particularly within the wood processing industry, there may be more sustainable approaches than relying on external hires. Instead, upgrading the skills and knowledge of existing employees must be emphasized. Offering training to individuals already acquainted with the company ensures optimal utilization of the existing technology and its vast potential for application. Additionally, involving affected employees in the process implementation fosters a sense of commitment towards adopting new technologies, effectively quelling fears related to potential workforce reductions that may complement automation initiatives. Ongoing training should accompany the entire transition, enabling employees to actively contribute to the implementation process, as their involvement is vital for building a firm foundation and dedication to new technologies.

6 Conclusions

The transition towards data-driven, smart manufacturing and adopting industry 4.0 strategies present lumber manufacturers with unprecedented opportunities and challenges. Embracing cutting-edge technologies such as IoT, big data analytics, cloud computing, robotics, and automation secures a smooth transition towards smart sawmilling by increasing productivity and recovery, minimizing waste, downtime, and environmental impact while practicing flexible manufacturing and market competitiveness. State-of-the-art highlights the tremendous opportunities sensing and machine vision technology combined with AI and machine learning can offer for optimal log breakdown and sawing strategies, sawing monitoring, smart kiln-drying, and lumber grading. While topics such as log CT scanning and sawing strategies are further studied and practiced, there is still a knowledge gap on using AI and big data analytics for intelligent sawing monitoring and smart quality control. It is worth mentioning that AI and data analytics could be applied to manufacturing processes such as sawing or the entire manufacturing system to study the production flow and bottleneck analysis. The need for more skilled workers emerges as the primary hindrance to implementing new technologies, with outdated facilities and uncertain financial benefits also being significant factors.

References

1. Affonso, C., Rossi, A. L. D., Vieira, F. H. A., & de Leon Ferreira, A. C. P. (2017). Deep learning for biological image classification. *Expert Systems with Applications, 85*, 114–122.
2. Avramidis, S., Lazarescu, C., & Rahimi, S. (2023) Basics of wood drying. In *Springer handbook of wood science and technology* (pp. 679–706). Springer International Publishing. https://doi.org/10.1007/978-3-030-81315-4_13
3. Ayanleye, S., Nasir, V., Avramidis, S., & Cool, J. (2021). Effect of wood surface roughness on prediction of structural timber properties by infrared spectroscopy using ANFIS, ANN and PLS regression. *European Journal of Wood and Wood Products, 79*, 101–115. https://doi.org/10.1007/s00107-020-01621-x
4. Belley, D., Duchesne, I., Vallerand, S., Barrette, J., & Beaudoin, M. (2019). Computed tomography (CT) scanning of internal log attributes prior to sawing increases lumber value in white spruce (*Picea glauca*) and jack pine (*Pinus banksiana*). *Canadian Journal of Forest Research, 49*(12), 1516–1524.
5. Berberovic, A., & Milota, M. R. (2011). Impact of wood variability on the drying rate at different moisture content levels. *Forest Products Journal, 61*(6), 435–442.
6. Berglund, A., Broman, O., Grönlund, A., & Fredriksson, M. (2013). Improved log rotation using information from a computed tomography scanner. *Computers and Electronics in Agriculture, 90*, 152–158.
7. Bharati, M. H., MacGregor, J. F., & Tropper, W. (2003). Softwood lumber grading through on-line multivariate image analysis techniques. *Industrial and Engineering Chemistry Research, 42*(21), 5345–5353.
8. Bond, B. H., & Espinoza, O. (2016). A decade of improved lumber drying technology. *Current Forestry Reports, 2*(2), 106–118.
9. Byrne, G., Dornfeld, D., Inasaki, I., Ketteler, G., König, W., & Teti, R. (1995). Tool condition monitoring (TCM)—The status of research and industrial application. *CIRP Annals, 44*(2), 541–567.
10. Cabardo, S. J. (2007). *Variability in physical and transport properties regarding drying behavior for regrowth and plantation blackbutt timber in New South Wales*. Ph.D. Dissertation, University of Sidney, South Wales, Australia.
11. Cronin, K. P., Norton, B., & Taylor, J. (1997). Drying lumber in kilns: A methodology for stochastic analysis using Markov modeling. *Drying Technology, 15*(3–4), 765–790.
12. Elustondo, D. M., & Avramidis, S. (2005). Comparative analysis of three methods for stochastic lumber drying simulation. *Drying Technology, 23*(1–2), 131–142.
13. Elustondo, D., Avramidis, S., & Oliveira, L. (2005). Industrial evaluation of re-dry strategy for softwood lumber. *Maderas Ciencia y Tecnologia, 7*(2), 65–78.
14. Elustondo, D., Oliveira, L., & Ananias, R. A. (2013). Visual method to assess lumber sorting before drying. *Drying Technology, 31*(1), 32–39.
15. Fathi, H., Kazemirad, S., & Nasir, V. (2020). A nondestructive guided wave propagation method for the characterization of moisture-dependent viscoelastic properties of wood materials. *Materials and Structures, 53*, 1–14. https://doi.org/10.1617/s11527-020-01578-6
16. Fathi, H., Nasir, V., & Kazemirad, S. (2020). Prediction of the mechanical properties of wood using guided wave propagation and machine learning. *Construction and Building Materials, 262*, 120848. https://doi.org/10.1016/j.conbuildmat.2020.120848
17. Fredriksson, M. (2014). Log sawing position optimization using computed tomography scanning. *Wood Material Science and Engineering, 9*(2), 110–119.
18. Fredriksson, M. (2016). Handling positioning errors when optimizing sawing of Scots pine and Norway spruce logs using CT scanning. *Journal of wood science, 62*(5), 400–406.
19. Fredriksson, M., Berglund, A., & Broman, O. (2015). Validating a crosscutting simulation program based on computed tomography scanning of logs. *European Journal of Wood and Wood Products, 73*, 143–150.
20. Fredriksson, M., Wadsö, L., Johansson, P., & Ulvcrona, T. (2016). *Wood Material Science and Engineering, 11*(4), 189–200.

21. Fujimoto, T., Kurata, Y., Matsumoto, K., & Tsuchikawa, S. (2010). Feasibility of near-infrared spectroscopy for on-line grading of sawn lumber. *Applied Spectroscopy, 64*(1), 92–99.
22. Fujimoto, T., Kobori, H., & Tsuchikawa, S. (2012). Prediction of wood density independently of moisture conditions using near-infrared spectroscopy. *Journal of Near Infrared Spectroscopy, 20*(3), 353–359.
23. Galligan, W. L., & McDonald, K. A. (2000). *Machine grading of lumber: Practical concerns for lumber producers*. US Department of Agriculture, Forest Service, Forest Products Laboratory.
24. Gazo, R., Wells, L., Krs, V., & Benes, B. (2018). Validation of automated hardwood lumber grading system. *Computers and Electronics in Agriculture, 155*, 496–500.
25. Gazo, R., Vanek, J., Abdul_Massih, M., & Benes, B. (2020). A fast pith detection for computed tomography scanned hardwood logs. *Computers and Electronics in Agriculture, 170*, 105107.
26. He, T., Liu, Y., Xu, C., Zhou, X., Hu, Z., & Fan, J. (2019). A fully convolutional neural network for wood defect location and identification. *IEEE Access, 7*, 123453–123462.
27. He, T., Liu, Y., Yu, Y., Zhao, Q., & Hu, Z. (2020). Application of deep convolutional neural network on feature extraction and detection of wood defects. *Measurement, 152*, 107357.
28. Hu, J., Song, W., Zhang, W., Zhao, Y., & Yilmaz, A. (2019). Deep learning for use in lumber classification tasks. *Wood Science and Technology, 53*, 505–517.
29. Ji, A., Cool, J., & Duchesne, I. (2021). Using X-ray CT scanned reconstructed logs to predict knot characteristics and tree value. *Forests, 12*(6), 720.
30. Kline, D. E., Surak, C., & Araman, P. A. (2003). Automated hardwood lumber grading utilizing a multiple sensor machine vision technology. *Computers and Electronics in Agriculture, 41*(1–3), 139–155.
31. Koponen, H. (1984). Dependences of moisture diffusion coefficients of wood and wooden panels on moisture content and wood properties. *Paperi ja puu, 66*(12), 740–745.
32. Kusiak, A. (2017). Smart manufacturing must embrace big data. *Nature, 544*(7648), 23–25.
33. Legg, B., Dorfner, B., Leavengood, S., & Hansen, E. (2021). Industry 4.0 implementation in US primary wood products industry. *Wood Industry/Drvna Industrija, 72*(2), 143–153.
34. Li, H., Wang, L., Wang, B. J., Wei, P., Yu, W., Fan, Z., & Du, G. (2021). Preliminary evaluation of a density-based lumber grading method for hem-fir CLT manufacturing. *European Journal of Wood and Wood Products, 79*, 967–975.
35. Licow, R., Chuchala, D., Deja, M., Orlowski, K. A., & Taube, P. (2020). Effect of pine impregnation and feed speed on sound level and cutting power in wood sawing. *Journal of Cleaner Production, 272*, 122833.
36. Merela, M., Oven, P., Serša, I., & Mikac, U. (2009). A single point NMR method for an instantaneous determination of the moisture content of wood. *Holzforschung, 63*(3), 348–351.
37. Mohammadpanah, A., & Hutton, S. G. (2015). Flutter instability speeds of guided splined disks: An experimental and analytical investigation. *Shock and Vibration*, 942141.
38. Mohammadpanah, A., & Hutton, S. G. (2016). Maximum operation speed of splined saws. *Wood Material Science and Engineering, 11*(3), 142–146.
39. Mohammadpanah, A., & Hutton, S. G. (2016). Dynamics behavior of a guided spline spinning disk, subjected to conservative in-plane edge loads, analytical and experimental investigation. *Journal of Vibration and Acoustics, 138*(4), 041005.
40. Mohammadpanah, A., & Hutton, S. G. (2017) Theoretical and experimental verification of dynamic behaviour of a guided spline arbor circular saw. *Shock and Vibration*.
41. Mohammadpanah, A., Lehmann, B., & White, J. (2019). Development of a monitoring system for guided circular saws: An experimental investigation. *Wood Material Science and Engineering, 14*(2), 99–106.
42. Nasir, V., & Cool, J. (2019). Optimal power consumption and surface quality in the circular sawing process of Douglas-fir wood. *European Journal of Wood and Wood Products, 77*, 609–617. https://doi.org/10.1007/s00107-019-01412-z
43. Nasir, V., & Cool, J. (2020). A review on wood machining: Characterization, optimization, and monitoring of the sawing process. *Wood Material Science & Engineering, 15*(1), 1–16. https://doi.org/10.1080/17480272.2018.1465465

44. Nasir, V., & Cool, J. (2020). Intelligent wood machining monitoring using vibration signals combined with self-organizing maps for automatic feature selection. *The International Journal of Advanced Manufacturing Technology, 108*, 1811–1825. https://doi.org/10.1007/s00170-020-05505-5
45. Nasir, V., & Cool, J. (2020). Characterization, optimization, and acoustic emission monitoring of airborne dust emission during wood sawing. *The International Journal of Advanced Manufacturing Technology, 109*(9–12), 2365–2375. https://doi.org/10.1007/s00170-020-05842-5
46. Nasir, V., & Cool, J. (2021). Cutting power and surface quality in sawing kiln-dried, green, and frozen hem-fir wood. *Wood Science and Technology, 55*, 505–519. https://doi.org/10.1007/s00226-020-01259-1
47. Nasir, V., & Sassani, F. (2021). A review on deep learning in machining and tool monitoring: Methods, opportunities, and challenges. *The International Journal of Advanced Manufacturing Technology, 115*(9–10), 2683–2709. https://doi.org/10.1007/s00170-021-07325-7
48. Nasir, V., Cool, J., & Sassani, F. (2019). Intelligent machining monitoring using sound signal processed with the wavelet method and a self-organizing neural network. *IEEE Robotics and Automation Letters, 4*(4), 3449–3456. https://doi.org/10.1109/LRA.2019.2926666
49. Nasir, V., Cool, J., & Sassani, F. (2019). Acoustic emission monitoring of sawing process: Artificial intelligence approach for optimal sensory feature selection. *The International Journal of Advanced Manufacturing Technology, 102*, 4179–4197. https://doi.org/10.1007/s00170-019-03526-3
50. Nasir, V., Kooshkbaghi, M., & Cool, J. (2020). Sensor fusion and random forest modeling for identifying frozen and green wood during lumber manufacturing. *Manufacturing Letters, 26*, 53–58. https://doi.org/10.1016/j.mfglet.2020.09.011
51. Nasir, V., Mohammadpanah, A., & Cool, J. (2020). The effect of rotation speed on the power consumption and cutting accuracy of guided circular saw: Experimental measurement and analysis of saw critical and flutter speeds. *Wood Material Science & Engineering, 15*(3), 140–146. https://doi.org/10.1080/17480272.2018.1508167
52. Nasir, V., Dibaji, S., Alaswad, K., & Cool, J. (2021). Tool wear monitoring by ensemble learning and sensor fusion using power, sound, vibration, and AE signals. *Manufacturing Letters, 30*, 32–38. https://doi.org/10.1016/j.mfglet.2021.10.002
53. Nasir, V., Kooshkbaghi, M., Cool, J., & Sassani, F. (2021). Cutting tool temperature monitoring in circular sawing: Measurement and multi-sensor feature fusion-based prediction. *The International Journal of Advanced Manufacturing Technology, 112*, 2413–2424. https://doi.org/10.1007/s00170-020-06473-6
54. Orlowski, K., Sandak, J., Negri, M., & Dzurenda, L. (2009). Sawing frozen wood with narrow kerf saws: Energy and quality effects. *Forest Products Journal, 59*(3), 79.
55. Rahimi, S. (2021). *Estimating the final moisture content variation in kiln-dried Pacific coast hemlock*. Doctoral dissertation, University of British Columbia.
56. Rahimi, S., & Avramidis, S. (2022). Predicting moisture content in kiln-dried timbers using machine learning. *European Journal of Wood and Wood Products, 80*(3), 681–692. https://doi.org/10.1007/s00107-022-01794-7
57. Rahimi, S., Faezipour, M., & Tarmian, A. (2011). Drying of internal-check prone poplar lumber using three different conventional kiln drying schedules. *Journal of the Indian Academy of Wood Science, 8*, 6–10. https://doi.org/10.1007/s13196-011-0016-5
58. Rahimi, S., Avramidis, S., & Lazarescu, C. (2022a) Estimating moisture content variation in kiln-dried Pacific coast hemlock. *Holzforschung, 76*(1), 26–36. https://orcid.org/0000-0001-6640-6626
59. Rahimi, S., Nasir, V., Avramidis, S., & Sassani, F. (2022b). Wood moisture monitoring and classification in kiln-dried timber. *Structural Control and Health Monitoring, 29*(4), e2911. https://doi.org/10.1002/stc.2911
60. Rahimi, S., Nasir, V., Avramidis, S., & Sassani, F. (2022c). Benchmarking moisture prediction in kiln-dried Pacific Coast hemlock wood. *International Wood Products Journal, 13*(4), 219–226. https://doi.org/10.1080/20426445.2022.2104212

61. Rahimi, S., Nasir, V., Avramidis, S., & Sassani, F. (2023). The role of drying schedule and conditioning in moisture uniformity in wood: A machine learning approach. *Polymers, 15*(4), 792. https://doi.org/10.3390/polym15040792
62. Rais, A., Pretzsch, H., & van de Kuilen, J. W. G. (2020). European beech log and lumber grading in wet and dry conditions using longitudinal vibration. *Holzforschung, 74*(10), 939–947.
63. Rosenkilde, A., & Glover, P. (2002). High resolution measurement of the surface layer moisture content during drying of wood using a novel magnetic resonance imaging technique. *Holzforschung, 56*, 312–317.
64. Rosner, S., Riegler, M., Vontobel, P., Mannes, D., Lehmann, E. H., Karlsson, B., & Hansmann, C. (2012). Within-ring movement of free water in dehydrating Norway spruce sapwood visualized by neutron radiography. *Holzforschung, 66*(6), 751–756.
65. Rummukainen, H., Makkonen, M., & Uusitalo, J. (2021). Economic value of optical and X-ray CT scanning in bucking of Scots pine. *Wood Material Science and Engineering, 16*(3), 178–187.
66. Salas, C., & Moya, R. (2014). Kiln-, solar-, and air-drying behavior of lumber of *Tectona grandis* and *Gmelina arborea* from fast-grown plantations: Moisture content, wood color, and drying defects. *Drying Technology, 32*(3), 301–310.
67. Schajer, G. S. (2001). Lumber strength grading using X-ray scanning. *Forest Products Journal, 51*(1), 43–50.
68. Shahverdi, M., Oliveira, L., & Avramidis, S. (2017). Kiln-drying optimization for quality pacific coast hemlock lumber. *Drying Technology, 35*(15), 1867–1873.
69. Tao, F., Qi, Q., Liu, A., & Kusiak, A. (2018). Data-driven smart manufacturing. *Journal of Manufacturing Systems, 48*, 157–169.
70. Thomas, E. (2017). An artificial neural network for real-time hardwood lumber grading. *Computers and Electronics in Agriculture, 132*, 71–75.
71. Thybring, E. E., & Fredrikson, M. (2023). Wood and moisture. *Springer handbook of wood science and technology* (pp. 355–398). Springer International Publishing.
72. van Blokland, J., Nasir, V., Cool, J., Avramidis, S., & Adamopoulos, S. (2021). Machine learning-based prediction of internal checks in weathered thermally modified timber. *Construction and Building Materials, 281*, 122193. https://doi.org/10.1016/j.conbuildmat.2020.122193
73. van Blokland, J., Nasir, V., Cool, J., Avramidis, S., & Adamopoulos, S. (2021). Machine learning-based prediction of surface checks and bending properties in weathered thermally modified timber. *Construction and Building Materials, 307*, 124996. https://doi.org/10.1016/j.conbuildmat.2021.124996
74. Wang, S. Y., Chen, J. H., Tsai, M. J., Lin, C. J., & Yang, T. H. (2008). Grading of softwood lumber using non-destructive techniques. *Journal of Materials Processing Technology, 208*(1–3), 149–158.
75. Wang, Y., Zhang, W., Gao, R., Jin, Z., & Wang, X. (2021). Recent advances in the application of deep learning methods to forestry. *Wood Science and Technology, 55*(5), 1171–1202.
76. Watanabe, K., Lazarescu, C., Shida, S., & Avramidis, S. (2012). A novel method of measuring moisture content distribution in timber during drying using CT scanning and image processing techniques. *Drying Technology, 30*(3), 256–262.
77. Watanabe, K., Hayashi, T., & Kobayashi, I. (2016). Probabilistic evaluation of the final moisture content of kiln-dried lumber using the bootstrap method. *Journal of Wood Science, 62*(6), 479–486.
78. Williston, E. M. (1988). *Lumber manufacturing*. Forest Products Society.
79. Wu, H., & Avramidis, S. (2006). Prediction of lumber kiln drying rates by neural networks. *Drying Technology, 24*(12), 1541–1545.
80. Zhang, B. G., & Liu, D. Y. (2006). Exploring a new developing way of wood drying technology in China. *China Forest Products Industry, 33*(4), 3–6.
81. Zhang, J., Cao, J., & Zhang, D. (2006). ANN-based data fusion for lumber moisture content sensors. *Transaction of the Institute of Measurement and Control, 28*(1), 69–79.

Polymer-Based Nanocomposites for Tissue Engineering: Surface to Bulk Structural Design

E. Tamjid

Abstract Materials design and development for the replacement of injured tissues has been a major challenge for decades. Polymer-based nanocomposites (PNCs) are the most appealing class of materials due to their biocompatibility, mechanical properties, hydrophilicity, bioactivity, biodegradability, and processability. Recent advances in the design and fabrication of PNCs rely on the manufacturing of fibrous structures and hierarchical architectures to mimic the structure of extracellular matrix (ECM) where all biochemical and biomechanical signals are transferred to the surrounding tissue. So far, various PNCs have been developed for the healing or regeneration of soft, hard, and semi-hard tissues. In this chapter, different polymeric biomaterials in the form of fibrous structures, three-dimensional scaffolds, injectable smart hydrogels, biomedical implant coatings, and antibacterial wound dressings are presented. The mechanical features that fulfill the requirement of natural tissues, as well as the most appealing biological responses relating to protein adsorption, cell attachment, proliferation, and differentiation, are discussed. The materials selection criteria and the role of nanofillers to generate nanotopography and bioactivity, as well as the future perspective of Machine Learning (ML) in optimization of materials composition and fabrication processes of PNCs in tissue engineering, especially for personalized medicine is stated.

1 Introduction

The aim of tissue engineering is to construct functional substitutes that replace, repair, or regenerate the damaged tissues or organs by providing an appropriate template for new tissue formation throughout the healing process. This goal is realized via implantation of suitable cells isolated from donor tissue and biocompatible scaffold materials [7]. To construct the organs or tissues it is necessary to have a holistic understanding of body as one of the most integrated systems in the universe. In this

E. Tamjid (✉)
Department of Nanobiotechnology, Faculty of Biological Sciences, Tarbiat Modares University, Tehran, Iran
e-mail: tamjid@modares.ac.ir

M.-R. Alam and M. Fathi (eds.), *Integrated Systems: Data Driven Engineering*,
https://doi.org/10.1007/978-3-031-53652-6_14

regard, it is important to note that the body is composed up of a hierarchical structure which is closely associated with its complicated functionality. Thus, to reconstruct such a complex system it is crucial to generate the similar hierarchy using the biomaterials in the same size scales. In other words, as living organisms are composed of organs, tissues, cells, organelles, molecules, in centi-, milli-, micro-, and nano scale, correspondingly, any synthetic biomaterial device acquired to resemble them ought to be composed of parts, composites, microstructure, higher order structure, and chemical primary structure in relevant size range. For instance, bone as the second most abundant part of body, is composed of osteons and blood vessels in the range of hundreds of microns, while the osteons themselves are constructed from osteocyte cells in micrometer range, and hydroxyapatite nanocrystals and collagen type I nanofibers in molecular level. Likewise, the articular cartilage at the end of each bone joint is composed of micrometric chondrocyte cells, and aggrecan biomolecules and collagen type II nanofibers in lower levels. Thus, any biomaterial intended to interact in cellular level have to be in the micron range, while, providing features in nanoscale to efficiently interact in molecular level (i.e., proteins), as well.

2 Biomaterial-Tissue Interactions

The success of any biomaterial implantation in the body, essentially relies on the interactions occurring in the interface of biomaterial and tissue. Hence, any host response arising in the physiological environment, including protein adsorption, cell attachment, and tissue remodeling, is strictly dependent on such interactions at the equivalent size and time scales. Therefore, any biomaterial intended to biomimic the living tissue should provide the structural hierarchy and biological functionality. In the recent years, the utilization of nanomaterials in tissue engineering has gained a significant attention due to their advantages over commercial materials, in terms of improved cytocompatibility, mechanical and electrical properties. Moreover, they stimulate unique nanotopography and surface chemistry which lead to increased protein adsorption, cell attachment, proliferation, and induced rapid differentiation in tissue engineering applications [11]. Accordingly, the development of polymer-based nanocomposites (PNCs) has been an attractive research area, and this chapter is focused on surface to bulk structural design of various polymeric and hydrogel-based biomaterials for tissue engineering applications. Briefly, the surface modified metallic implants, antibacterial coatings and textiles, 3D scaffolds for tissue engineering, as well as hydrogel-based bioinks for regenerative medicine have been described. Furthermore, the current trends and the future prospects are addressed.

3 Surface Modification of Metallic Implants

In the recent years, the interest in tissue remodeling using the intrinsic healing potential of body has led to studies on biodegradable metals, and magnesium alloys are on top of this trend. Despite the extreme similarity of density, mechanical strength and elastic modulus of magnesium to natural bone, its high biodegradation rate is the main drawback. It is worthy to note that the biodegradation mechanism of magnesium is the corrosion. Accordingly, many studies have been focused on hindering the biodegradation of magnesium via (i) alloying, and (ii) surface modification and coating to inhibit the electrochemical reactions [16]. On the other hand, cell attachment as a very first stage in biomaterial-tissue interactions is extremely dependent on the surface topography, roughness, chemistry, and stiffness [22]. Therefore, any technology leading to surface modification accordingly alters the cellular behavior. Many researches have focused on utilizing different alloying [6] or coating technologies [27] to overcome the corrosion issue, while, enhancing the surface favorability for cell attachment and subsequent tissue remodeling. For instance, Hadavi et al. [10], deposited an electrospun polycaprolactone (PCL) coating on a magnesium-calcium alloy to reduce the corrosion rate, as well as enhancing the bone regeneration and mineralization by alloying elements. The results showed that corrosion current density was reduced by an order of magnitude, due to the surface protection by polymeric coating and corrosion by-products. Meanwhile, the cell viability and attachment were notably increased on coated samples in comparison with the uncoated ones. In another study [20], an electrospun PNC based on PCL and bioactive glass (BG) nanoparticles was developed to coat AZ91, a biomedical grade magnesium alloy, to reduce the biodegradation rate in simulated body fluid (SBF). The average thickness of the PNC film and the fibers were 30 μm $\pm$ 5 and 300 $\pm$ 31 nm, respectively. Electrochemical studies and potentiodynamic polarization tests in SBF indicated the decrease of the degradation rate by two orders of magnitude, and reduced degradation current density in surface modified samples, respectively. In vitro bioactivity tests showed that the morphology and the precipitated phases formed upon immersion of the magnesium alloy in SBF varied with the polymer and composite coating. Magnesium hydroxide, zinc hydroxide, magnesium dihydrogen phosphate, and HA particles, as well as more HA particles were formed in the presence of BG NPs. The surface modification resulted in improved in vitro cytocompatibility, predominantly on PCL/BG composite coating (Fig. 1a and b).

Although, the reduced degradation rate of magnesium is favorable, there are concerns about the by-products, mainly the hydrogen bubbles, and magnesium hydroxide, which results in the loss of mechanical integrity as the outcome of local corrosion [31], and the local alkalization [4], respectively. To resolve this issue, an electrospun PNC coating containing a metal organic framework nanoparticle, ZIF-8, with a specific surface area of 80,000 m^2/g, was utilized to capture the hydrogen bubbles during AZ91 alloy corrosion. The results (Fig. 1c and d) indicated that the bubble evolution has decreased by 87%, in addition to reduction in corrosion by an order of magnitude [14].

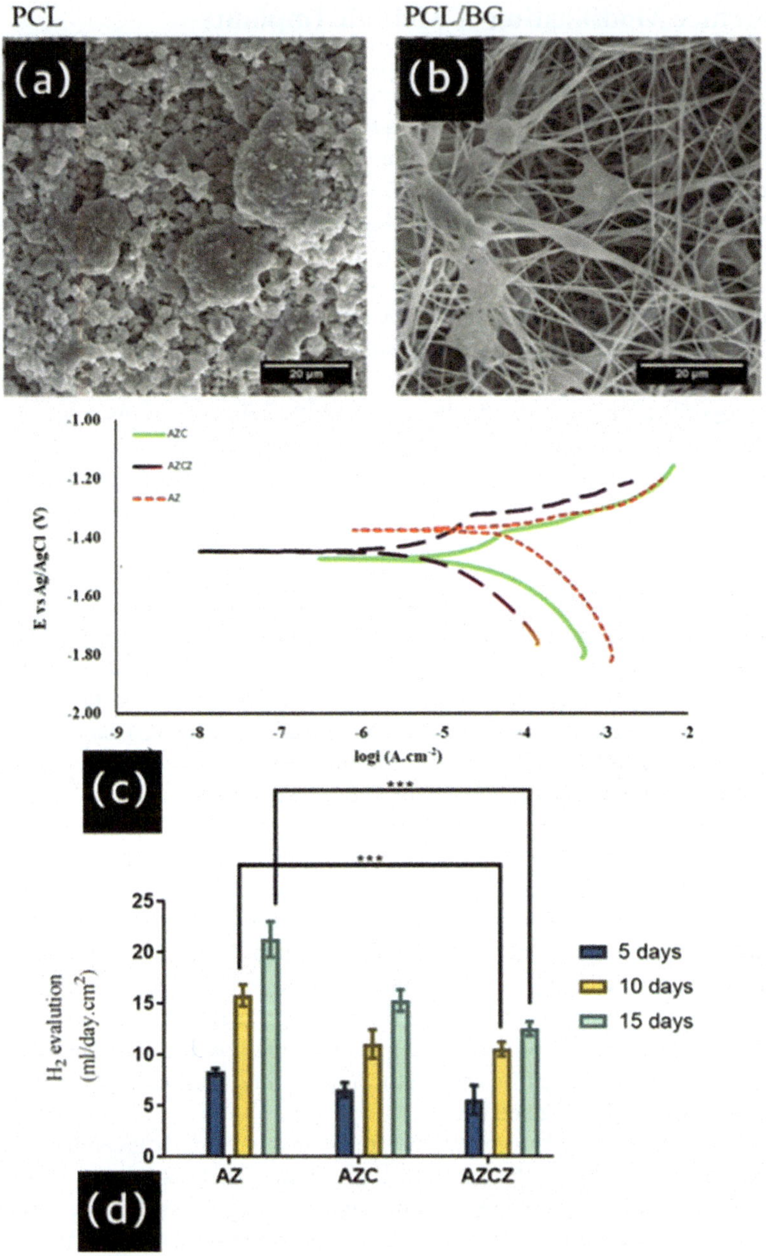

Fig. 1 SEM images showing the surface of **a** AZ91 Mg and **b** Mg/PCL-BG after MG63 cell incubation for 7 days [20], Electrochemical evaluations of the biodegradation of AZ91 alloy in SBF at 37 °C: **c** Tafel plots exhibiting the influence of surface coating on the corrosion rate, **d** The average rate of hydrogen gas evolution at different time intervals [14]

4 Antibacterial Coatings for Textiles and Implants

The importance of surface is not restricted to the corrosion and biodegradation. Many important phenomena occur on the surface, namely the microbial infection which is one of the main obstacles in implantation of biomaterials. Antibacterial textiles are largely used in biomedical applications, in the form of implantable fabrics, filters and surgical dressings. Hence, the development of antibacterial textiles has been of interest by many researchers [9]. For instance, Abedi Mohtasab et al. [2], have studied the efficiency of a PCL electrospun mat in biofilm prevention on S. Mutans, S. Aureus, E. Coli, and P. Aeruginosa bacteria species. The results showed that the nanofibrous scaffold revealed a significant antimicrobial and anti-biofilm formation effect on all of the studied bacterial strains, however, the efficiency was mainly dependent on the bacteria species, i.e., the maximum inhibition was achieved on Pseudomonas Aeruginosa, as revealed by SEM observations and microdilution assay. Moreover, the cell viability studies showed that the most significant effect was shown on human fibrosarcoma cancerous cell line (HT1080) with a drastic decrease by 40% after 48 h as compared to the control (Fig. 2). In a study on antibiotic eluting 3D-printed highly porous PNC scaffolds, cumulative drug release, bacterial growth inhibition zone, and cell viability was studied in vitro. It was shown that without application of any additional coating, sustainable release for up to 72 h was attained due to chemical interactions between functional groups of tetracycline hydrochloride (TCH) and the nanoparticles. All drug-eluting composite scaffolds exhibited tolerable hemocompatibility, in addition to acceptable in vitro cellular and antibacterial effect on gram positive (Staphylococcus aureus) and gram negative (Escherichia coli) bacteria with preserving the cell viability over 95% [29]. These findings may pave the way for designing structurally engineered composite scaffolds with sustainable drug release profile by additive manufacturing techniques for tissue engineering applications.

5 Three-Dimensional Scaffolds for Tissue Engineering

The functionality of tissue as a three-dimensional compartment of the body is closely related to its architecture. Therefore, one of the main necessities in tissue engineering is to mimic the natural three-dimensional architecture of the ECM. Thus, despite modifying the surface, it is crucial to design the bulk of the biomaterial similar to the target tissue. Such an architecture should mimic the tissue in terms of exterior dimensions, interior porosity, interconnection of the pores, and chemical composition. Rapid prototyping techniques (RP) are versatile methods to precisely fabricate bulk products with complex geometries and fine structures through computer-aided design (CAD) drawings and manufacture automatically three-dimensional objects layer-by-layer according to the virtual design. These techniques are appropriate to manufacture a wide range of biomaterials with controlled bulk structure fulfilling

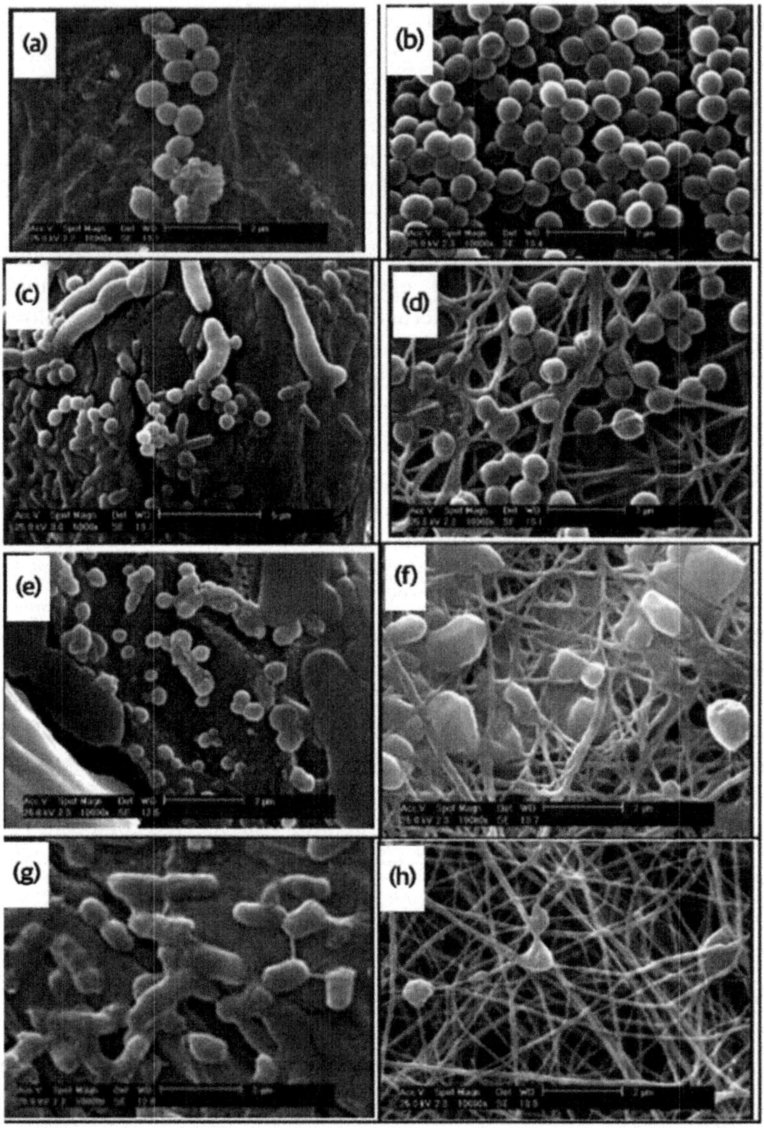

Fig. 2 SEM images of **a, b** Streptococcus mutans, **c, d** Pseudomonas aeruginosa, **e, f** Escherichia coli, and **g, h** Staphylococcus aureus bacterial strains, on blank disc (left), and nanofibrous PCL scaffold (right); [2]

the requirements of tissue engineering scaffolds. Meanwhile, surface properties of such 3D bulk structures are of significant importance. Decorating the biomaterial-tissue interface in 3D scaffolds with micro and nanometer features improves the biomechanical properties of the scaffold in terms of surface topography, roughness, and stiffness, as well as, better cell adhesion to the scaffold surface. Thus, rapid prototyping can lead to customized constructs based on the data acquired from the medical scans to better match individual needs. RP enables the control of the scaffold porosity with desired structural integrity for tissue engineering applications. Up to date, several RP techniques have been developed and applied in TE. The most prominent RP techniques for TE Scaffolds fabrication are Extrusion-based RP techniques, Electrospinning (ES), Three-Dimensional Printing (3D Printing), Selective Laser Sintering (SLS), Stereolithography, Fused Deposition Modeling (FDM), Stereolithography (SL), Electron Beam Melting (EBM), Selective Laser Melting (SLM) [1, 21].

However, each RP process has its own advantages and disadvantages in fabricating tissue engineering scaffolds. First and foremost, ES is a convenient method which uses a spinneret under a high-voltage electric field to produce nanometric fibers through charging and ejecting a polymeric solution or melt which forms a filament in the range of micro to nanofibers after solidification or coagulation. It is beneficial in terms of controllable size, density, orientation, and composition [35]. The high surface to volume ratio of the fibers is of interest for many tissue engineering applications, i.e., the enhancement of cell attachment as an essential step in tissue remodeling. Although, the cell attachment is of crucial significance, the proliferation, and furthermore, the differentiation to the target tissue plays a key role in this process. Accordingly, some researchers [8] have utilized ES to generate nanofibers in order to not only improve the cell attachment due to the high specific surface area, but also promote the differentiation via adding osteogenic components into polymeric solutions. An example is the research performed by Rostami et al. [26] on a PNC nanofibrous scaffold based on PCL and graphene oxide (GOx) nanosheets as a carrier for osteogenic drugs to stimulate the differentiation of mesenchymal stem cells to bone tissue. The results showed that sustainable drug release from dexamethasone-loaded PCL/GOx nanocomposite electrospun scaffolds was achieved, resulting in dominance of the cell differentiation over the proliferation, as well as the significant mineralization through Alizarine Red staining.

In another study, 3D-printing was applied to fabricate a PCL-based nanocomposite scaffold containing bioactive glass (BG) in the size range of micro and nano, and the TiO_2 nanoparticles. It was shown that nanotopographical features in the range of 50 nm effectively enhanced the osteogenic differentiation in Alkaline phosphatase (ALP) assay, moreover, the penetration of the cells into the pore channels was observed in the depth of 250 μm inside the scaffold. Giemsa staining as an indicator of ECM approved the tissue formation within 4 weeks [29].

Cell differentiation, as a process leading to specialized cells is another critical phase in tissue engineering (Rosenberg and [25]), which relies on biochemical and physicochemical properties of biomaterial-tissue interface. Additionally, as the scaffold is attributed to biomimic the ECM, design of the porous structure is critical in controlling the differentiation pathways. It has been shown that differentiation of mesenchymal stem cells (MSCs) to several linages is feasible through manipulation of pore size [17]. Tamjid et al., fabricated PCL-based nanocomposite scaffolds with gradient pore structure (GPS) through 3D printing. In vitro osteogenic and chondrogenic differentiation of bone marrow-derived mesenchymal stem cells (MSCs) on GPS scaffolds was seen within the larger pores in the range of 450 μm to smaller pores of about 200 μm, respectively (Fig. 3a and b). The pore size gradient was recognized to be responsible through curvature-assisted biomineralization, cell density, local nutrient availability, and the extent of hypoxic conditions [28].

To biomimic any natural tissue, it is necessary to not only simulate its biological functionality, but also replicate its mechanical performance. To achieve this goal, PNCs are of great attention, as they provide the biological functionality via their polymeric matrix and retain the mechanical endurance through addition of nanoparticles. To this point, many researches have utilized various ceramic bioactive nanoparticles to take advantage of increased strength, alongside with the bioactivity enhancement. In case of bone remodeling, rather than the mechanical strength, the Young's modulus (E) is of great concern, due to the stress shielding phenomenon which leads to osteolysis and subsequent implant loosening. As the E modulus of the sponge bone is in the range of 1.5 to 45 GPa [19], it is always crucial to achieve the same range of E in bone tissue engineering scaffolds. Recently, silk fibroin as a natural fibrous protein has attracted a great attention in 3D-Printing for its biocompatibility [15]. A silk fibroin-based PNC scaffold using an indirect 3D-printing technique was developed, where BG particles of micro and nano-scale were incorporated to increase the bioactivity, while enhancing the compressive strength [24]. The E modulus of the nano and microcomposite scaffolds in dry condition were 0.94 and 1.2 MPa, respectively. These findings (Fig. 3c and d) were in agreement with previous studies [5]. Notably, the E was reduced by an order of magnitude at wet conditions, falling into 0.22 MPa. In the meantime, the difference between the polymeric, composite and the nanocomposite scaffolds was negligible under wet circumstances. This finding reveals that the mechanical durability of scaffolds in physiological environments might be far altered in comparison with many in vitro studies performed in dry conditions.

Mainly, tissue engineering and wound healing are complicated processes based on signaling molecules controlling the dynamic interactions in the interface of the engineered scaffold and the newly formed tissue. As each tissue has its own specific architecture, it is essential to regenerate the same design while healing is happening. It is pertinent to note that a successful design of an engineered scaffold for healing of full-thickness skin wounds depends on the entirely mimicking the natural ECM structure and providing a proper microenvironment for a successful tissue regeneration. Obviously, any RP technique has shortcomings in fabricating tissue engineering scaffolds. Hence, the future research should be focused on development of RP machines designed specifically for fabrication of tissue engineering scaffolds. Another approach to address this gap is to combine different RP techniques and biomaterials fabrication methods to generate hierarchical structures resembling a specific tissue. For example, in a study by Zandi et al., biomimetic bilayer scaffolds composed of gelatin nanofibers (GFS) and photocrosslinkable composite hydrogels

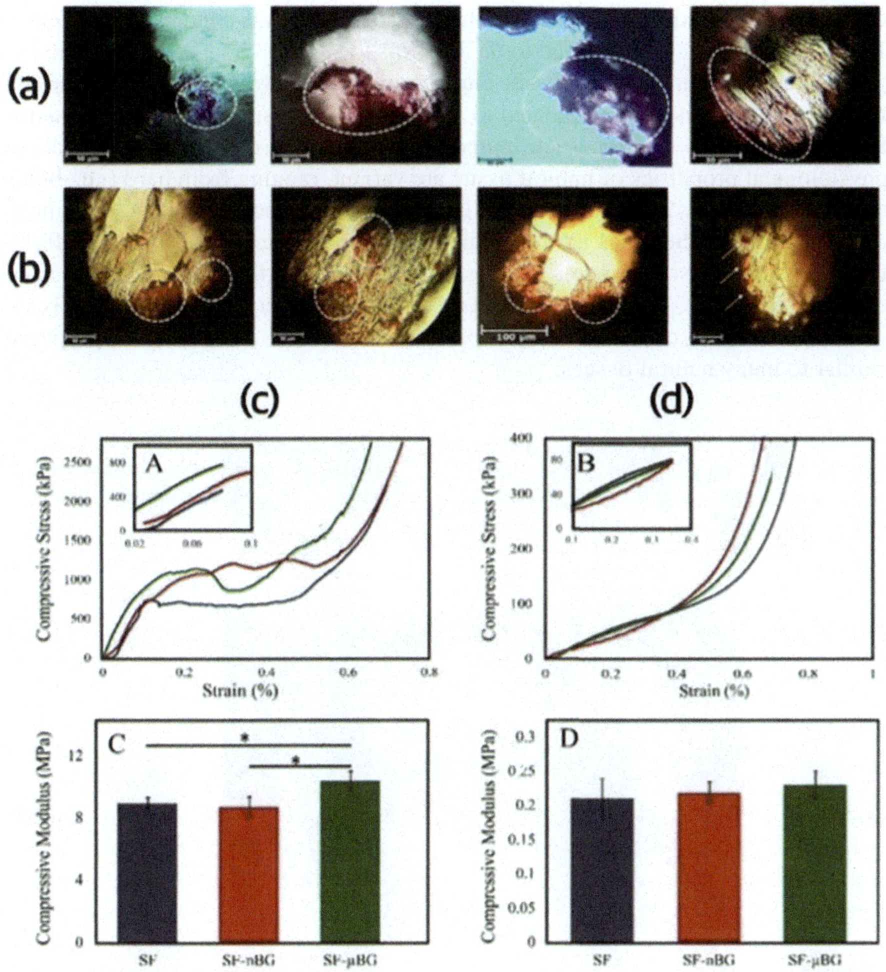

Fig. 3 In vitro **a** chondrogenic and **b** osteogenic differentiation of bone marrow-derived mesenchymal stem cells (MSCs) on GPS scaffolds, within the smaller pores of about 200 μm and larger pores in the range of 450 μm, respectively [28], (A, B) compressive stress–strain curve, and (C, D) compressive modulus at **c** dry condition and **d** wet condition; in all plots (blue) silk fibroin, (red) silk fibroin-nano bioglass and (green) silk fibroin-micro bioglass ($p^* < 0.05$) [24]

loaded with epidermal growth factors (EGF) were presented. The nanofibers served as the dermis layer, while, EGF-loaded nanocomposite hydrogels based on gelatin metacryloyl (GelMA) modified with silicate nanoplatelets (Laponite) represented the epidermis matrix for the full-thickness wound healing (Fig. 4a). Such a bilayer construct as a full-thickness wound model demonstrated a significant improvement in in vitro wound closure potential, where the adhesion of the scaffolds to the native tissue was increased by 3.5 folds, and the wound closure was enhanced up to 93.1 ± 1.5% after 14 days, as compared to the controls [34, 33].

Besides, the main hurdle in tissue engineering is the development of biomaterials which can be precisely manipulated to the anticipated designs through the existing RP techniques. To develop such materials, it is notable that the mechanical and physiological properties of natural tissue are variant, ranging from hard (i.e., bone), and semi-hard (i.e., cartilage) to soft tissues (e.g., skin, nerve, kidney, liver, heart, …). Certain characteristics are not achievable by utilizing bulk PNCs as scaffolds. Examples are the soft or semi hard tissues like skin, cartilage, and cardiac tissue engineering, where, porous, softer, and more flexible materials are required. Hydrogels are functional candidates which provide mechanical and structural properties similar to many natural tissues.

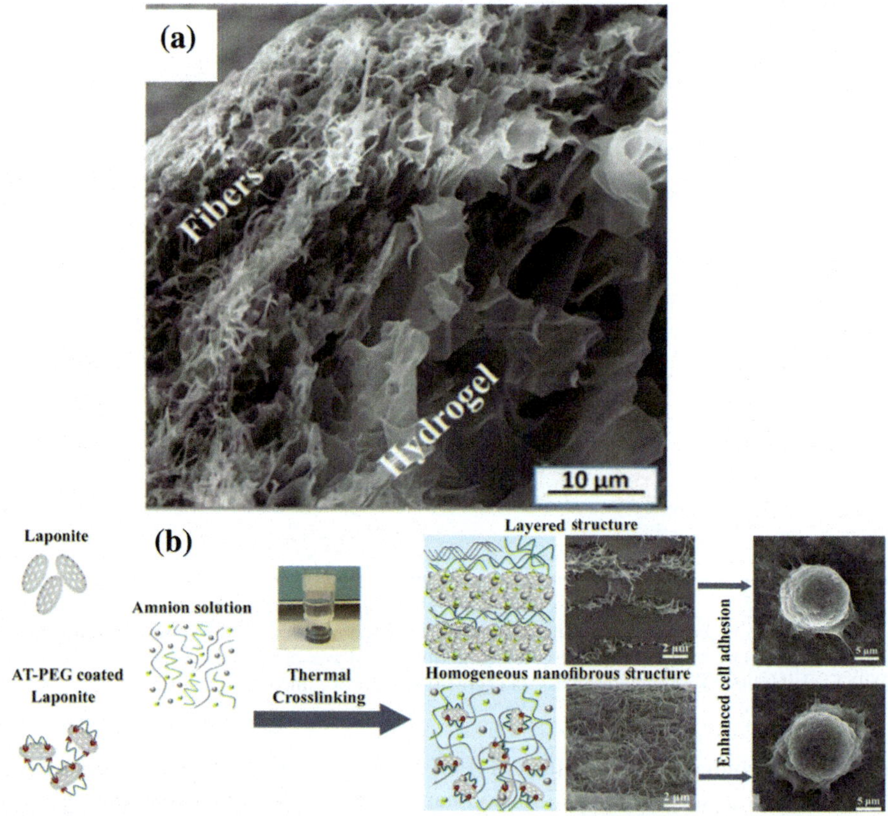

Fig. 4 **a** Biomimetic bilayer scaffolds composed of gelatin nanofibers (GFS) and photocrosslinkable composite hydrogels loaded with epidermal growth factors (EGF) [34, 33] **b** effect of bare (top) and coated (bottom) laponte on architecture of crosslinked amnion on enhancement of cell attachment [12]

6 Hydrogel-Based Bioinks for Regenerative Medicine

The key requirement for tissue engineering is the retention of 3D structure together with the high amount of water content to stimulate the native ECM, facilitate the cell immobilization and growth factor delivery, and provide mechanical support for the cells. To realize this requisite, cell-laden hydrogels are appropriate candidates, which have gained great attention in the last two decades [18]. Hydrogels as 3D polymeric networks composed of crosslinked water-soluble polymers, can retain a high amount of water content in their network while maintaning their original structure [3]. To spatially-control the placement of cells in a designed 3D microenvironment, 3D bioprinting technique is an opportunity. Hydrogels are the best nominee to support cells to be printed in well-defined shapes and subsequently generate 3D tissue, forming an ideal printable media known as bioinks. Development of such hydrogel-based bioinks is an interdisciplinary field, which combines the materials science and cell biology, and offers a variety of solutions to simplify the 3D bioprinting technology. The main concern in elaboration of printable bioinks is the retention of cell survival and function amide the shear forces during the printing. The other issue is attaining appropriate rheological behavior which is of crucial importance in extrusion-based 3D-printing devices. Furthermore, the bioinks should be printed facilely into shape-persistent and free-standing structures. To realize these requirements, shear-thinning hydrogels has attracted considerable attention in fabricating 3D scaffolds with high structural reliability and resolution. For instance, in a recent study by Zandi et al., shear-thinning and printable nanocomposite hydrogels based on laponite and glycosaminoglycan nanoparticles (GAGNPs) were developed for bioprinting applications. The rapid formation of nanocomposite hydrogels was accomplished within seconds due to the interactions between the negatively charged groups of GAGNPs and the positively charged edges of laponite. Moreover, it was shown that the encapsulated cells were protected from aggressive shear stresses through shear-thinning behavior of the hydrogel during bioprinting. Rheological studies demonstrated fast recovery of GLgels over multiple strain cycles. In vitro studies confirmed the ability of GLgels to support cell growth, proliferation, and spreading, as well as osteogenic differentiation of pre-osteoblastic bone marrow stromal cells encapsulated inside the GLgels through ALP activity assay and calcium deposition analysis. Likewise, in vivo biocompatibility and biodegradability was confirmed by the subcutaneous implantation of the GLgel in rats [34, 33].

As mentioned earlier, the aim of scaffold-based tissue engineering is to provide a suitable microenvironment which completely resembles physicochemical properties of ECM and facilitates the cellular vital processes. ECM as a complex media is composed of macromolecular components such as proteins, growth factors, and glycosaminoglycans (GAGs) secreted by various cells in the body. Despite the recent advances in development of PNCs based on synthetic materials, hydrogels derived from decellularized extracellular matrices (dECMs) derived from various tissues or organs are beneficial for regenerative medicine (TERM) applications due to their superior biocompatibility as a rich source of protein and growth factor ingredients.

To this end, numerous studies have shown that dECM scaffolds derived from various resources play regulating role in cell survival, cell proliferation, cell morphogenesis, and differentiation. Recently, Kafili et al., employed injectable hydrogels based on human amniotic membrane and polyethyleneglycol-modified nanosilicates for tissue engineering (Fig. 4b). A thermosensitive gelation behavior at the physiological temperature, as well as the shear-thinning, i.e., decreased storage modulus (up to 80%), swelling ratio (up to 50%), and gelation rate (up to 16%) were attained [12].

Consequently, the engineered shear-thinning hydrogels with cell differentiation characteristics facilitate the development of functional bioinks for 3D printing of constructs for tissue engineering applications [13]. To efficiently mimic a natural tissue, the last step is to generate stimuli-responsive biomaterials with properties variable with time, which similar to the natural tissue. Different stimuli are of interest, i.e., temperature, pH, electromagnetic radiation, magnetic field, and biological factor. Stimuli-responsive hydrogels are beneficial in advancements of tissue engineering, remarkably in development of bioinks applicable in four-dimensional printing (4D-printing). Bioinks are capable of changing dimensions, properties, and functions versus time, which fully resembles the behavior of natural tissues in real physiological conditions, and might be useful in generating reliable interactive tissues leading to viable organ printing.

7 Future Perspectives

To realize the viable organ printing, it is necessary to produce an ideal tissue regenerating biomaterial, which fulfils all requirements addressed above. Designing such an integrated system is impossible unless biomaterials' composition, scaffold fabrication methods, and cell/tissue modeling are optimized. Machine learning (ML) and artificial intelligence (AI) are beneficial to accelerate such scientific innovation, improve clinical trials, and extend fundamental understanding of many biological phenomena. In the near future ML technologies can be utilized to various areas of tissue engineering research and development. Recently, some studies have utilized AI-based software to optimize the tissue engineering scaffold design and technology. For instance, Rafieyan et al. have developed an ML software to confront cardiovascular tissue engineering (CTE) issues. CTE specifications collected from the literature were used to create a dataset, involving different materials, cell lines, and fabrication methods commonly used in CTE scaffold fabrication. The specifications were used as variables in the study to rate the CTE scaffolds based on a scale of 0–3, according to cell responses, i.e., cell viability, proliferation, and differentiation. Applying twenty-eight ML algorithms to determine the most effective one for predicting cell behavior on various CTE scaffolds, the high performance of XGBoost with an accuracy of 87% was indicated, while, by implementing ensemble learning algorithms and using five algorithms with the best performance, an accuracy of 93% with the AdaBoost Classifier and Voting Classifier was achieved.

Such studies clarify that the emergence of ML in tissue engineering, though relatively modern, has already shown to be potentially promising in integrated and multivariate analyses of the relationships between biological, chemical, and physical parameters in design and development of tissue regenerative products, and highlights the innovative methodologies, emerging strategies, and areas of potential advancement within this rapidly growing fields of research [23].

8 Conclusions

The development of ideal biomaterials for tissue engineering relies on promoting injured tissue to heal through endogenous regeneration, i.e., reconstruct the hierarchical design, preserve its complicated functions, and retain the vital interactions occurring on biomaterial-tissue interface. To achieve this goal, various approaches has been used so far. As discussed in this chapter, the surface design methodologies via polymeric nanocomposites include creation of 2D constructs in the form of coatings, films, textiles, etc. via different techniques specially electrospinning. Electrospinning is a These 2D electrospun mats are advantageous in stimulating the biomaterial-tissue interactions on their surface via providing high surface area which is favorable for protein adsorption and cell attachment. Moreover, bioactive nanoparticles are beneficial in cell proliferation and differentiation. Nevertheless, the 3D structure of natural tissue necessitates to design Three-dimensional scaffolds which better resembles the ECM architecture and retains its vital functions. Thus, a paradigm shift from surface to bulk design has led to application of RP technologies, especially, 3D printing of polymeric nanocomposites in scaffold manufacturing to attain the required external dimensions, as well as, internal hierarchical architecture. Recently, novel interactive biomaterials based on stimuli-responsive nanocomposite hydrogels compatible to 4D printing have gained significant attention, which are promising in future tissue regeneration advances. Such printable cell-laden bioinks might pave the way to functional living tissues leading to forthcoming organ printing.

References

1. Abdelaal, O. A. M., & Darwish, S. M. H. (2013). Review of rapid prototyping techniques for tissue engineering scaffolds fabrication. In *Characterization and development of biosystems and biomaterials*. Springer. *Advanced Structured Materials, 29*, 33–54.
2. Abedi Mohtasab, T. P., Tamjid, E., & Haji-Hosseini, R. (2019). A study on biofilm formation, antibacterial properties and cell viability of poly(ε-Caprolactone)-based electrospun nanofibrous scaffold. *Modares Journal of Biotechnology, 10*(3), 373–380.
3. Ahmed, E. M. (2015). Hydrogel: Preparation, characterization, and applications: A review. *Journal of Advanced Research, 6*(2), 105–121.
4. Cai, L., et al. (2023). Biodegradation mechanisms of pure Mg in presence of glucose, vitamin C, and citric acid. *Smart Materials in Manufacturing, 1*, 100014.

5. Chandrasekaran, A., et al. (2016). Physico-chemical and biological studies on three-dimensional porous silk/spray-dried mesoporous bioactive glass scaffolds. *Ceramics International, 42*(12), 13761–13772. Elsevier
6. Deng, M., et al. (2021). Approaching "stainless magnesium" by Ca micro-alloying. *Materials Horizons, 8*, 589–596.
7. Di Silvio, L. (2007). Bone tissue engineering and biomineralization. In *Tissue engineering using ceramics and polymers* (pp. 319–331). Woodhead Publishing.
8. Ghasemvand, F., et al. (2023). Chitosan, polyethylene oxide/polycaprolactone electrospun core/shell nanofibrous mat containing rosuvastatin as a novel drug delivery system for enhancing human mesenchymal stem cell osteogenesis. *Frontiers in molecular biosciences, 10*, 1220357.
9. Gulati, R., Sharma, S., & Sharma, R. K. (2022). Antimicrobial textile: Recent developments and functional perspective. *Polymer Bulletin, 79*, 5747–5771. Springer.
10. Hadavi, S. H., et al. (2021). In vitro corrosion behavior and biological properties of magnesium zinc-calcium alloy coated with polycaprolactone nanofibers. *Journal of Ultrafine Grained and Nanostructured Materials, 54*(1), 93–100.
11. Hasan, A., et al. (2018). Nanoparticles in tissue engineering: Applications, challenges and prospects. *International journal of nanomedicine, 13*, 5637–5655. Dovepress.
12. Kafili, G., et al. (2022). Development of injectable hydrogels based on human amniotic membrane and polyethyleneglycol-modified nanosilicates for tissue engineering applications. *European Polymer Journal, 179*, 111566.
13. Kafili, G., et al. (2023). Development of printable nanoengineered composite hydrogels based on human amniotic membrane for wound healing application. *Journal of Materials Science, 58*(30).
14. Khalili, M. A., & Tamjid, E. (2021). Controlled biodegradation of magnesium alloy in physiological environment by metal organic framework nanocomposite coatings. *Scientific Reports, 11*(1), 8645. Nature.
15. Kim, S. H., et al. (2018). Precisely printable and biocompatible silk fibroin bioink for digital light processing 3D printing. *Nature Communications, 9*, 1620.
16. Kumar, S., & Katyal, P. (2022). Factors affecting biocompatibility and biodegradation of magnesium based alloys. *Materials Today: Proceedings, 52*(3), 1092–1107. Elsevier.
17. Leach, J. K., & Whitehead, J. (2018). Materials-directed differentiation of mesenchymal stem cells for tissue engineering and regeneration. *ACS biomaterials science & engineering, 4*(4), 1115–1127.
18. Mantha, S., et al. (2019). Smart hydrogels in tissue engineering and regenerative medicine. *Materials (Basel), 12*(20), 3323.
19. Morgan, E. F., Unnikrisnan, G. U., & Hussein, A. I. (2018). Bone mechanical prop-erties in healthy and diseased states. *Annual review of biomedical engineering, 20*, 119–143.
20. Panahi, Z., Tamjid, E., & Rezaei, M. (2020). Surface modification of biodegradable AZ91 magnesium alloy by electrospun polymer nanocomposite: Evaluation of in vitro degradation and cytocompatibility. *Surface and Coatings Technology, 386*, 125461.
21. Peltola, S. M., et al. (2008) A review of rapid prototyping techniques for tissue engineering purposes. *Annals of Medicine, 40*(4), 268–280 Taylor & Francis.
22. Rafiee, K., Naffakh-Moosavy, H., & Tamjid, E. (2020). The effect of laser frequency on roughness, microstructure, cell viability and attachment of Ti6Al4V alloy. *Materials Science and Engineering: C, 109*, 110637.
23. Rafieyan, S., et al. (2023). MLATE: Machine learning for predicting cell behavior on cardiac tissue engineering scaffolds. *Computers in Biology and Medicine, 158*, 106804.
24. Razaghzadeh Bidgoli, M., et al. (2019). Fabrication of hierarchically porous silk fibroin-bioactive glass composite scaffold via indirect 3D printing: Effect of particle size on physico-mechanical properties and in vitro cellular behavior. *Materials Science and Engineering: C, 103*, 109688. Elsevier.
25. Rosenberg, L. E., & Rosenberg, D. D. (2012). Chapter 15-birth defects. In *Human genes and genomes* (pp. 241–258). Academic Press.

26. Rostami, F., Tamjid, E., & Behmanesh, M. (2020). Drug-eluting PCL/graphene oxide nanocomposite scaffolds for enhanced osteogenic differentiation of mesenchymal stem cells. *Materials Science and Engineering: C, 115*, 111102. Elsevier
27. Rouein, Z., et al. (2022). In vitro bioactivity and biocompatibility of magnesium implants coated with poly(methyl methacrylate)-bioactive glass composite. *Materials Today Communications, 33*, 104872.
28. Tamjid, E., et al. (2022). Three-dimensional gradient porous polymeric composites for osteochondral regeneration. *Journal of Polymer Research, 29*, 163.
29. Tamjid, E., et al. (2013). Tissue growth into three-dimensional composite scaffolds with controlled micro-features and nanotopographical surfaces. *Journal of biomedical materials research Part A, 101*(10), 2796−2807. Wiley.
30. Tamjid, E., et al. (2020). Sustainable drug release from highly porous and architecturally engineered composite scaffolds prepared by 3D printing. J*ournal of Biomedical Materials Research Part A, 108*(6), 1426–1438. John Wiley & Sons Inc.
31. Tesař, K., & Balík, K. (2020). Nucleation of corrosion products on H2 bubbles: A problem for biodegradable magnesium implants? *Materials Today, 35*, 195–196.
32. Williams, D. F. (2022). Biocompatibility pathways and mechanisms for bioactive materials. *The bioactivity zone, Bioactive Materials, 10*, 306–322.
33. Zandi, N., et al. (2021). Nanoengineered shear-thinning and bioprintable hydrogel as a versatile platform for biomedical applications. *Biomaterials, 267*, 120476.
34. Zandi, N., et al. (2021). Biomimetic nanoengineered scaffold for enhanced full-thickness cutaneous wound healing. *Acta Biomaterialia, 124*, 191-204. Elsevier.
35. Zhong, W. (2016). 3-Nanofibres for medical textiles. In L. van Langenhove (Eds.), *Woodhead Publishing series in textiles, advances in smart medical textiles* (pp. 57–70). Woodhead Publishing.

Knowledge-Based Design and Fabrication of Advanced Composites and Hybrid Materials for Nanomedicine

A. Simchi and P. Kianfar

Abstract Medical applications of nanotechnology in tissue engineering, biomedical devices, biosensors, and biological machines have significantly enhanced over the past decade. Advanced composite and hybrid materials (ACHMs) have become the foundation of nanomedicine research because biological systems are complex in nature and require materials with specific structures and functionality. Knowledge-based design of new materials, development of advanced manufacturing techniques, and engineering of the material/biological system interfaces are the key driving forces to expand the applications of ACHMs in nanomedicine. Merging existing knowledge and experience with the implementation of artificial intelligence can further improve the design and manufacture of advanced materials, especially when guided by a bio-inspired approach. Herein, we present recent advances in ACHMs in nanomedicine. Concepts for the design, challenges, and fabrication of materials for biomedical applications are discussed. From a materials perspective, polymer-matrix nanocomposites in the form of hydrogels, nanofibers, and architecturally designed structures are elaborated. Advanced electrospinning and 3D printing techniques are highlighted. Applications of ACHMs in the design and fabrication of functional biosensors and wearable devices are also demonstrated. Moreover, the use of ACHMs in pharmaceuticals, in vivo imaging, cancer diagnosis, and therapy is addressed. Finally, the challenges for the nanotechnology implementation concerning regulatory issues are discussed and prospects are presented.

1 Introduction

Rapid advances in materials science and engineering are enabling new discoveries in a variety of disciplines, particularly the application of nanotechnology to medicine [1]. The complex nature of biological systems requires advanced materials with desired structure and functionality at both micro- and nanoscales. ACHMs provide the desired functionalities by altering the building blocks through a synergistic combination of

A. Simchi (✉) · P. Kianfar
Department of Materials Science and Engineering, Sharif University of Technology, Tehran, Iran
e-mail: simchi@sharif.edu

M.-R. Alam and M. Fathi (eds.), *Integrated Systems: Data Driven Engineering*,
https://doi.org/10.1007/978-3-031-53652-6_15

components [2]. Therefore, these are a promising class of materials with tailored properties and structures [3]. With exceptional features, such as tunable physical, chemical, and mechanical properties, ACHMs offer great potential for use in biological environment thanks to their biocompatibility, biodegradability, and bioactivity. To this end, knowledge-based design (KBD), which relies on existing data and current technologies, optimizes and improves material cycle development by understanding the relationship between structure, material properties, and material function [4]. In this article, we introduce the basic design concepts and advanced fabrication methods of ACHMs for nanomedicine. Herein, the incorporation of nanotechnology in the design of ACHMs by a knowledge-based approach and the application of ACHMs in nanomedicine are the main guiding principles. The various applications of these materials where nanoscale features play a role and add value are briefly discussed. Particular attention is given to tissue engineering, cancer diagnosis, cancer therapy, in vivo imaging, and biosensing. The challenges and future prospects for materials design and applications in nanomedicine are also outlined.

2 Knowledge-Based Design

The interdisciplinary nature of nanomedicine requires a systemic and knowledge-based approach to integrate different disciplines, existing knowledge, and experimental data to meet initial design requirements and ultimate goals and functions. Therefore, the design and fabrication of ACHMs must be viewed as an integrated process that is fundamentally different from traditional sequential fabrication. In this regard, bioinspiration is the most promising approach, where design mimics the way nature works, and develops concepts, processes, and technologies [5, 6]. This strategy uses approaches from various disciplines, including life sciences, physics, engineering, and medicine, to advance technologies in harmony with nature. Herein, the chemistry, molecular structure, and nano-bio interface [7] should be carefully designed to achieve biocompatibility and biodegradability. The integration of computational modeling and simulations into the process design is important to optimize development efforts. The designed and fabricated material is then tested under in vitro and in vivo conditions, and the development cycle continues until the product meets all requirements for the intended use [8].

In addition to material properties, the development of ACHMs for nanomedicine also requires a well-designed structural architecture in two and three dimensions. Various advanced fabrication technologies such as self-assembly, 3D bioprinting, advanced electrospinning, and template-assisted techniques are available to precisely manufacture designed structures [1]. Additionally, a deep understanding of the nano-bio interface is essential because once the nanomaterial enters the physiological environment, the dynamic physicochemical interactions can drastically change the biological activity of the designed nanostructure and the fate of the nanomedicine [7]. The desired interaction at the interface can be influenced by the properties of the nanostructure such as size, shape, surface charge, topography, chemical composition, intrinsic properties of the components, architecture, the composition of surrounding biological fluids, their electrostatic interactions, hydrogen bonds, salt bridges, ligand-receptor interactions, hydrophobic interactions, and more [7].

3 Novel Biofabrication Techniques

One of the key driving forces in the development of ACHMs used in nanomedicine is the emerging and novel fabrication techniques that can prepare well-designed structures that mimic the complex architecture of living tissues or organs. The most useful techniques are 3D bioprinting, advanced electrospinning, self-assembly, and multilayer techniques (Fig. 1), which are briefly discussed below.

The 3D printing technology is one of the most promising techniques to fabricate complex structures and architectures [12]. This technique uses a computer-aided design (CAD) model to print building blocks or bio-inkslayer-by-layer in a detailed

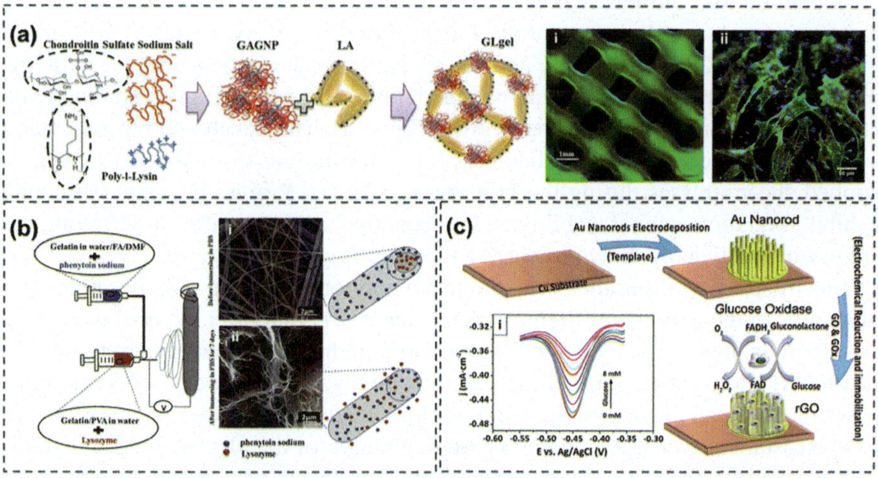

Fig. 1 **a** Schematic representation of the synthesis process of printable glycosaminoglycan nanoparticles/laponite hydrogels formed by self-assembly, fluorescence microscopy images of the printed GAG /laponite hydrogel scaffold (the scale bar is 1 mm), and the F-actin fluorescence image of W-20-17 stromal cells encapsulated in the printed scaffold on day 5 of the cell culture (scale is 50 μm) [9]. **b** Schematic representation of the preparation of core-sheath structure for dual drug delivery by the electrospinning method. The core is loaded with lysozyme and the sheath is loaded with phenytoin sodium. SEM images show the morphology of fibers (b.i) before and (b.ii) after incubation in PBS (37 °C) (the scale is 2 μm) [10]. **c** Schematic representation of the preparation of the glucose biosensor, in which the gold nanorods were obtained by template-assisted electrochemical growth on a polycarbonate membrane. Glucose oxidase/GO nanosheets were electrochemically immobilized on the gold nanorods. The differential pulse voltammetry shows the response of the biosensor at different glucose concentrations [11]

arrangement f [13]. Printable inks and building blocks can be enriched with nanostructures to improve properties such as thermal and electrical conductivity, mechanical strength, and sensitivity to stimuli. Various types of 3D printing technologies have emerged, including vat photopolymerization, powder bed fusion, material extrusion, material jetting, binder jetting, which have made their way into medicine and nanomedicine applications [14]. The choice of 3D printing technique depends mainly on the material properties as well as print resolution, size, speed, throughput, precision and also how the layers are bonded together [14]. Depending on 3D printing technique and material properties, the printable ink can be in the form of solid, liquid, melts, and resin. Accordingly, 3D bioprinting is the automated fabrication of cellularized 3D constructs to create complex tissue structures by localizing different cell types and materials to mimic the anatomical complexity of a tissue [15]. The technique has widely been utilized for tissue engineering and regenerative medicine and all correlated industries [16].

Hydrogels are the most promising building blocks for 3D bioprinting. They typically contain 70 to 99% water, which makes them highly porous and permeable [17] and suitable for the rapid diffusion of nutrients while facilitating cell migration [18]. Natural and bio-based polymers (e.g., proteins and polysaccharides) are the most commonly used raw materials for the synthesis of hydrogels [19] to mimic the cell niche [20]. Recently, Zandi et al. [9] have developed a novel hydrogel-based bioink composed of silicate nanomaterials, laponite, and glycosaminoglycan nanoparticles for bone tissue engineering (Fig. 1a). The developed bioink exhibits a well-tuned shear-thinning behavior, protecting the encapsulated cells from the applied shear stresses during the bioprinting process. While the printed structures exhibit good dimensional stability as free-standing structures after bioprinting, they are osteoinductive, as shown by the ALP assay.

Electrospinning is another versatile technique for producing polymer fibers with diameters in the nanometer to micrometer range [21] that offer exceptional properties, such as high surface area to volume ratio, high porosity and interconnectivity, flexibility, and lightweight. During electrospinning, a high voltage is applied in to order to create an electric field. The liquid polymer (the polymer solution or melt) is electrostatically charged and at a certain strength of the electric field, it expands and accelerates in the form of a jet towards a collector of opposite charge due to the repulsive electric forces. During the movement toward the collector, the polymer jet passes through various stages until the solidified polymer reaches the collector and is deposited in the form of a solid fiber. A recent development is the fabrication of fibrous nanostructures with tunable surface morphology and composition via concurrent electrospinning, electrospray/electrospinning, emulsion electrospinning, and direct writing techniques [22, 23]. Centrifugal spinning, which is characterized by high yield and low cost, is another nanofiber production technology that relies on the centrifugal force due to high rotational speed to stretch polymer jets and produce fine fibers. The main challenge in this technique is the uncontrolled morphology and bead defects. These drawbacks can be overcome by applying a high voltage through a new technique named electrocentrifugal spinning [24].

Hybrid electrospun nanofibers have made their way into a variety of nanomedicine applications including drug delivery systems, wound dressings, biosensors, tissue engineering, and more [25]. Recently, Zandi et al. [10] have designed a core-sheath electrospun mat for the simultaneous delivery of hydrophilic and sensitive therapeutic agents using coaxial electrospinning technology. The sheath is based on water-resistant cross-linked gelatin and the core is based on water-soluble gelatin/polyvinyl alcohol (Fig. 1b). The core fibers protect the hydrophilic drug, i.e., lysozyme as a model protein, from denaturation, and the shell contains the therapeutic agent of phenytoin sodium. In the release studies, sustained and prolonged release of both therapeutic agents was observed in PBS, demonstrating the potential of the core/shell structure as a suitable platform for drug delivery.

In self-assembly, a pre-existing disordered system of components spontaneously forms an organized pattern using non-covalent bonds involving hydrogen bonds, electrostatics, host–guest interactions, or a combination of non-covalent interactions [1, 26, 27]. Due to the crucial role of self-assembly in biological systems where supermolecules are formed, this technique is of particular interest. The non-covalent nature of the bonds allows for adaptability through dissociation and reconstruction. Self-assembled hydrogels are formed by the spontaneous assembly of monomers into polymeric forms without the formation of covalent bonds [9, 28]. Correspondingly, peptide-directed hydrogels have attracted considerable interest due to their outstanding properties such as biocompatibility, self-healing properties, biodegradability, and ease of tuning. These hydrogels consist of a linear hydrophilic polymer with self-assembling peptides as branches. The aggregation of these peptides in an organized manner through reversible physical interactions formed the peptide self-assembly. Zandi et al. [9] used the self-assembly technique to prepare a hydrogel based on glycosaminoglycan nanoparticles and laponite for bone tissue engineering applications (Fig. 1a).

The physicochemical properties of composite and hybrid nanostructures can be tailored by various synthesis methods, such as surface modification, physical blending, and in situ deposition [1]. Surface modification of a nanostructure can be achieved by physical (e.g., adsorption of small molecules by non-covalent bonds) and chemical (e.g., grafting by covalent bonds) methods. In-situ deposition is another efficient approach to uniformly incorporate nanoparticles into polymer matrices or prepared nanostructures, which can be achieved either via the gas or liquid phase [29]. Shen et al. [30] prepared silver–graphene oxide (Ag–GO) polylactide acid nanocomposite by in situ polymerization technique for biomedical and tissue engineering applications. They showed improved flexural properties and antibacterial properties compared to counterparts prepared by mechanical blending technique. In another work, Mazaheri et al. [11] developed a hybrid structure for enzymatic detection of glucose (Fig. 1c). Reduced GO nanosheets were functionalized by the glucose enzyme, i.e., glucose oxidase, and immobilized on the gold nanorod electrodes by in situ deposition as an excellent electrochemical sensing platform. The enzymatic biosensor exhibited high electrochemical activity and selectivity due to the large available active surface area and promoted electron transfer from the enzyme to the gold sensing platform.

4 Biocompatible Hybrid Materials

Nanocomposites based on natural polymers are becoming increasingly promising for nanomedicine due to their excellent properties such as biocompatibility, biodegradability, high water retention capacity, and similarity to the extracellular matrix [31]. Figure 2a shows a wide variety of natural polymers that can be utilized for biomedical applications. Despite all the advantages of natural polymers, they face some challenges. These include low mechanical properties, rapid degradability, risk of contamination by microbial species, difficulty in tailoring properties, and large-scale synthesis [32, 33]. Therefore, the development of hybrid materials is necessary, where the specified functionality and properties are increasingly demanding. To this end, recent advances in materials design and manufacturing have focused on the development of bioactive nanomaterials that can be used to treat, regenerate, evaluate, recognize, or replace a tissue, organ, or function [34]. These materials have tailored properties and possess interactions with biomolecules, cells, and tissues at the interface that resemble bioactivity, including cell adhesion, proliferation, differentiation, biomimetic cell or tissue activities, the release of bioactive compounds or ions, targeted drug delivery, catalytic, and diagnostic activities [35]. The bioactivity of a nanostructure is influenced by its physical structure (e.g., size, shape…), surface properties (surface charge, hydrogen bonding, hydrophobicity, and chemical modifications), nanotopography (nanopatterns, nanopores or nanochannels and surface roughness) and chemical composition [34]. Figure 2b shows different types of nano-enabling materials, including metals, carbon-based materials, ceramics, and metal–organic materials, which have widely been used in nanomedicine. Besides, biological compounds (i.e., enzymes, antibodies, proteins, peptides) and therapeutic agents (i.e., anticancer drugs, anti-inflammatory drugs, antibiotics, etc.) should be mentioned as another class of materials that can provoke and induce bioactivity such as diagnostic, targeting, and triggering activity (Fig. 2c).

5 Applications of Biocomposites and Hybrid Materials

Nanomedicine is the medical application of nanotechnology for diagnostic, preventive, and treatment purposes. ACHMs have remarkable potential in nanomedicine, for example, in tissue engineering, biosensing, drug delivery, bioimaging, and cancer diagnosis and therapy [35, 34]. For tissue engineering, the development of advanced scaffolds that mimic the functionality of the target tissue and enable regeneration is essential. The outstanding properties of ACHM scaffolds provide better cell-scaffold interactions, cell migration, differentiation, and proliferation [1]. For bone tissue engineering, for example, an ideal scaffold requires an advanced design that mimics the complicated physicochemical environment and properties of bone to stimulate

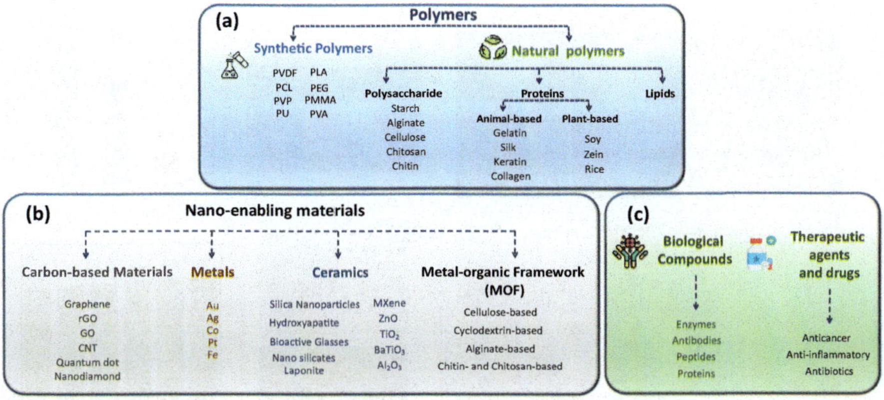

Fig. 2 Wide variety of **a** natural polymers and **b** nanomaterials that can be utilized for biomedical applications. **c** Examples of biological an therapeutic agents.

host tissue for regenerative purposes [36, 37]. To date, numerous 3D hybrid scaffolds based on bioceramics (for example, tricalcium phosphate (TCP), hydroxyapatite (HA), and bioglass) and polymers (e.g., chitosan, collagen, alginate, fibrin, silk, poly-L-lactic acid (PLLA), polyethylene glycol) have been proposed for bone tissue regeneration [38]. The hybrid nanostructures of organic and inorganic compounds are of particular interest due to the hybrid intrinsic structure of bone tissue. Tamjid et al. [39] have applied an indirect 3D printing method to fabricate 3D composite scaffolds with tailored internal architecture, including well-designed micro- and macro-porosity (Fig. 3a). The scaffold was based on polycaprolactone (PCL) decorated with TiO_2 and bioactive glass nanoparticles as osteoconductive and bioactive agents. They showed that the designed structure promoted adhesion, proliferation, and differentiation of osteoblasts.

For wound healing and skin regeneration, the development of novel materials that resemble and mimic the function of skin tissue is essential. While the self-healing properties of the skin are adequate for minor wounds, crevice wounds require a longer time to heal and carry the risk of infection, inflammation, and scarring [43]. Mahmoudi et al. [40] prepared electrospun nanocomposites based on chitosan/poly(vinyl pyrrolidone) (PVP)/GO with adjustable pore structure, improved elastic modulus and tensile strength, and controlled water permeability for skin tissue engineering. They showed that the nanofibrous composite structure promoted cell attachment and viability and accelerated wound closure rate compared with sterile gauze sponges (Fig. 3b). In another work, Zandi et al. [41] proposed bilayered scaffolds resembling full-thickness skin. The hybrid scaffolds consisted of an epidermis-like layer of a growth factor loaded GelMA/laponite nanocomposite hydrogel and a dermis-like layer of gelatin nanofibrous structure. In vivo investigations, including wound closure rate, histopathological analysis, and collagen metabolism, have shown the great potential of the bilayer composite scaffold for wound healing application (Fig. 3c).

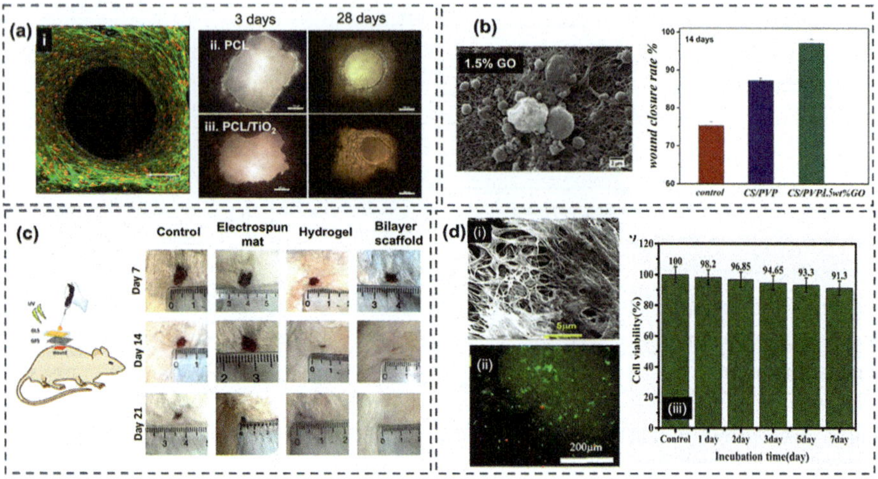

Fig. 3 **a** Confocal laser scanning microscopy image of stained osteoblast cells deposited on the pore structure of a 3D-printed scaffold made of PCL decorated with TiO_2 and bioglass (a.i). (a.ii and (a.ii) Phase contrast microscopy images of MC3T3-E1 cell in-growth in the pore channels of PCL PCL/TiO_2 after 3 and 28 days [39]. **b** SEM images of mesenchymal cells on the chitosan/PVP/GO electrospun composite surface at (b.i) 1.5% GO and (b.ii) the wound closure rate after 14 days compared to chitosan/PVP and a control [40]. **c** Schematic representation of the preparation of the bilayer scaffold consisting of an electrospun gelatin layer and a GelMA/laponite hydrogel layer and in vivo wound healing process [41]. **d** FESEM image of 3D nanofiber network of bacterial nanocellulose fibers (d.i). Fluorescence microscopy images of H9C2 cells encapsulated in a chitosan/poloxamer hydrogel containing Au-decorated bacterial cellulose after 7 days of incubation (d.ii). The green and red colors indicate live and dead cells, respectively. (d.iii) Viability of the encapsulated H9C2 cells at different incubation times [42]

Cartilage is a connective tissue composed of sporadically distributed chondrocytes in a highly organized extracellular matrix (ECM). However, due to the lack of chondrocyte mobility, vascular network, and progenitor cells, cartilage tissue has a low self-healing capacity. Eslahi et al. [44] have developed an innovative thermosensitive injectable hydrogel based on keratin, Pluronic, and chitosan incorporated with a mineral nano-clay, namely laponite. The results showed that the pore structure and viscoelastic properties can be tailored by the degree of crosslinking and the content of laponite. In-vitro assays with chondrocyte cells have shown high cell attachment and viability as a function of hydrogel formulation and structure, giving this hydrogel a high capacity for application in cartilage tissue engineering. Similarly, the intrinsic inability of cardiomyocytes to regenerate makes cardiac tissue engineering a real challenge [45]. The major and well-known issue in cardiac tissue engineering is to design a complex construct that has the key characteristics of native cardiac muscle, i.e., sufficient density of myocytes and supporting cells, the ability to exchange oxygen between blood and cells, and to propagate the electrical signal for harmonic contractions. Recently, Tohidi et al. [42] have developed a thermo-responsive and injectable chitosan/poloxamer-based hydrogel for cardiac tissue repair (Fig. 3d).

Bacterial nano-cellulose fibers decorated with gold nanoparticles were added to the hydrogel to provide electrical signal transmission and mechanical stiffness near to myocardium. The porous 3D nano-enabled structure exhibited a sol–gel transition at 15–22 °C depending on Au concentration. Based on descent cytocompatibility, biodegradability, and electrical conductivity, it was demonstrated that the hybrid nanomaterial could be a potential candidate for an injectable biomaterial for cardiac tissue engineering.

Recently, ACHMs have gained massive recognition in biosensing. Several examples of the application of these nanomaterials in biosensors are reviewed by [46–48]. Recently, Saiedi et al. [49] have presented a novel electrochemical immunosensor for the detection of one of the diagnostic biomarker for cardiovascular disease, namely cardiac troponin I (Fig. 4a). A screen-printed electrode (SPE) was modified with ZIF-67 MOF decorated with Au–Ag nanoparticles to attain excellent sensitivity, specificity, reproducibility, and stability. Hatami et al. [50] have prepared a free-standing three-dimensional graphene nanostructure with high porosity (90%) decorated by rose-like yttrium hexacyanoferrate particles to modify a gold screen-printed electrode (SPE). The fabricated electrode was employed as a redox mediator for the detection of ascorbic acid, an important reducing agent in the immune system, without interfering with other electrically active compounds such as glucose (Fig. 4b). In another work, Angizi et al. [51] have developed a biosensor based on SPE modified with gold and 2D boron nitride quantum dots (BNQD) with high electrochemical catalytic activity for the oxidation of ascorbic acid. Rahmati et al. [52] have proposed an innovative electrochemical sensor based on macroporous graphene networks decorated with gold nanoparticles for melatonin detection. Highly porous 3D graphene nanostructures were prepared through template-assisted chemical vapor deposition and decorated with Au nanoparticles by electrodeposition (Fig. 4c). The macroporous nanostructure exhibits a high active surface area and sensitivity.

ACHMs play a promising role in other nanomedicine applications, including diagnostics (e.g., dual bioimaging), therapeutics (e.g., cancer therapy and targeted drug delivery), and monitoring (e.g., wearable devices). Recently, Ramedani et al. [53] have developed hybrid theranostic liposomes based on a liposome nanocarrier composed of poly(vinylidene fluoride) (PVDF), graphene quantum dots, and silibinin for in vivo imaging and therapy of cancer tumors. The hybrid liposomes are acoustically sensitive and release the therapeutic agent into the target tissue area in a controlled manner. Labeling the nanocarriers with the antibody PD-1 results in targeted delivery of the therapeutic agent to the cancer cell with low distribution to other tissues. The developed ultrasound-activated piezoelectric nanostructure is a promising platform for the sustained release of a hydrophobic therapeutic agent for breast cancer therapy, especially when the ability to label and monitor cells is intended. In another study, Mohandes et al. [54] have developed a novel hybrid magnetic-fluorescent nanostructure for dual bioimaging in vivo. Polymer-modified magnetic Mn ferrite nanoparticles were coupled with nitrogen-doped carbon dots to gain the advantages of both photoluminescence and magnetization for dual magneto-fluorescence imaging. No toxicity was observed during in vitro (up to a concentration

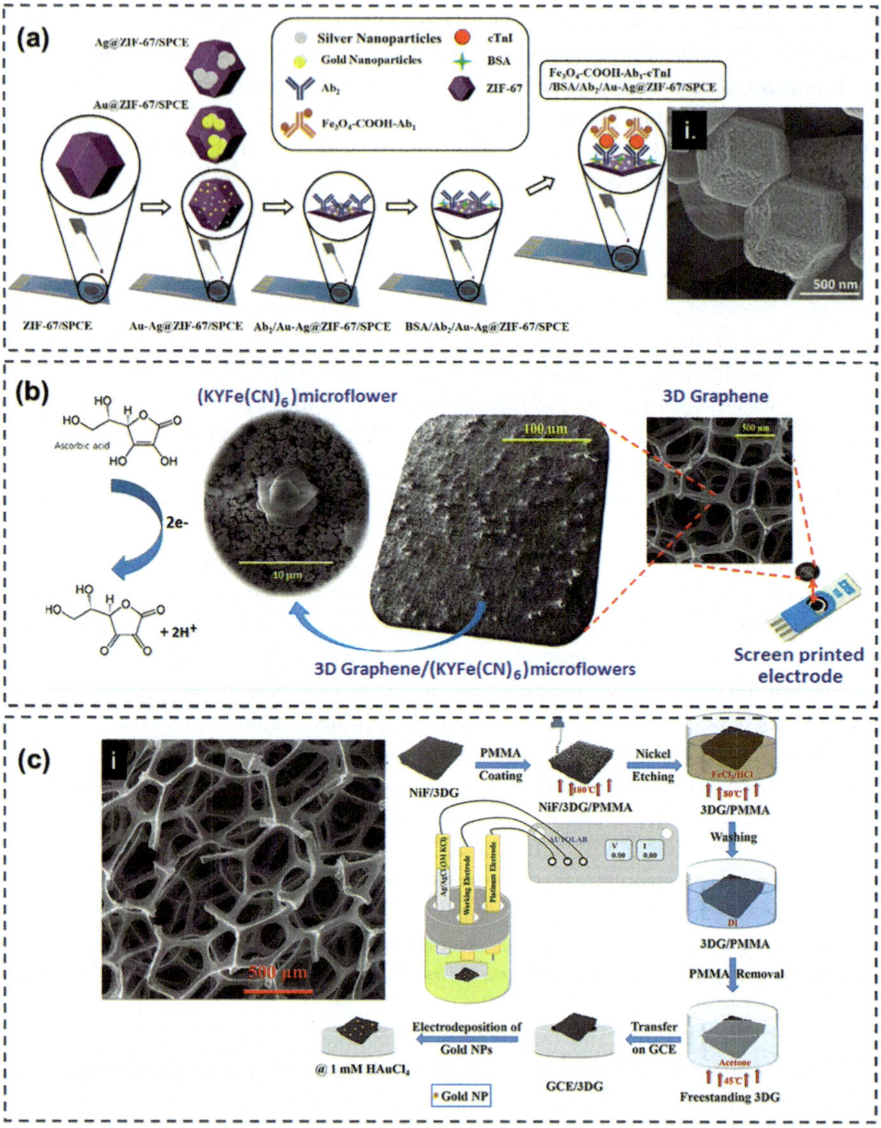

Fig. 4 **a** Schematic illustration of the fabrication of an immunosensor for the detection of cardiac troponin I based on a multilayer nanostructure consisting of a modified SPE with ZIF-67 MOF decorated with Au–Ag nanoparticles and loaded with anti-cardiac troponin I polyclonal antibody. (a.i) FESEM image of ZIF-67 MOF decorated with Au–Ag nanoparticles. **b** The illustration of a modified gold SPE with a highly porous 3D graphene structure decorated with yttrium hexacyanoferrate micro-flowers for the detection of ascorbic acid [50]. **c** (c.i) FESEM images of 3D graphene synthesized by CVD process and schematic representation of the preparation method of the graphene-gold sensor for melatonin detection (the scale bar is 500 µm) [52]

of 125 μg/mL) and in vivo (dosage of 5 mg/kg) studies. Another example is the fabrication of a self-powered, wearable piezoelectric sensor for medical pulse monitoring [55]. Nanofibrous PVDF membranes were reinforced with zirconium-based MOF nanoparticles for an enhanced piezoelectric response. The developed wearable sensor exhibited an output voltage of 568 mV, which is among the highest-performance self-powered wearable sensors.

6 Challenges and Perspective

The development of an ACHM presents numerous challenges that can be divided into two key aspects: the material and design aspect and the implementation aspect. The material aspect includes all aspects related to the design and development of the material, such as synthesis of a complex composition, hybridization, process design, interface design, and recycling. Implementation and regulatory challenges include issues, such as lack of test methods for safety and risk assessment, potential risks of derived products, accumulation and bio-persistence, lack of an accurate predictive model for human responses, incomplete assessment and analytical approaches, lack of global standardization, lack of standardized nomenclature, translation of research results into practical applications, and reproducible large-scale manufacturing [56, 57]. Therefore, the development of ACHMs should be seen in a circular exchange between researchers, stakeholders, and practitioners, rather than in a linear science-to-practice approach [58]. An effective consortium involving all research, industry, and clinician institutions working together can overcome many challenges. Moreover, complications in many research efforts can be avoided if all research approaches keep in mind the final and ultimately clinical practice [4]. Furthermore, overcoming these challenges may lie in the convergence of artificial intelligence (AI) and nanomedicine. Indeed, the convergence of AI and nanomedicine is a promising approach in several directions. By rapidly analyzing patient data, AI can play a role in diagnosis, disease progression assessment, and prediction, moving one step closer to precision medicine. AI can also optimize and accelerate the design and development process of ACHMs. Therefore, nanomedicine can benefit from AI by predicting and optimizing the interaction of developed materials with the immune system or cell membrane [59, 60].

7 Conclusions

The development of ACHMs for nanomedicine requires an integrated system design and a knowledge-based approach that incorporates theoretical, computational, and experimental methods in a coordinated framework. To this end, the selection of materials, synthesis, and hybridization techniques to achieve favorable properties and functionalities are of great importance. On the other hand, advanced fabrication techniques should be employed to integrate the desired functionality into the on-demand structural architectures by controlling the building blocks at the finest level, structural aspects, and the nano-bio interface. The result of the harmony between smart material design and smart fabrication methods is a variety of applications that address unresolved problems in healthcare. In the meantime, addressing the current challenges through a collaborative consortium involving all research, industry, and clinical entities is paramount. Undoubtedly, the application of artificial intelligence in the healthcare system is leading to a faster and more accurate approach to nanomedicine. Meanwhile, more knowledge about all features of hybrid materials and production methods is still required for smarter design and fabrication processes. It is also important to consider cost analysis and market evaluation for the full implementation of ACHMs from economic perspectives.

References

1. Park, W., et al. (2020). Advanced hybrid nanomaterials for biomedical applications. *Progress in Materials Science, 114*, 100686.
2. Das, P. P., et al. (2021). Advancement in hybrid materials, its applications and future challenges: A review. *Materials Today: Proceedings, 47*, 3794–3801.
3. Taheri-Ledari, R. (2022). Classification of micro and nanoscale composites. In: *Heterogeneous micro and nanoscale composites for the catalysis of organic reactions* (pp. 1–21). Elsevier.
4. Lendlein, A., & Trask, R. S. (2018). Multifunctional materials: Concepts, function-structure relationships, knowledge-based design, translational materials research. *Multifunctional Materials, 1*(1), 10201.
5. Chen, A. Y., Zhong, C., & Lu, T. K. (2015). Engineering living functional materials. *ACS Synthetic Biology*, 8–11.
6. Nagel, J. K. S., Schmidt, L., & Born, W. (2018). Establishing analogy categories for bio-inspired design. *Designs, 2*(4), 47.
7. Zhao, H., et al. (2022). Engineering Nano-Bio Interfaces from Nanomaterials to Nanomedicines. *Accounts of Materials Research, 3*(8), 812–829.
8. Vajna, S. (2002). Approaches of knowledge-based design. In: *DS 30: Proceedings of DESIGN 2002, the 7th international design conference Dubrovnik* (pp. 93–100).
9. Zandi, N., Sani, E. S., et al. (2021). Nanoengineered shear-thinning and bioprintable hydrogel as a versatile platform for biomedical applications. *Biomaterials, 267*, 120476.
10. Zandi, N., et al. (2020). Core-sheath gelatin based electrospun nanofibers for dual delivery release of biomolecules and therapeutics. *Materials Science and Engineering: C, 108*, 110432.
11. Mazaheri, M., Simchi, A., & Aashuri, H. (2018). Enzymatic biosensing by covalent conjugation of enzymes to 3D-networks of graphene nanosheets on arrays of vertically aligned gold nanorods: Application to voltammetric glucose sensing. *Microchimica Acta, 185*, 1–9.
12. Zhang, J., et al. (2021). 3D Bioprinting of human tissues: Biofabrication, bioinks, and bioreactors. *International Journal of Molecular Sciences, 22*(8), 3971.
13. Di Marzio, N., et al. (2020). Bio-fabrication: Convergence of 3D bioprinting and nano-biomaterials in tissue engineering and regenerative medicine. *Frontiers in Bioengineering and Biotechnology, 8*, 326.
14. Ahmad, N., Packirisamy, G., & Dutta, R. (2019). *3D printing technology in nanomedicine.* Elsevier.
15. Murphy, S. V., & Atala, A. (2014). 3D bioprinting of tissues and organs. *Nature Biotechnology, 32*(8), 773–785.

16. Moroni, L., et al. (2018). Biofabrication strategies for 3D in vitro models and regenerative medicine. *Nature Reviews Materials, 3*(5), 21–37.
17. Ullah, F., et al. (2015). Classification, processing and application of hydrogels: A review. *Materials Science and Engineering: C, 57*, 414–433.
18. Chimene, D., Kaunas, R., & Gaharwar, A. K. (2020). Hydrogel bioink reinforcement for additive manufacturing: A focused review of emerging strategies. *Advanced Materials, 32*(1), 1902026.
19. Varghese, S. A. et al. (2020). Natural polymers and the hydrogels prepared from them. In: *Hydrogels based on natural polymers* (pp. 17–47). Elsevier.
20. Slaughter, B. V., et al. (2009). Hydrogels in regenerative medicine. *Advanced Materials, 21*(32–33), 3307–3329.
21. Teo, W. E., & Ramakrishna, S. (2006). A review on electrospinning design and nanofibre assemblies. *Nanotechnology, 17*(14), R89.
22. Buzgo, M. et al. (2018). Blend electrospinning, coaxial electrospinning, and emulsion electrospinning techniques. In *Core-shell nanostructures for drug delivery and theranostics* (pp. 325–347). Elsevier.
23. Chen, H., et al. (2017). Direct writing electrospinning of scaffolds with multidimensional fiber architecture for hierarchical tissue engineering. *ACS Applied Materials & Interfaces, 9*(44), 38187–38200.
24. Xu, H., et al. (2023). A review on current nanofiber technologies: Electrospinning, centrifugal spinning, and electro-centrifugal spinning. *Macromolecular Materials and Engineering, 308*(3), 2200502.
25. Abadi, B. et al. (2022). Electrospun hybrid nanofibers: Fabrication, characterization, and biomedical applications. *Frontiers in Bioengineering and Biotechnology, 10*, 986975.
26. Radvar, E., & Azevedo, H. S. (2019). Supramolecular peptide/polymer hybrid hydrogels for biomedical applications. *Macromolecular Bioscience, 19*(1), 1800221.
27. Rivas, M., et al. (2019). Peptide self-assembly into hydrogels for biomedical applications related to hydroxyapatite. *Gels, 5*(1), 14.
28. Zhang, Y., et al. (2020). Self-assembled organic nanomaterials for drug delivery, bioimaging, and cancer therapy. *ACS Biomaterials Science & Engineering, 6*(9), 4816–4833.
29. Kango, S., et al. (2013). Surface modification of inorganic nanoparticles for development of organic–inorganic nanocomposites—A review. *Progress in Polymer Science, 38*(8), 1232–1261.
30. Shen, X.-J., et al. (2019). Improved mechanical and antibacterial properties of silver-graphene oxide hybrid/polylactid acid composites by in-situ polymerization. *Industrial Crops and Products, 130*, 571–579.
31. Aravamudhan, A. et al. (2014). Natural polymers: polysaccharides and their derivatives for biomedical applications. In *Natural and synthetic biomedical polymers* (pp. 67–89). Elsevier.
32. Green, J. J., & Elisseeff, J. H. (2016). Mimicking biological functionality with polymers for biomedical applications. *Nature, 540*(7633), 386–394.
33. Kalirajan, C., et al. (2021). A critical review on polymeric biomaterials for biomedical applications. *Polymers, 13*(17), 3015.
34. Zhao, Y. et al. (2021). Advanced bioactive nanomaterials for biomedical applications. In: *Exploration* (p. 20210089). Wiley Online Library.
35. Abd Elkodous, M. et al. (2019). Therapeutic and diagnostic potential of nanomaterials for enhanced biomedical applications. *Colloids and Surfaces B: Biointerfaces, 180*, 411–428.
36. Abdelaziz, A. G., et al. (2023). A review of 3D polymeric scaffolds for bone tissue engineering: Principles, fabrication techniques, immunomodulatory roles, and challenges. *Bioengineering, 10*(2), 204.
37. Pina, S., Oliveira, J. M., & Reis, R. L. (2015). Natural-based nanocomposites for bone tissue engineering and regenerative medicine: A review. *Advanced Materials, 27*(7), 1143–1169.
38. Koons, G. L., Diba, M., & Mikos, A. G. (2020). Materials design for bone-tissue engineering. *Nature Reviews Materials, 5*(8), 584–603.

39. Tamjid, E., et al. (2013). Tissue growth into three-dimensional composite scaffolds with controlled micro-features and nanotopographical surfaces. *Journal of Biomedical Materials Research Part A, 101*(10), 2796–2807.
40. Mahmoudi, N., & Simchi, A. (2017). On the biological performance of graphene oxide-modified chitosan/polyvinyl pyrrolidone nanocomposite membranes: In vitro and in vivo effects of graphene oxide. *Materials Science and Engineering: C, 70*, 121–131.
41. Zandi, N., Dolatyar, B., et al. (2021). Biomimetic nanoengineered scaffold for enhanced full-thickness cutaneous wound healing. *Acta Biomaterialia, 124*, 191–204.
42. Tohidi, H., et al. (2022). An electroconductive, thermosensitive, and injectable chitosan/pluronic/gold-decorated cellulose nanofiber hydrogel as an efficient carrier for regeneration of cardiac tissue. *Materials, 15*(15), 5122.
43. Chaudhari, A. A., et al. (2016). Future prospects for scaffolding methods and biomaterials in skin tissue engineering: A review. *International Journal of Molecular Sciences, 17*(12), 1974.
44. Eslahi, N., et al. (2016). Hybrid cross-linked hydrogels based on fibrous protein/block copolymers and layered silicate nanoparticles: tunable thermosensitivity, biodegradability and mechanical durability. *RSC Advances, 6*(67), 62944–62957.
45. Vunjak-Novakovic, G. et al. (2010). Challenges in cardiac tissue engineering. *Tissue Engineering Part B: Reviews*. Mary Ann Liebert, Inc. 140 Huguenot Street, 3rd Floor New Rochelle, NY 10801 USA, *16*(2), 169–187.
46. Hatamie, A. et al. (2020). Textile based chemical and physical sensors for healthcare monitoring', *Journal of the Electrochemical Society, 167*(3), 37546.
47. Naresh, V., & Lee, N. (2021). A review on biosensors and recent development of nanostructured materials-enabled biosensors. *Sensors, 21*(4), 1109.
48. Saeidi, M. et al. (2023). Electrochemical wearable biosensors and bioelectronic devices based on hydrogels: Mechanical properties and electrochemical behavior. Preprints.
49. Saeidi, M., et al. (2022). Multilayered mesoporous composite nanostructures for highly sensitive label-free quantification of cardiac troponin-I. *Biosensors, 12*(5), 337.
50. Hatamie, A., et al. (2019). Yttrium hexacyanoferrate microflowers on freestanding three-dimensional graphene substrates for ascorbic acid detection. *ACS Applied Nano Materials, 2*(4), 2212–2221.
51. Angizi, S., et al. (2018). Mechanochemical green synthesis of exfoliated edge-functionalized boron nitride quantum dots: Application to vitamin C sensing through hybridization with gold electrodes. *ACS Applied Materials & Interfaces, 10*(34), 28819–28827.
52. Rahmati, R., et al. (2020). Sensitive voltammetric detection of melatonin in pharmaceutical products by highly conductive porous graphene-gold composites. *ACS Sustainable Chemistry & Engineering, 8*(49), 18224–18236.
53. Ramedani, A., Sabzevari, O., & Simchi, A. (2022). Hybrid ultrasound-activated nanoparticles based on graphene quantum dots for cancer treatment. *International Journal of Pharmaceutics, 629*, 122373.
54. Mohandes, F., et al. (2022). Magneto-fluorescent contrast agents based on carbon Dots@ Ferrite nanoparticles for tumor imaging. *Journal of Magnetism and Magnetic Materials, 561*, 169686.
55. Moghadam, B. H., Hasanzadeh, M., & Simchi, A. (2020). Self-powered wearable piezoelectric sensors based on polymer nanofiber–metal–organic framework nanoparticle composites for arterial pulse monitoring. *ACS Applied Nano Materials, 3*(9), 8742–8752.
56. Allan, J., et al. (2021). Regulatory landscape of nanotechnology and nanoplastics from a global perspective. *Regulatory Toxicology and Pharmacology, 122*, 104885.
57. Metselaar, J. M., & Lammers, T. (2020). Challenges in nanomedicine clinical translation. *Drug Delivery and Translational Research, 10*, 721–725.
58. Pokrajac, L. et al. (2021). Nanotechnology for a sustainable future: Addressing global challenges with the international network4sustainable nanotechnology.

59. Das, K. P., & Chandra, J. (2022). 'Nanoparticles and convergence of artificial intelligence for targeted drug delivery for cancer therapy: Current progress and challenges. *Frontiers in Medical Technology, 4.*
60. Liao, X., Ma, H., & Tang, Y. J. (2022). Artificial intelligence: A solution to involution of design–build–test–learn cycle. *Current Opinion in Biotechnology, 75*, 102712.

TExKG in Health Domain: The Application of Knowledge Graph Based Framework for Explainable Recommendations in the Contexts of Elderly Care, Mental Health, and Emergency Responses

Hasan Abu-Rasheed, Mubaris Nadeem, Mareike Dornhöfer, Johannes Zenkert, Christian Weber, and Madjid Fathi

Abstract Generating explainable output of intelligent algorithms is subject to several requirements, which originate from the domain of application, involved stakeholders, and the goal of the algorithm. Industrial and health domains differ in multiple aspects, such as the level of criticality, in terms of generating wrong explanations. Those differences challenge explainability approaches to account for the requirements from each of those domains. In this paper, we investigate the ability of the knowledge graph (KG) based explainability framework (TExKG) in accommodating the explanation requirements from both industrial and health domains. We implement the framework in three different use cases in the health domain: elderly care, mental health, and health-emergency responses. We then evaluate the framework based on its ability to accommodate the requirements from these cases, in comparison to its ability of corresponding to requirements from the industrial domain. Our findings highlight the flexibility of the knowledge graph-based framework towards the different domains of application. We trace this flexibility back to the open-box nature of the KG, and the specialized components that the framework offers to integrate domain experts in the explanation generation task.

Keywords Explainable AI · Knowledge graphs · Recommendation systems · Industry 4.0 · Elderly care · Mental health · Rescue operations

H. Abu-Rasheed (✉) · M. Nadeem · M. Dornhöfer · J. Zenkert · C. Weber · M. Fathi
Institute of Knowledge Based Systems & Knowledge Management, University of Siegen, 57076 Siegen, Germany
e-mail: hasan.abu.rasheed@uni-siegen.de
URL: https://www.eti.uni-siegen.de/ws

M.-R. Alam and M. Fathi (eds.), *Integrated Systems: Data Driven Engineering*,
https://doi.org/10.1007/978-3-031-53652-6_16

1 Introduction

The application of artificial intelligence (AI) is becoming more important in many domains of daily life and work. The focus is not only on more technical- and industrial-oriented domains, but also on health-oriented ones, in order to support humans in situations of mental- and information-overload. The need for intelligent algorithms in those domains is not limited to generating an accurate prediction, but also to explaining the logic behind that prediction. This is due to the sensitivity of the decision making step in a domain like the health one. Generating explainable output of intelligent algorithms has witnessed a rising attention in the recent literature [1]. The explanation approach, shape and content, however, differs considerably based on the nature of each domain. When examining the differences between industrial and health domains, it becomes evident that health-related applications place a significant emphasis on tasks including interaction with other humans, e.g. during elderly care situations or psychological consultations. The acquisition of data and the role of the experts is different in such scenarios, in comparison to industrial and manufacturing scenarios.

Therefore, health-related domains are not inherently prepared to apply industrial approaches for explaining intelligent algorithms in daily tasks, due to their focus on sensitive, human-oriented, interaction. This style of interaction reveals several requirements for applying explainability approaches to AI-based algorithms. Two major requirements from those are: (1) the sensitivity of human-health topic leaves little margin for error in generating predictions and their explanations from the intelligent algorithm. (2) the interaction in health domain requires flexible decision making that is usually conducted by domain experts, and based on concrete, unified, terminology. This means that an explanation that is generated automatically for a patient must be directly influenced, and controlled by, the domain experts. To better understand the second point, one can think of the organizational structure of the industrial domain, in which each organizational unit may have its own style and vocabulary in describing and explaining, e.g., a system failure [2, 3], even if the failure is the same in all units. In the health domain, on the other hand, the description of, e.g., a mental disorder, or explaining its characteristics, does not allow the same level of flexibility in the selection of vocabulary or the documentation styles by different experts.

To address those special requirements from the health domain, explainability approaches need to correspond to those requirements, allowing more expert control on the explanation generation, as well as less error margin in the explained output.

In this paper, we propose and evaluate a solution for providing explanations of intelligent algorithms, such as recommender systems, based on semantic knowledge graphs (KG) as an explicit, human- and machine-readable knowledge representation. This solution is based on the authors' graph-based explanation framework (TExKG) [4], which was evaluated on two use cases in Industry 4.0 and job-market data. This paper shows the differences in requirements and implementation between these industrial-oriented and the health-oriented domains. We aim to prove the applicability of TExKG framework in health domains, complying with the differences in data acquisition, the approaches of expert involvement, and explanation requirements. We phrase the research question to elaborate how the TExKG framework is applicable in health domains, and we approach answering it based on three use cases from three research projects, two in Germany and one in a European collaboration, considering their requirements and differences from a knowledge integration perspective.

The paper is structured as follows: Section two discusses the background behind the TExKG framework and similar literature. Section three gives an overview of the three health-related use cases and their requirements. Section four documents the implementation of TExKG in each use case and the subsequent adaptation steps conducted, along with discussing the results. Section five summarizes the paper and gives a future perspective.

2 Background

Recent years have witnessed an increasing attention towards the importance of explaining AI algorithms, especially on those that have such complex structures that the interpretation of their predictions is challenging [1]. The present survey by Clement et al. [1] shows that the health domain forms 23% of the overall application domains of explainable AI (XAI) literature that the authors surveyed. While industry 4.0 comes second with 6%, the requirements from XAI from both domains are quite different, as the authors show, due to the difference in stakeholders and processes involved in each domain. A wide range of approaches was used for generating explanations of black-box algorithms, depending on the domain of application, stakeholders and users involved, characteristics of required explanations, context of explanation, complexity of algorithms, and most importantly the effect of uncertainty that explanations may have on users. From these approaches, KGs are considered one of the most promising, due to their capability to integrate heterogeneous information sources and represent the different aspects of the domain on a multidimensional level [4, 5]. The use of KGs in XAI grew rapidly in the recent years in multiple domains, especially the health domain. 43% of all cases where KGs were implemented for XAI took place within the health domain as Rajabi and Etminani [5] show in a recent survey.

2.1 Explainable AI Based on KG

Knowledge graphs (KG) are a semantic form of knowledge representation, where knowledge items are represented as nodes and relations. Oftentimes they are domain-specific, covering or excluding a certain aspect of a knowledge domain [6]. Hogan et al. (2022) define a KG as "*a graph of data intended to accumulate and convey knowledge of the real world, whose nodes represent entities of interest and whose edges represent potentially different relations between these entities*" [7].

A KG can be generated either manually from (semi-)structured or unstructured data sources [9], or automatically with the help of data- or text-mining pipelines, as in [8], where a text mining pipeline (TMP) for the enrichment of the KG based on textual information sources was implemented.

An essential application scenario of KGs is the generation of KG-based recommendations, where KGs provide additional context to support a recommendation, whether a path-based recommendations or embedding-based one [10]. Xu et al. (2021) argue that a disadvantage of using just collaborative filtering (CF) for creating recommendations is the lack of context-awareness and that a combination between KGs and collaborative filtering might bridge this gap [11]. Huang et al. (2021) follow the path-based approach in KGs to create explainable recommendations, as their main point of interest are accurate and explainable recommendations, while arguing that existing solutions, based solely on embeddings, lack explainability. Their solution incorporates path-enhanced recurrent networks [12]. Other references utilize reinforcement learning (RL) as a method for creating recommendations. Xian et al. (2019) propose a policy-guided path reasoning, incorporating Markov decision making and RL methods, where an agent learns "*the potential items of interest, such that the path history can serve as a genuine explanation for why the item is recommended to the user.*" [13]. Song et al. (2019) [14] incorporate RL as well, with the specific aim to find meaningful paths within the KG and generating fitting explanations. A previous work by the authors [15] follows a similar approach, where a Markov decision process is utilized in a light-RL implementation. In this approach, the proposed path exploration and ranking algorithm is based on a new reward function, which corresponds to the learner's context to generate explainable learning-path recommendations.

As these works show, KGs provide a transparent data structure, which can be queried for information that supports the construction of explanations [16, 17]. The combination of KGs and machine learning algorithms, provides a better potential for producing transparent and interpretable algorithms, which utilize the prediction capabilities of ML and the transparency and comprehensiveness of KGs to generate an explainable prediction [4, 5]. Moreover, KGs enable a better integration of domain experts, as their creation phase may be designed to be directly influenced by the expert-based requirements. This allows a flexible design which accommodates the differences between industrial and health related domains. In a previous work [4], we have taken this concept a step further and designed a comprehensive, graph-based explainabilty framework (TExKG), to integrate KGs and black-box algorithms for generating explanations for recommender systems. While the framework proved its potential in industry 4.0 and job-market domains, we show in this paper its adaptability to the health domain, within three use cases in the fields of elderly care, mental health, and emergency responses.

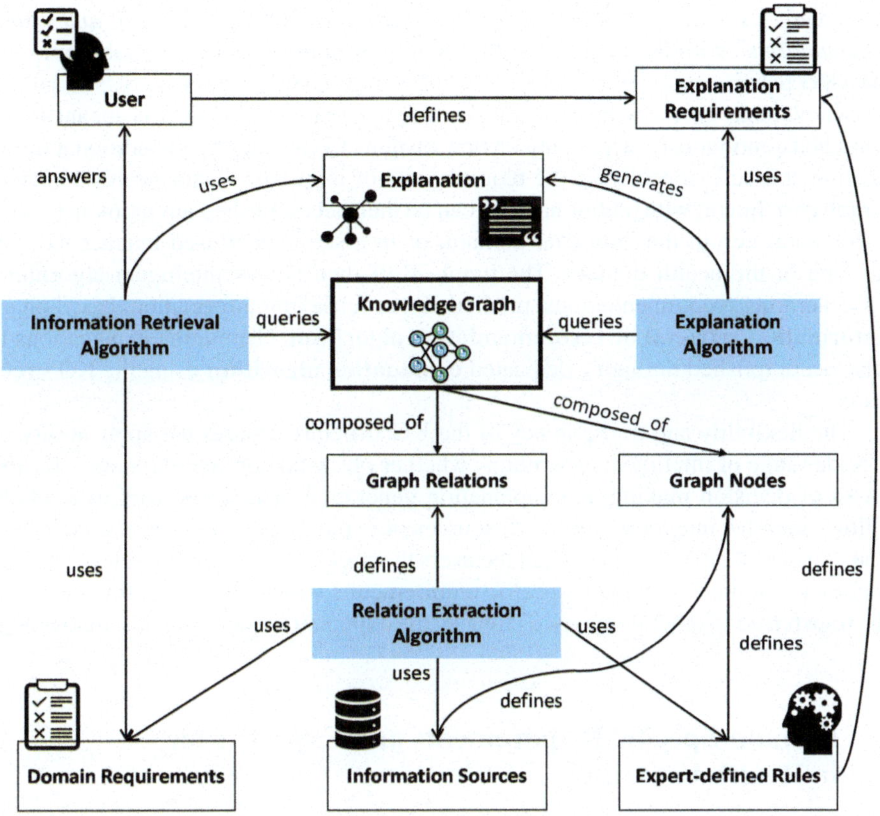

Fig. 1 Transferable knowledge graph framework, published in [4]

2.2 *TExKG: Transferable, Knowledge Graph Based Framework for Explainable Recommendations*

In a previous publication, the authors introduced a transferable, knowledge graph based framework for recommendation and explanation (TExKG), showing its validity from the results of two use cases in the domains of Industry 4.0 and job market [4], which form more structured and data-intensive domains. Figure 1 shows the published concept image of the framework and its concepts from an ontological point of view. The core idea of the TExKG is to establish domain specific knowledge graphs which are created and enriched with the help of text mining algorithms for relation extraction. Relation extraction and graph-node definition in the proposed framework are based on the consideration of **information sources**, **domain requirements**, as well as **expert-defined rules**. This is meant to allow for accommodating different levels of expert-, domain-, and data-dependency in the overall system, which in turn corresponds to a wider range of explanation requirements. The knowledge graph is modelled "*as a single source of truth*" to "*generate high-level, textual, and visual explanations*" with the help of the different algorithms that are integrable into the TExKG [4].

Knowledge grows with time, the knowledge graph in TExKG is meant to be enriched continuously, depending on new insights from experts and additional information sources. Considering the requirements of industrial versus health domains, graph enrichment with additional data can be then accomplished autonomously with more data, i.e., in the industrial domain, or in a semi-supervised manner, i.e., by experts in the health domain. The graph itself then allows intelligent algorithms of generating recommendations, predictions, or retrieving information, based on an **information retrieval or recommendation algorithm**. Generating explanations is also accomplished through a dedicated **explanation algorithm** using the KG structure.

The flexibility and transparency of the KG structure ensures the applicability of a wide range of intelligent algorithms, whether black-box or open-box ones, for the tasks of decision making and explanation generation. The framework as a whole allows then for integrating different sources of explanation requirements, including the domain, databases, experts and the users themselves. This, in turn, offers a potential to accommodate domain-specific requirements, which allows the framework to be transferred to other domains, as long as the explanation requirements are defined.

3 Domain-Specific Requirements and Experts Role in Health Use Cases

In order to investigate the ability of the TExKG framework in accommodating the explanation requirements from the health domain, we analyse three different health-related use cases, and apply the framework in each use case, to define its potential and limitations. Our focus in this research is on the unique requirements of the health domain, regarding: (1) the level of tolerance towards errors in explanations, and (2) the flexibility in representing domain and expert knowledge in an autonomous manner.

As discussed in Sect. 1, the health domain requires a sensitive implementation that does not allow for a margin of error. A wrong explanation generated for a patient about a potential diagnosis of their mental state, for instance, can lead to severe consequences regarding the acceptance of the diagnoses or commitment to therapy. This is not the case in an industrial setting, where the recipient of the explanation is usually an experienced professional, who is already aware, partially or completely, of the potential outputs of the machine.

Additionally, the industrial domain allows for a more autonomous expansion of the data sources [18]. This is due to the highly structured nature of the domain, and the data it includes, unlike health domains, whose data is not only based on, e.g., sensor readings, but rather on qualitative feedback, open-ended answers, and other forms of semi- of un-structured data. The nature of health-related applications requires a higher level of expert supervision on the analysis of data, definition of vocabularies, and the extension of the knowledge bases available about the use case. A framework that is meant to enable explainable output of intelligent algorithms used in the health domain is then supposed to allow for a direct expert influence and control over the knowledge base and the explanation generated from it. The framework is also supposed to offer the necessary means to minimize or eliminate the chances of generating a wrong explanation in a sensitive application use case.

We introduce in this section three use cases from projects that have been conducted on a European level. Two projects were conducted in Germany, and one project was a European collaboration between Germany, Ireland, Portugal, the Netherlands, and France. The three use cases reflect different aspects of health-domain applications. Therefore, they are utilized to evaluate the adaptability of TExKG towards the requirements from each use case. The collected requirements are then used to create project-related versions of the TExKG framework, and indicate the differences in how the framework is interpreted and used in different cases.

3.1 Emergency Response in Health Care—Project KIRETT

The objective of KIRETT[1] project [19, 20] is to enhance the initial care, provided during rescue operations, through the implementation of a wearable device. The resulting wearable aims to enable computer-assisted situation recognition and provide the rescue personnel with contextually-relevant recommendations for actions. The primary goals of this use case are: minimizing the incidence of late-stage complications resulting from improper treatment, and augmenting the likelihood of survival. The requirements of this use case are oriented towards fast and reliable action-recommendations. Reliability here is defined by: (1) a zero-margin for errors, and (2) a full expert-supervision of the creation of the knowledge base for the action-recommendations. Actions in this use case are defined by expert-created manuals of action sequences that are followed in the response to a medical emergency situation.

3.2 Elderly Care—Project WBsmart

The project WBsmart[2] targets the domain of vocational education and training (VET) of employees working in elderly care. This project aims to establish an educational space with intelligent learning recommendations and explanations for the workers in elderly homes. Learning recommendations in this use case serve the goal of enabling the employees of solving complex problems that they might face in their daily work

[1] www.eti.uni-siegen.de/ws/projekte/kirett.

[2] www.wbsmart.eu.

with the elderly, such as managing cases of conflict or special situations of residents having dementia or Alzheimer. The age of the target group requires a high level of awareness, which reflects on the sensitivity of the explanations that the workers need when receiving a recommendation on how to handle a certain problem. Moreover, the learning recommendation in vocational education demand a strong pedagogical foundation, which only experts can define. The explanations in this use case are then goal oriented and are meant to allow a level of reflection on the daily problems that a care employee may have.

3.3 *Mental Health—Project OSCAR*

OCSAR[3] project is a collaborative project that aims to support mental health and professional development of early career researchers. This use case focuses on providing personalized recommendations, accompanied with mental health oriented mentoring. An essential point that this use case reveals is the great emphasis on avoiding any diagnosis or therapy-oriented recommendations. The domain of mental health and the experts in that domain set clear boundaries between recommendations that *enhance the knowledge on mental health problems and their potential solutions*, and recommendations that *suggest mental health diagnosis and corresponding actions to follow accordingly*.

The explanations in this use case are provided to the researchers seeking more understanding of how to enhance their skills in managing their mental health. Therefore, the explanation requires to follow two main rules: (1) it must not include any therapeutic dimension, and (2) its terminology and formulation has to be strictly controlled by domain experts. Table 1 provides an overview on the three use cases and defines the corresponding stakeholders, i.e., users and experts, along with the requirements from those experts and the domain of application.

4 Use Case-Based Implementation

After the analysis of the requirements from the three use cases, we conducted the experiment in this research through implementing the TExKG framework [4] in each use case, to evaluate its ability to accommodate the corresponding requirements, and fulfil the use case goals. Implementing the explainability framework was conducted by the technical staff of the corresponding project to each use case. Data sources, as well as the information from the domain and experts have been collected from the project teams, to make decisions on how to use the framework's components

[3] www.oscar-ai.eu.

Table 1 Requirements

	Use case 1	Use case 2	Use case 3
Domain	Emergency response and rescue operations	Elderly Care	Mental Health
User	First-aid rescue operators and health professionals	Employees in elderly care homes	Scientific researchers
Experts	Medical directors, rescue service directors	Managers of elderly care homes, as well as VET pedagogical experts	Psychologists, working in the field of mental health consultation
Domain requirements	– Fast responses (in real-time) – Accuracy (controlled treatment procedures) – Reliability (controlled explanation content) – Wearable device should support mass casualty incident (MCI) scenarios	– Complex work-related problems require holistic decision-making and problem-solving – Connections between topics from multiple fields is necessary for an detailed overview of a problem solution	– Relations between the learning materials are directional – High relevance thresholds are to be defined due to domain sensitivity – Accuracy in the formulation and content of the explanation
Expert-defined rules	– The visual display of node and relation information should be understandable – Reasoning from the situation detection algorithms is rule-based and can therefore be explained by experts – Graph needs to include associated information that supports the explainability of a recommended treatment operation	– For the project, mandatory VET courses such as Hygiene in elderly care homes or interaction with dementia patients shall be used – The topics shall be interconnected and not isolated from each other, to reach a higher level of cognition and understanding – Reflection-level explanations take the form of a question – Training content should be semantically connected based on a pre-defined abstraction level (i.e., the foundational dimensions of learning)	– Priority in recommendation and explanation is given to the order of relations created by the experts, in the form of relation weights – Relation extraction is needed from longer texts, i.e., the description properties – Explanation and recommendation must not include treatment information – Extending the knowledge base has to be curated by experts in a semi-supervised manner

for the KG construction, decision making algorithm, and the explanation generation approach. In the following sub-sections, implementation steps in each use case are explained in detail.

4.1 TExKG Implementation for Emergency Response

The goal of this use case is to apply an intelligent algorithm to support ambulances in emergency and first aid situations. The KG is the core of the technology map in this use case, as it covers the semantic knowledge representation of treatment and first aid instructions for paramedic care. Search functions are deployed on the KG and their results are shown as instructions on a wearable device [19]. As the central medical treatment knowledge base, the KG is constructed based on the "Treatment Groups and Standard Operating Procedures" [21]. These are predefined and concrete guidelines for steps to follow in an emergency situation. Within this use case, the task of the KG is to support health professionals by providing: (1) next treatment steps, (2) additional information, and (3) medical intakes for patients. It supports keeping track of standardized medical treatments and uses multiple intelligence algorithms to detect the right treatment point to provide the right medical action-groups for that treatment. Final decision is always made by medical experts and rescue operators.

Explanation requirements have a crucial role within this use case to generate a KG-based clarification of the detected treatment points. It influences the explanation shape and content, and relies directly on the situation detection algorithm and the KG itself. Based on the situation detection algorithm, the machine learning algorithms used are meant to be explained, along with their output presented on the wearable device. The construction of the KG is based on a process-oriented manual, describing health care and decision procedures provided by health professionals. Based on the high level of accuracy needed in this application, the explanation is generated considering multiple sources, to enable a "rule based interpretation" of the output of the situation detection algorithm.

An automated approach for extending the knowledge base is not possible, which means that the construction of the KG has to be fully supervised by the domain experts. To correspond to this requirement, the relation extraction approach, see Fig. 2, is a manual one that is based on the expert curated treatment manuals and the expert-defined rules for the types of nodes and relations of the KG.

4.2 TExKG Implementation for Elderly Care

The second application scenario for the TExKG framework focuses on the elderly care domain. To apply the framework, a process oriented view is presented in Fig. 3 to illustrate how the KG is created and queried for the recommendation and explanation purposes of this use case. The basis for creating and enriching the KG are domain

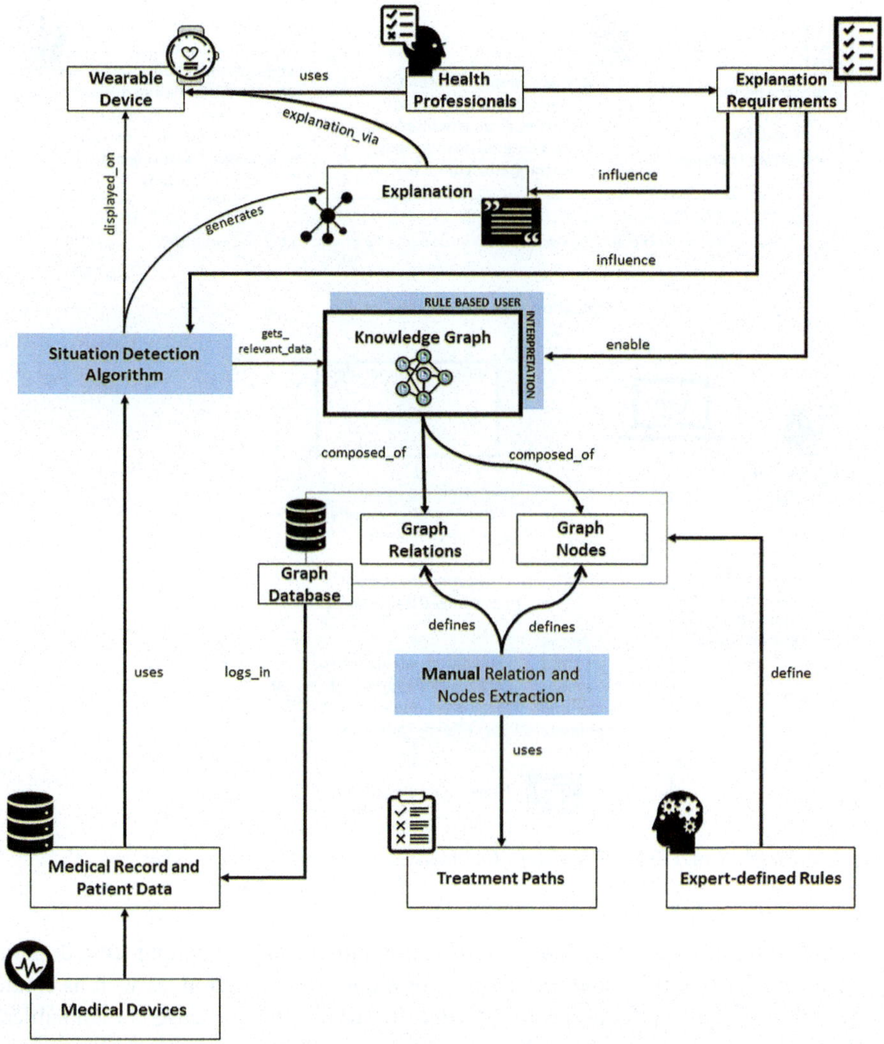

Fig. 2 Implementation of TExKG components in the use case of emergency reasons

requirements regarding VET for elderly care employees, which represent an on-the-job training approach for supporting complex problem-solving.

To achieve the use case objectives, the KG is constructed from educational materials, represented by graph nodes. Relations among those materials are created through a semantic relation extraction approach. This is also the same approach followed to enrich the KG, where domain specific information sources are analyzed with the help of NLP to find semantic similarities between the topics. Information resources in this use case are training handbooks, care-giving guidelines and existing curricula

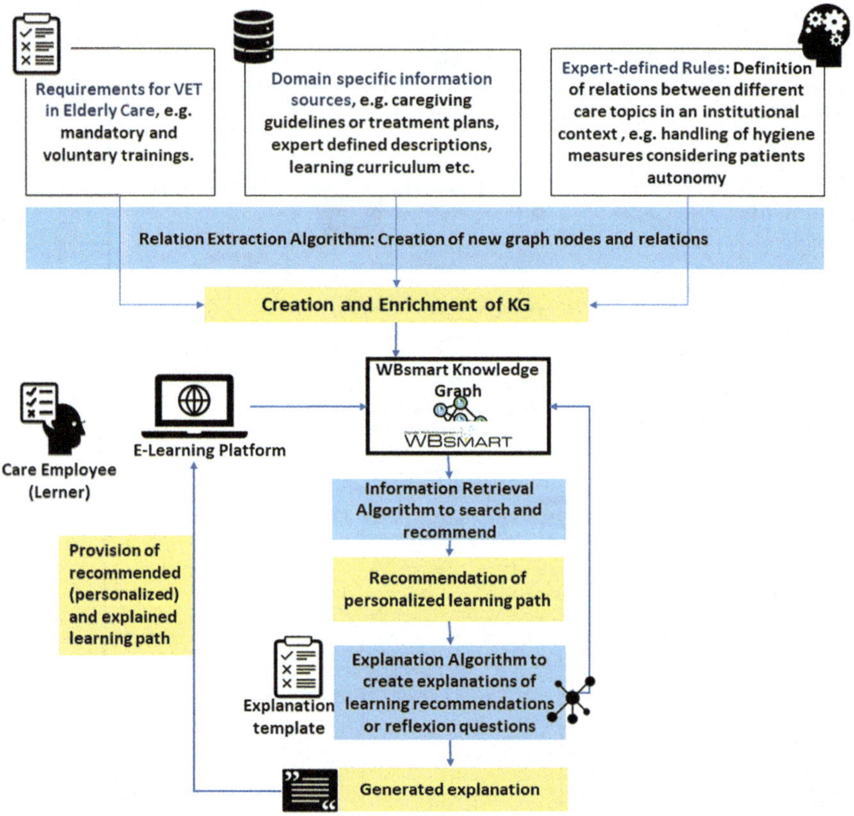

Fig. 3 Adaptation of TExKG framework for elderly care use case – Process view

descriptions. Relation extraction is additionally influenced by the input from domain experts, who define rules and restrictions for the KG construction, as well as create descriptions to enrich the nodes in the KG. The KG in this use case is then subject to further extension with time, through adding more content and extracting more relations. The enrichment of the KG follows a semi-supervised task, revealing a level of flexibility in comparison to the emergency response task, but not a complete flexibility since the domain of elderly care still requires controlled content of the recommended materials.

Based on the KG, the recommendation algorithm is implemented to generate personalized learning recommendations for care employees. The recommendation algorithm in this use case is a graph-based exploring algorithm based on RL. It is used to create a) direct, short-term learning recommendations, which are tailored to solving an urgent problem, and b) exploratory, long-term recommendations, which provide more learning material to reach a deeper level of knowledge on the problem in hand.

The explainability approach in this use case is important to ensure that elderly care workers understand why they receive certain learning recommendations. This not only serves providing clarifications, but also motivating the employee to commit to following the recommendation, and reflect it on the complex problem they are facing. Explanations are generated in the form of a semi-structured textual template, which acquires information from the KG relations about the similarities between recommended materials, and their relevance to the complex problem in hand. The template also utilizes information provided by the experts, in the form of reflection questions, which allows the employee to better comprehend the exact effect of each learning material. The hybrid approach used to construct the explanation template is meant to reduce the potential for generating wrong explanations.

4.3 TExKG Implementation for Mental Health

Being in the domain of mental health, this use case features a high level of sensitivity. Therefore, the implementation of the TExKG framework depends, to a large extent, on the expert defined rules, to avoid generating incorrect or insufficient explanations. The use case is described in terms of offering learning path recommendations to users, who would like to learn more and enhance their mental well-being. This is similar in some aspects to the previous elderly care use case, where the workers request learning path recommendations as well. The learning materials in this use case are also curated by domain experts. A KG structure is created to connect materials from different topics and courses to each other. The KG is then used to generate the learning path recommendations through an Markov decision process (MDP) algorithm.

The need of explaining the recommendation to the users in the mental health domain is indicated by the experts in the domain, and it plays, to some extent, a supporting role that a human mentor usually provides to the learners. This is of course needed in situations where the contact to the human mentor is not easy or feasible. Therefore, the implementation of the TExKG framework focuses on clarifying the prediction of the MDP algorithm, based on the information available in the KG.

We create the KG in the mental health use case, starting from the educational resources that were defined by domain experts, see Fig. 4. TExKG's information sources in this use case are databases of educational content on the mental health and well-being. These sources include a fixed taxonomy, based on which learning materials are created. For each educational resource, a textual content is defined by "Title" and "Description" metadata properties. This becomes the first requirement for the relation extraction algorithm in TExKG for finding semantic similarities between the learning materials.

The requirements from the mental health domain have indicated in this use case that the relations between learning materials should be directional ones, meaning that the order of the materials is important in this domain. These requirements from the domain align with the expert defined rule that the order defined by content curators has to be prioritized over the order defined by the relation extraction algorithm. In

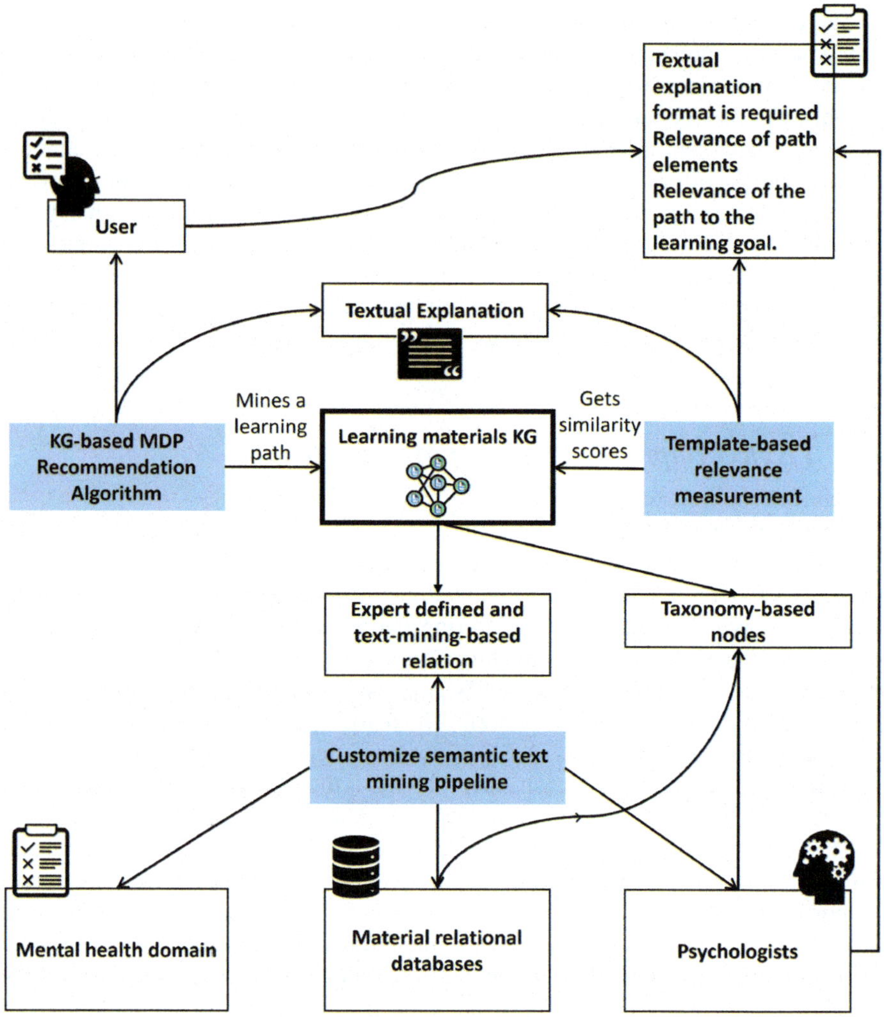

Fig. 4 Adaptation of TExKG framework for mental health use case

other words, if new directed relations are to be added to the KG, their weights are to be set less than those of the expert created relations. The relation extraction algorithm in this case is similar to the one in the elderly-care domain, and is realized through a customized text mining pipeline for semantic relation extraction from material titles and descriptions. The explainability algorithm in the mental health use case is a template-based one, which justifies to the user: (1) the similarity between the recommended elements, and (2) the relevance of the recommendation for the learning goal. In this use case, we defined the template content with domain experts, and extracted the information for the template slots from the KG. A textual format of the explanation is utilized, as experts did not identify a considerable necessity for visualizing the explanation, especially that visualizations will present an additional cognitive load on understanding the recommendation, which is not preferred in cases of, e.g., mental stress. User requirements from the explanations in this use case were aligned with the expert ones, which indicated that a justification of the relevance of the recommended path to the learning goal is essential. Users did not specify a certain shape of the explanation.

5 Evaluation Results and Discussion

To evaluate the ability of the TExKG framework to correspond to the requirements of the health domain and adopt to the different implementation scenarios in Sect. 4, we follow a use case-based evaluation procedure, similar to the one in [22, 23]. We define for that purpose a set of competency-evaluation questions, which are meant to validate the framework for achieving certain tasks required in the implementation use cases. We focus in this process on the two main requirements that separate the health domain from the industrial one, namely: (1) the required level of correctness in the explanation, based on the sensitivity of the domain, and (2) the flexibility of KG enrichment and explanation formulation, in terms of being fully-, semi-, or un-supervised task.

As a result of the evaluation design phase, a set of four competency questions were identified as follows:

- Q1: What components of the framework are used to represent the different algorithms needed in each use case?
- Q2: How did the framework enable the integration of domain experts in recommendation and explanation tasks?
- Q3: Was the KG-based explanation capable of corresponding to the sensitivity level of each use case?
- Q4: Did the construction approach of the KG correspond to the requirements of knowledge base extension in each use case?

In the following sub-sections, we discuss the results found from the implementation experiment for each of the previous evaluation questions.

5.1 Framework's Ability to Cover Algorithmic Requirements (Q1)

In the three evaluation use cases, graph search algorithm was implemented for emergency responses, while a recommendation algorithm was required for the elderly case and mental health use cases. In TExKG framework the dedicated algorithm components, shown in blue color in Fig. 1, are not limited to a certain algorithmic approach or method. They are principally designed to accommodate any open-box or black-box algorithm. This is because the framework is designed with the KG as the main source of data and information, that a decision-making algorithm needs to generate a prediction, recommendation, or otherwise.

The construction of the KG can itself be based on an algorithm, as revealed in the elderly care and mental health use cases. This task is also covered by the relation extraction component, which was capable of accommodating the NLP-based algorithm in these use cases, as well as the manual approach needed for the emergency response use case. Relation extraction component is designed to utilize multi-source information, which is not only acquired from the databases, but also from the experts and the domain itself. This means that other relation extraction approaches, such as machine learning based link-prediction, are also applicable in the framework since they have a direct connection to the required information needed to generate that prediction, while allowing at the same time an observable and controlled method to guide and supervised the relation extraction of the KG.

It is important to point out that the KG, as a graphical database, has the potential to offer the functionality of a relational database, in terms of retrieving data about the individuals in the database. Despite being the core of the framework, which is directly connected to the decision-making and explainability algorithms, those algorithms do not necessarily need to be graph-based ones. For example, a simple user-user collaborative-filtering (CF) recommendation algorithm, is still fully capable of querying the needed information from the KG about the user profiles and the similarities between the users, to generate the final recommendation. What would be essential in this regard is a proper querying approach, which works with the KG itself.

5.2 Domain-Expert Integration in the Frameworks Tasks (Q2)

In the health domain, the sensitivity of the use cases towards generating wrong predictions, recommendations, or explanations is on a high level. In the three implementation use cases, the level of sensitivity varied, but remained an essential requirement for the framework to allow a full- or semi-supervised approach for KG construction, recommendation and explanation generation. The framework component *Expert-defined Rules*, becomes a key component in the implementation of TExKG in the health domain, as it allows the experts of having a direct control over all parts of the system, including:

- The ability to define the type and content of graph nodes and relations. This allows the expert to set the information attached to each element in the database, as well as the information about the relations between the elements. In the emergency response use case, this took the form of a sequencing task, to reflect the sequences defined in treatment guidelines. In the elderly care domain, it took the form of including specific reflection questions to each recommended material, to allow self-reflection. In the mental health use case, the experts defined higher weights to the recommendation orders that originated from the manual definition of the content.

- The ability to control the recommendation generated for the users. Although there is no direct link between the *Expert-defined Rules* and the decision making algorithm, the indirect connection through the KG itself is sufficient to allow the expert of defining what is to be recommended (node content), as well as what information the algorithm uses to generate a prediction of a recommendation (node content and relation types). In the implementation use cases, this integration was clear through the rules that experts defined to set fixed directional relations between treatment steps for the emergence response, and adjust the recommended content through the additional, manually set, properties of the learning materials in the elderly care use case.
- The control of the explainability algorithm, and the resulting explanations, through defining the shape, content, and limitations of the explanation. In the emergency response use case, the use of a wearable device allows the experts to define visual and audible formats of the explanation. In the elderly care domain, the explanation was mainly textual and utilized the reflection questions to a great extent. In the mental health domain, the explanation algorithm was prevented from generating any therapeutic information to the explanation, which was accomplished through the expert controls over the explanation templates.

It is evident from the implementation use case that the ability of the TExKG framework to accommodate experts in the overall system implementation was an essential feature.Without this feature, the framework could not have been utilized in the health domain.

5.3 *Explanation Sensitivity in the Health Domain (Q3)*

The high impact that a wrong explanation poses in the health domain is a serious motivation to ensure that the explainable system is observable and controllable when it generates the explanation. From the three implementation use cases, we identify two factors that are essential to accomplish this objective: (1) the algorithm in which the explanation is generated, and (2) the source of information that is used for the explanation content. The control of these two factors is then a key to avoid generating a wrong or misleading explanation. To account for these two factors, the TExKG framework relies considerably on two pillars: (1) the control of experts over the overall explanation generation phase (algorithm, shape, and content), and (2) the KG structure as a transparent data source, which serves as a factual information source, once its construction is well observed and controlled. Explanations in the emergency response use case were the highest sensitive ones among the three use cases. While the experts defined concrete shape and content of the explanation, the most important factor in ensuring the correctness of the explanation in this use case was the KG itself. This is because the relation extraction step was conducted manually for this use case, which meant that each relation in the KG was fully controlled and factually evaluated, before the explanation algorithm queried the KG to generate the explanation. In the elderly care and mental health use case, the role of the KG as a factual source on information was supported, to a large extent, by the expert definition of the content of the explanation templates. This control over the information that is queried from the KG to fill the template slots, balanced the semi-supervised manner of KG construction in these two use cases, without compromising the correctness level of the explanation.

An important result of the implementation experiment in this research was the conclusion reached about the **compromise** needed to balance the KG construction transparency, with the time and effort needed to validate the KG structure, in order to assure a factual information source for the explainability algorithm. This, in turn, is also applied to the explanation structure, whether a visual or textual one, and the amount of information needed to build, e.g., the explanation template. Our implementation shows that the TExKG framework was able to accommodate both factors and allows for a solid control over the KG structure and the explanation algorithm, in addition to the essential role of the KG itself as an open-box, transparent, data representation method, which is the core of the overall framework.

5.4 Extending the Knowledge Base in the Health Domain (Q4)

In the three implementation use cases, the knowledge base is subject to natural extension with time, and due to the existence of new content to be recommended. In contrast to the flexibility of the industrial domain in automating the extension of the knowledge base, e.g., through regular addition of new data to the database, the health domain and its use cases are less flexible to an automated extension of the data base. This is not only due to the sensitivity of the domain, but also due to the strictness it shows in terms of describing and interpreting the data that is mostly of a qualitative and open-ended nature, e.g., interview scripts in a mental health consultation session. This level of strictness has to be accounted for in any explainabiltiy framework, to ensure that the databases used can be properly extended with the semi- of full-supervision of experts.

In TExKG, the knowledge graph structure and its construction steps have shown the ability to correspond to the requirements from each of implementation use case. In the emergency response scenario, the extension of the data base was completely conducted based on pre-defined manuals for the treatments steps and standard procedures. This was possible due to the expert defined rules and the information source components in the framework. In the elderly care use case, new materials that correspond to new work-related problems are added to the KG in a semi-supervised manner, where the experts define new content, and then validate samples of the automatically extracted relations by the NLP algorithm. In the mental health use case, experts validated all the relations extracted, as they defined the weights and directions of manually created relations in the KG. It was evident from the implementation experiment that a wider range of control and observation was both essential and possible due to the structure of the TExKG framework, and the composition of different components that balanced the roles of machine- and human-based decision making, to ensure more transparent data sources and their utilization for the recommendation and explanation tasks.

6 Conclusion

In this paper, we investigated the ability of the KG-based explainability framework TExKG to correspond to the explainability and decision-making requirements of the health domain. In comparison to a previous evaluation of the framework in the industrial domain, we defined the main requirements of the health domain through three different use cases, in emergency response, elderly care, and mental health. For each use case, experts defined the corresponding goals, stakeholders, and the requirements to fulfill the explanation and decision-making tasks. Then, the TExKG was implemented in each use case individually to evaluate its capabilities to fulfill the use case objectives and requirements.

We evaluated the implementations in the three use cases following the competency question method, where a set of four questions on the main tasks that the framework is meant to fulfill were defined. Competency questions were answered in the light of the results from each use case implementation. Our results showed high adaptation capability that our KG-base explainability framework offered in the health domain. The role of KG as a transparent data structure, and the integration of experts in the framework has proven to be a key factor for accommodating the strict requirements of the health domain, correspond to the sensitivity of its applications, and account for semi- to full-supervised extensions of the knowledge bases used for generating the recommendations and their explanations.

References

1. Clement, T., Kemmerzell, N., Abdelaal, M., & Amberg, M. (2023). *XAIR: A systematic metareview of explainable AI (XAI) aligned to the software development process* (pp. 78–108). Machine Learning and Knowledge Extraction: Multidisciplinary Digital Publishing Institute.
2. Zenkert, J., Weber, C., Dornhöfer, M., Abu-Rasheed, H., & Fathi, M. (2023). Knowledge integration in smart factories. In *Encyclopedia of engineering* (pp. 153–172). MDPI Books. https://doi.org/10.3390/encyclopedia1030061

3. Abu-Rasheed, H., Weber, C., Zenkert, J., Czerner, P., Krumm, R., & Fathi, M. (2021). A text extraction-based smart knowledge graph composition for integrating lessons learned during the microchip design. In K. Arai, S. Kapoor & R. Bhatia (Eds.), *Proceedings of the intelligent systems and applications* (pp. 594–610). Springer International Publishing: Cham. https://doi.org/10.1007/978-3-030-55187-2_43
4. Abu-Rasheed, H., Weber, C., Zenkert, J., Dornhöfer, M., & Fathi, M. (2022). *Transferrable framework based on knowledge graphs for generating explainable results in domain-specific, intelligent information retrieval. Informatics*, *9*, (6). https://doi.org/10.3390/informatics9010006
5. Rajabi, E., & Etminani, K. (2022). Knowledge-graph-based explainable AI: A systematic review. *Journal of Information Science*. https://doi.org/10.1177/01655515221112844
6. Fensel, D., Şimşek, U., Angele, K., Huaman, E., Kärle, E., Panasiuk, O., Toma, I., Umbrich, J., & Wahler, A. (2020). *Knowledge graphs: Methodology, tools and selected use cases*. Cham: Springer International Publishing. https://doi.org/10.1007/978-3-030-37439-6
7. Hogan, A., Blomqvist, E., Cochez, M., D'amato, C., Melo, G. D., Gutierrez, C., Kirrane, S., Gayo, J. E. L., Navigli, R., Neumaier, S., Ngomo, A.-C.N., Polleres, A., Rashid, S. M., Rula, A., Schmelzeisen, L., Sequeda, J., Staab, S., & Zimmermann, A. (2022). *Knowledge Graphs. ACM Computer Service, 54*, 1–37. https://doi.org/10.1145/3447772
8. Abu-Rasheed, H., Dornhöfer, M., Weber, C., Kismihók, G., Buchmann, U., & Fathi, M. (2023). Building contextual knowledge graphs for personalized learning recommendations using text mining and semantic graph completion. In *The 23rd IEEE international conference on advanced learning technologies (ICALT 2023), 10-13 July 2023*. Utah, USA: IEEE Computer Society Publications, IEEEXplore. https://doi.org/10.1109/ICALT58122.2023.00016
9. Ege, B., & Paschke, A. (Eds.). (2021). *Semantische Datenintelligenz im Einsatz*. Wiesbaden: Springer Vieweg.
10. Syed, M. H., Huy, T. Q. B., & Chung, S.-T. (2022). Context-aware explainable recommendation based on domain knowledge graph. *Big Data and Cognitive Computing, 6*, 11. https://doi.org/10.3390/bdcc6010011
11. Xu, G., Jia, G., Shi, L., & Zhang, Z. (2021). Personalized course recommendation system fusing with knowledge graph and collaborative filtering. *Computational Intelligence and Neuroscience, 2021*, 1–8. https://doi.org/10.1155/2021/9590502
12. Huang, Y., Zhao, F., Gui, X., & Jin, H. (2021). Path-enhanced explainable recommendation with knowledge graphs. *World Wide Web, 24*, 1769–1789. https://doi.org/10.1007/s11280-021-00912-4
13. Xian, Y., Fu, Z., Muthukrishnan, S., de Melo, G., & Zhang, Y. (2019). Reinforcement knowledge graph reasoning for explainable recommendation. In Proceedings of the 42nd international ACM SIGIR conference on research and development in information retrieval (pp. 285–294). Paris, France: Association for Computing Machinery. 10/gg8xrh
14. Song, W., Duan, Z., Yang, Z., Zhu, H., Zhang, M., & Tang, J. (2019). Explainable knowledge graph-based recommendation via deep reinforcement learning. arXiv:1906.09506 [cs]
15. Abu-Rasheed, H., Weber, C., Dornhöfer, M., & Fathi, M. (2023). Pedagogically-informed implementation of reinforcement learning on knowledge graphs for context-aware learning recommendations. In *The 18th European conference on technology enhanced learning EC-TEL 2023, 4-8 September, Aveiro, Portugal*. https://doi.org/10.1007/978-3-031-42682-7_35
16. Upadhyay, C., Abu-Rasheed, H., Weber, C., & Fathi, M. (2021). Explainable job-posting recommendations using knowledge graphs and named entity recognition. In *2021 IEEE international conference on systems, man, and cybernetics (SMC)* (pp. 3291–3296). Melbourne, Australia. https://doi.org/10.1109/SMC52423.2021.9658757
17. Abu-Rasheed, H., Weber, C., Zenkert, J., Krumm, R., & Fathi, M. (2022). Explainable graph-based search for lessons-learned documents in the semiconductor industry. In K. Arai (Ed.), *Intelligent computing*. Lecture notes in networks and systems (Vol. 283). Cham: Springer. https://doi.org/10.1007/978-3-030-80119-973
18. Weber, C., Abu-Rasheed, H., & Fathi, M. (2022). Adding context to industry 4.0 analytics: A new document driven knowledge graph construction and contextualization approach. In *IEEE*

international conference on electro information technology (eIT), Mankato, MN, USA (pp. 550–555). https://doi.org/10.1109/eIT53891.2022.9813992
19. Zenkert, J., Weber, C., Nadeem, M., Bender, L., Fathi, M., Ahammed, A. S., Ezekiel, A. M., Obermaisser, R., & Bradford, M. (2022). KIRETT - A wearable device to support rescue operations using artificial intelligence to improve first aid. In *2022 IEEE international smart cities conference (ISC2)* (pp. 1–4). Pafos, Cyprus: IEEE. https://doi.org/10.1109/ISC255366.2022.9922361
20. Nadeem, M., Zenkert, J., Weber, C., Fathi, M., & Hamza, M. (2023). Smart UX-design for rescue operations wearable - a knowledge graph informed visualization approach for information retrieval in emergency situations. In *2023 IEEE International Conference on Electro Information Technology (eIT)* (pp. 180–185). Romeoville, IL, USA. https://doi.org/10.1109/eIT57321.2023.10187320
21. Arbeitsgruppe: Aerztlicher Leiter Rettungsdienst and Feuerwehr Siegen, DRK Rettungsdienst Siegen-Wittgenstein: Behandlungspfade und Standardarbeitsanweisungen fuer den Rettungsdienst im Kreis Siegen-Wittgenstein, Germany (2020).
22. Ilkou, E. et al. (2021). EduCOR: An educational and career-oriented recommendation ontology. In A. Hotho et al. (Eds.), *The semantic web – ISWC 2021. ISWC 2021*. Lecture notes in computer science (Vol. 12922). Cham: Springer. https://doi.org/10.1007/978-3-030-88361-432
23. Chari, S., Seneviratne, O., Gruen, D. M., Foreman, M. A., Das, A. K., & McGuinness, D. L. (2020). Explanation ontology: A model of explanations for user-centered AI. In J.Z. Pan et al. (Eds.), *The semantic web – ISWC 2020. ISWC 2020*. Lecture notes in computer science (Vol. 12507). Cham: Springer. https://doi.org/10.1007/978-3-030-62466-815

Intelligent Sustainability

Data-Driven Identification and Control of Positive Systems

Yueyang Wang and Bahram Shafai

Abstract This paper considers the data-driven identification and control of positive dynamic systems. Since such systems appear in diverse applications whereby analytical models may not be easily available, the input-output data are collected for identification and control. Direct application of subspace system identification (SSID) to collected data of positive system does not guarantee the positivity of the resulting state space parameters. The proposed data-driven identification offers a procedure to obtain the state space representation of a positive system by applying nonnegative matrix factorization (NMF) to the Hankel matrix. The identified positive system is used for subsequent positive stabilization and performance requirements. This approach resembles the indirect adaptive control strategy with the understanding that the classical system identification based adaptive control cannot guarantee positivity constraint. Subsequently, we also address a direct approach in which data-driven positive control can be achieved. Using formulas for data-driven control, we formulate the problem of positive stabilization of linear systems and provide a solution for it directly from collected input output data. This allows to obtain state feedback gain matrices without the intermediate subspace identification. With the aid of available stability results of positive systems, the solution of data-driven positive stabilization can be obtained using data-dependent LMI. Finally, numerical examples are provided to support the theoretical results.

1 Introduction

System identification is an experimental modeling of dynamical systems using the measurements of its input output data. One can perform this process by frequency-domain and time-domain methods. Subspace system identification (SSID) is an

Y. Wang (✉) · B. Shafai
Department of Electrical and Computer Engineering, Northeastern University, Boston, MA 02115, USA
e-mail: wang.yueya@northeastern.edu

B. Shafai
e-mail: shafai@ece.neu.edu

M.-R. Alam and M. Fathi (eds.), *Integrated Systems: Data Driven Engineering*,
https://doi.org/10.1007/978-3-031-53652-6_17

important class of time-domain methods, which has been used by researchers for many years [9, 11, 20, 27]. These methods apply an extension of classical realization technique, known as Ho and Kalman algorithm [8]. Different algorithms exist depending on the measurement data such as Eigensystem realization algorithm (ERA) and 4SID algorithm. They all benefit from reliable numerical algorithms of the QR factorization and SVD. In subspace methods, one may also construct the state estimates from input-output data using available techniques, and subsequently the state space model is obtained by solving a least-squares problem.

Unfortunately none of the SSID methods can directly be applied to nonnegative data collected from a positive system. The main reason is that the structure of the resulting identified system does not guarantee positivity. Thus, the challenging problem is how to extract nonnegative system parameters from the Hankel matrix of a positive system. Positive systems appeared in diverse applications and many papers published on the stability and control of this important class of systems, see [4, 10, 21–23, 25] and the references therein. However, the problem of system identification for positive systems has not been fully investigated, except the results reported in [7, 12]. Consequently, we propose a new approach based on the Nonnegative matrix factorization (NMF) of Hankel matrix for positive system identification. NMF has found application in several areas, see [6, 28, 29] and the references therein. Therefore, it is also important to investigate the applicability of NMF to the problem of nonnegative SSID.

This paper provides a novel subspace system identification approach as applied to positive systems. The collected input and output data, which captured in the Hankel matrix are used in conjunction with NMF to obtain the state space parameters. Then we solve the state feedback control problem using data dependent LMI with or without SSID step. Finally, we provide numerical examples to support our theoretical results.

2 Preliminaries

2.1 Discrete-Time Positive Systems

Consider the LTI discrete-time system represented in the state space form

$$x(k+1) = Ax(k) + Bu(k) \tag{1}$$

$$y(k) = Cx(k) + Du(k) \tag{2}$$

where $x \in R^n$, $u \in R^m$ and $y \in R^p$ are state, input and output vectors, respectively. The coefficients matrices $\{A, B, C, D\}$ are of appropriate sizes, and it is assumed that the system is controllable and observable.

Given the initial condition $x(0)$ and the input sequence $u(k)$, the state response can be derived recursively using (1) and is given by

$$x(k) = A^k x(0) + \sum_{i=0}^{k-1} A^{k-1-i} Bu(i) \tag{3}$$

Assuming $x(0) = 0$, the direct relationship between input and output is obtained from (2) as

$$y(k) = Du(k) + \sum_{i=0}^{k-1} CA^{k-1-i} Bu(i) \tag{4}$$

which defines the associated transfer function

$$H(z) = C(zI - A)^{-1} B + D \tag{5}$$

and the corresponding impulse response matrices $h(0) = D, h(k) = CA^{k-1}B, k = 1, 2, \ldots,$ where $H(z) = \mathscr{Z}[h(k)]$.

Now, we give a brief overview of positive systems. The following results are known from the theory of positive systems [4, 10, 21, 23, 26].

Definition 1 The system (1), (2) is internally positive if for every initial condition $x_0 \in R_+^n$ and any positive input $u(k) \in R_+^m$, the state and output responses remain in positive orthant of state space i.e. $x(k) \in R_+^n$ and $y(k) \in R_+^p$. Similarly, the system (1), (2) represented by its input-output description (4) is externally positive if for any positive input $u(k) \in R_+^m$, the output response remains positive i.e. $y(k) \in R_+^p$.

Theorem 1 *The system (1), (2) is internally positive if and only if $A \in R_+^{n\times n}$, $B \in R_+^{n\times m}$, $C \in R_+^{p\times n}$ and $D \in R_+^{p\times m}$. Furthermore, the system (1), (2) represented by its input-output description (4) is externally positive if and only if its impulse response matrices are positive i.e. $h(k) \in R_+^{p\times m}$.*

Lemma 1 *Let the system (1), (2) be positive. Then it is asymptotically stable if and only if any one of the following equivalent conditions is satisfied:*

1. *There exists a diagonal positive definite matrix $P \succ 0$ such that $A^T PA - P \prec 0$ or equivalently the following LMI is feasible*

$$\begin{bmatrix} -P & A^T P \\ PA & -P \end{bmatrix} \prec 0$$

2. *There exists a vector $z \in R_+^n$ such that $(A - I)z < 0$*

Theorem 2 *There exists a state feedback control law $u = v + Kx$ for the system (1), (2) such that the closed-loop system*

$$x(k+1) = A_c x(k) + Bv \tag{6}$$

where $A_c = A + BK$, becomes positive stable if and only if

1. *The following LMI has a feasible solution for variables Y and Z*

$$\begin{bmatrix} -Z & ZA^T + Y^T B^T \\ AZ + BY & -Z \end{bmatrix} \prec 0 \tag{7}$$

$$AZ + BY \succeq 0 \tag{8}$$

where $Z \succ 0$ *is a positive definite diagonal matrix and* K *is given by* $K = YZ^{-1}$, *or*

2. *The following LP has a feasible solution for the variables* $z = [z_1 z_2 \ldots z_n]^T \in R^n$ *and* $y_i \in R^m, \forall i = 1, \ldots, n$

$$\begin{aligned} (A - I)z + B\sum_{i=1}^{n} y_i &< 0 \\ y_i &\geq 0 \;\; for \;\; i = 1, \ldots, n \\ a_{ij}z_j + b_i y_i &\geq 0 \;\; for \;\; i, j = 1, \ldots, n \end{aligned} \tag{9}$$

Furthermore, the feedback gain matrix K *is obtained by*

$$K = [\frac{y_1}{z_1} \;\; \frac{y_2}{z_2} \;\; \ldots \;\; \frac{y_n}{z_n}] \tag{10}$$

Note that the proof of (1) is based on condition (1) of Lemma 1 and the proof of (2) can be established based on condition (2).

2.2 Nonnegative Matrix Factorization

For a nonnegative matrix $H^{p \times q} = \left[h_1, h_2, \ldots, h_q\right]$ of size $p \times q$, we can find a nonnegative matrix $V \in R^{p \times r}$ and a nonnegative $U \in R^{r \times q}$ with respective sizes $p \times r$ and $r \times q$ such that:

$$H^{p \times q} \approx V^{p \times r} \times U^{r \times q} \tag{11}$$

where $V^{p \times r} = \left[v_1, \ldots, v_r\right]$, $U^{r \times q} = \left[u_1, \ldots, u_q\right]$, r are the basis matrix, the coefficient matrix, and the rank of factorization, respectively.

Usually, the factorization rank r is chosen such that $(p + q)r < pq$ [14]. Thus, VU can be seen as a compressed form of the data represented in H.

To evaluate the quality of the approximation, the cost function needs to be defined first. One useful measure is the square of the Euclidean distance between H and VU. Then, the formulation of NMF can be considered as an optimization problem [15].

$$\begin{aligned} \min \quad & f(V, U) = \sum_{i=1} \sum_{j=1} (H_{ij} - (VU)_{ij})^2 \\ \text{subject to} \quad & V \geq 0, U \geq 0 \end{aligned} \tag{12}$$

It has been recognized that the following "multiplicative update rule" is a reliable solution of the above optimization problem in terms of speed and ease of implementation.

$$\begin{aligned} U_{a\mu} &\leftarrow U_{a\mu} \frac{(V^T H)_{a\mu}}{(V^T V U)_{a\mu}} \\ V_{ia} &\leftarrow V_{ia} \frac{(H U^T)_{ia}}{(V U U^T)_{ia}} \end{aligned} \tag{13}$$

The Euclidean distance $||H - VU||$ is non-increasing under these updates and is invariant if and only if $V$and U are at a stationary point of the distance.

3 Subspace System Identification

3.1 Least-Square Formulation and Realization Algorithm

Consider the discrete-time system (1), (2) and incorporate the residues terms as,

$$x(k+1) = Ax(k) + Bu(k) + w(k) \tag{14}$$

$$y(k) = Cx(k) + Du(k) + v(k) \tag{15}$$

Note that $w(k)$ and $v(k)$ can also be considered as noise terms if one formulates the problem in stochastic framework of system identification. The above equations can be rewritten in a standard LS format as shown below

$$\underbrace{\begin{bmatrix} x(k+1) \\ y(k) \end{bmatrix}}_{\tilde{y}(k)} = \underbrace{\begin{bmatrix} A & B \\ C & D \end{bmatrix}}_{\Theta} \underbrace{\begin{bmatrix} x(k) \\ u(k) \end{bmatrix}}_{\phi(k)} + \underbrace{\begin{bmatrix} w(k) \\ v(k) \end{bmatrix}}_{e(k)} \tag{16}$$

where Θ represents the unknown parameter matrices, $\phi(k)$ is the regressor vector, and $\tilde{y}(k)$ can be regarded as an augmented output vector. Once the data are collected from input, output, and the states, the least square estimate of the parameters can be obtained.

Another approach is related to system realization problem since the input-output description of the system and its state space representation are connected by the Hankel matrix. Thus, the Hankel matrix is the most important part of subspace system identification methods.

The starting point is the input-output description of discrete-time LTI system

$$y(k) = \sum_{i=0}^{k} h(i)u(k-i) \tag{17}$$

where $h(i)$ represents the pulse response matrix or weighing sequence matrix. Note that the notation $h(i)$ is also used for weighting sequence of SISO system.

If we let $u_j(0) = 1$ for all $j = 1, 2, \ldots, m$ and $u_j(k) = 0$ for all $k = 1, 2, \ldots$ and substitute in (1), (2) we obtain

$$y(k) = \begin{cases} Du(0) & \text{for} \quad k = 0 \\ CA^{k-1}Bu(0) & \text{for} \quad k > 0 \end{cases} \tag{18}$$

The output response to the pulse input in (17) gives $y(k) = h(k)u(0)$ and comparing it with (18) leads to $h(0) = D$ and $h(k) = CA^{k-1}B$ for $k > 0$ known as Markov parameter matrices. System realization begins by constructing the generalized Hankel matrix of order $\alpha \times \beta$ by using $h(k)$ as follows,

$$H(\alpha, \beta) = \begin{bmatrix} h(1) & h(2) & \ldots & h(\beta) \\ h(2) & h(3) & \ldots & h(\beta+1) \\ \vdots & \vdots & \ddots & \vdots \\ h(\alpha) & h(\alpha+1) & \ldots & h(\alpha+\beta-1) \end{bmatrix} \tag{19}$$

Note that $h(0) = D$ is not included in $H(\alpha, \beta)$. It is not difficult to show that if $\alpha \geq n$ and $\beta \geq n$, the matrix $H(\alpha, \beta)$ is of rank n. In fact, one can state the following results.

Lemma 2 *The proper transfer function matrix* $H(z) = C(zI - A)^{-1}B + D$ *has degree n if and only if rank* $H(n, n) =$ *rank* $H(n+l, n+q) = n$ *for all* $l, q = 1, 2, \ldots$

The fundamental solution strategy of the realization algorithm is based on the decomposition of Hankel matrix. So, let $\alpha = \beta = l > n$ and construct (19) with $h(k) = CA^{k-1}B$. Thus, it is not difficult to observe that $H(l, l) = V_l U_l$, where V_l and U_l are extended observability and controllability matrices, respectively. Applying the singular value decomposition of the Hankel matrix gives

$$H = RST^T = \begin{bmatrix} R_1 & R_2 \end{bmatrix} \begin{bmatrix} S_1 & 0 \\ 0 & 0 \end{bmatrix} \begin{bmatrix} T_1^T \\ T_2^T \end{bmatrix} = R_1 S_1 T_1^T \tag{20}$$

where $S_1 = \text{diag}\{\sigma_i, i = 1, \ldots, n\}$, with singular values $\sigma_i > 0$. By obtaining the positive square root of matrix S_1, one can easily construct

$$V_l = R_1 S_1^{1/2} = \begin{bmatrix} C \\ CA \\ \ldots \\ CA^{l-1} \end{bmatrix} \qquad U_l = S_1^{1/2} T_1^T = \begin{bmatrix} B & AB & \ldots & A^{l-1}B \end{bmatrix} \tag{21}$$

Thus, the first p rows of V_l specifies C and the first m columns of U_l determines B. Finally, the matrix A can be obtained by writing

$$V_l A = \begin{bmatrix} CA \\ CA^2 \\ \vdots \\ CA^l \end{bmatrix} \triangleq \bar{V}_l \tag{22}$$

and solving for A by using the generalized inverse of the matrix V_l as

$$A = (V_l^T V_l)^{-1} V_l^T \bar{V}_l \tag{23}$$

3.2 Data-Based Subspace System Identification

So far, we constructed Hankel matrix from impulse response and applied Ho-Kalman based realization to obtain $\{A, B, C, D\}$. However, the goal of subspace system identification is to use large input-output data generated from persistently exciting inputs and then to obtain $\{A, B, C, D\}$. So, we let $\{u(k), y(k) : k = 0, 1, \ldots\}$ and construct the block Hankel matrices with collected data as

$$U_{0/l-1} = \begin{bmatrix} u(0) & \ldots & u(N-1) \\ \vdots & \ddots & \vdots \\ u(l-1) & \ldots & u(l+N-2) \end{bmatrix}_{ml \times N}, \quad Y_{0/l-1} = \begin{bmatrix} y(0) & \ldots & y(N-1) \\ \vdots & \ddots & \vdots \\ y(l-1) & \ldots & y(l+N-2) \end{bmatrix}_{pl \times N} \tag{24}$$

The expression of collective input-output Hankel matrices can be derived by repeated substitution of system equation (1), (2) starting with $\{u(k), y(k)\}$ and going onward l steps into future [3]. This leads to

$$x(k+l) = A^l x(k) + W_l u_l(k) \tag{25}$$

$$y_l(k) = V_l x(k) + T_l u_l(k) \tag{26}$$

where

$$u_l(k) = \begin{bmatrix} u(k) \\ \vdots \\ u(k+l-1) \end{bmatrix}_{ml \times 1}, \quad y_l(k) = \begin{bmatrix} y(k) \\ \vdots \\ y(k+l-1) \end{bmatrix}_{pl \times 1} \tag{27}$$

$k = 0, 1, \ldots, N-1$

correspond to the column vectors of $U_{0/l-1}$ and $Y_{0/l-1}$ respectively, and

$$W_l = \left[A^{l-1}B \dots AB \ B\right], \quad V_l = \begin{bmatrix} C \\ CA \\ \vdots \\ CA^{l-1} \end{bmatrix}, \quad T_l = \begin{bmatrix} D & 0 & \dots & 0 & 0 \\ CB & D & \dots & 0 & 0 \\ \vdots & \vdots & \ddots & \vdots & \vdots \\ CA^{l-2}B & CA^{l-1}B & \dots & CB & D \end{bmatrix} \tag{28}$$

are reverse extended controllability matrix, extended observability matrix, and Toeplitz matrix.

Thus, we have

$$\begin{aligned} U_{0/l-1} &= \left[u_l(0) \dots u_l(N-1)\right], Y_{0/l-1} = \left[y_l(0) \dots y_l(N-1)\right] \\ X_0 &= \left[x(0) \dots x(N-1)\right] \end{aligned} \tag{29}$$

and (26) in collective format can be written as

$$Y_{0/l-1} = V_l X_0 + T_l U_{0/l-1} \tag{30}$$

One can also define the state sequence X_l as

$$X_l = \left[x(l) \ x(l+1) \dots x(l+N-1)\right] \tag{31}$$

and relate it to X_0 by

$$X_l = A^l X_0 + W_l U_{0/l-1} \tag{32}$$

Now, let us combine $U_{0/l-1}$ and $Y_{0/l-1}$ using (30) to define the collective data matrix $Z_{0/l-1}$, which can be expressed by

$$Z_{0/l-1} \triangleq \begin{bmatrix} U_{0/l-1} \\ Y_{0/l-1} \end{bmatrix} = \begin{bmatrix} I_{lm} & 0_{lm \times n} \\ T_l & V_l \end{bmatrix} \begin{bmatrix} U_{0/l-1} \\ X_0 \end{bmatrix} \tag{33}$$

Under the assumption of "persistent excitation", $U_{0/l-1}$ is of full row rank ml and we have $rank(Z_{0/l-1}) = ml + n$.

The satisfaction of the above rank assumptions guarantees the existence of vectors $\{q^i : i = 1, \dots, (m+p)l\}$ such that

$$Z_{0/l-1} \underbrace{\left[q^1 \ q^2 \dots q^{(m+p)l}\right]}_{Q} = \begin{bmatrix} L_{11} & 0 \\ L_{21} & L_{22} \end{bmatrix} \tag{34}$$

This reveals the LQ decomposition of the collective data matrix $Z_{0/l-1}$ whereby L_{22} represents the Hankel matrix associated with generalized controllability and observability matrices i.e. $L_{22} = V_l U_l$. Once L_{22} is computed, the rest of the procedure to obtain $\{A, B, C, D\}$ can be performed by realization algorithm.

Alternatively, one can use (30) and (32) to obtain the state sequence X_l by eliminating X_0 which results in

$$X_l = \begin{bmatrix} W_l - A^l V_l^g T_l & A^l V_l^g \end{bmatrix} \begin{bmatrix} U_{0/l-1} \\ Y_{0/l-1} \end{bmatrix} \tag{35}$$

The state sequence X_{l+1} can also be obtained analogously, and the system matrices are estimated by least squares method applied to

$$\begin{bmatrix} X_{l+1} \\ Y_l \end{bmatrix} = \begin{bmatrix} \hat{A} & \hat{B} \\ \hat{C} & \hat{D} \end{bmatrix} \begin{bmatrix} X_l \\ U_l \end{bmatrix} \tag{36}$$

The above two approaches can be used for SSID without requiring positivity. However, if one imposes nonnegativity constraint, the constrained nonnegative least square methods should be applied to (36). This will be elaborated in the next section based on two separate constraint nonnegative least square procedures. Furthermore, a direct procedure based on the data Hankel matrix will be given in Sect. 5 using NMF.

4 Subspace Identification of Positive Systems

4.1 Data-Driven Positive Systems and Least Square Formulation

As we elaborated in Sect. 3, if the input signal to the system (1) is persistently exciting of order $n + l$, then

$$rank \begin{bmatrix} U_{0/l-1} \\ X_0 \end{bmatrix} = rank \begin{bmatrix} U_{0/l-1} \\ Y_{0/l-1} \end{bmatrix} = ml + n \tag{37}$$

Furthermore, any l-long input-output trajectory of (1) can be expressed as

$$\begin{bmatrix} u_l(k) \\ y_l(k) \end{bmatrix} = \begin{bmatrix} U_{0/l-1} \\ Y_{0/l-1} \end{bmatrix} q \tag{38}$$

where $q \in R^N$ [30]. The right-hand side of (38) represents any linear combination of the columns of the augmented input-output data matrices, $U_{0/l-1}$ and $Y_{0/l-1}$, which we represented by $Z_{0/l-1}$.

In the subsequent development, we consider a persistently exciting input sequence $u_l(k)$ of order $n + l$ with $l = 1$.

Theorem 3 *Let the condition (37) be satisfied with $l = 1$. Then the system (1) with positivity assumption has the following equivalent data-based representation*

$$x(k+1) = X_{1,N} \begin{bmatrix} U_{0/N} \\ X_{0/N} \end{bmatrix}^g \begin{bmatrix} u(k) \\ x(k) \end{bmatrix} \tag{39}$$

where we modified the notation of (29) and (31) for $l = 1$ as follows

$$U_{0/0} \triangleq U_{0/N} = \left[u(0) \ldots u(N-1)\right], X_0 \triangleq X_{0/N} = \left[x(0) \ldots x(N-1)\right]$$
$$X_1 \triangleq X_{1/N} = \left[x(1) \ldots x(N)\right]$$

Proof Assuming the system (1) is positive, [1] guarantees the existence of a positive generalized inverse term in (39). To derive (39), let us define

$$M = \begin{bmatrix} U_{0/N} \\ X_{0/N} \end{bmatrix} \quad , \quad v(k) = \begin{bmatrix} u(k) \\ x(k) \end{bmatrix} \tag{40}$$

Then for any $v(k)$, the algebraic equation

$$v = Mq \tag{41}$$

has a general solution

$$q = M^g v + (I - M^g M)w \tag{42}$$

If we write (1) compactly and substitute (41) we get

$$x(k+1) = \left[B \ A\right] \begin{bmatrix} u(k) \\ x(k) \end{bmatrix} = \left[B \ A\right] Mq(k) \tag{43}$$

Since $X_{0,N}$ and $X_{1,N}$ are related by

$$X_{1,N} = AX_{0,N} + BU_{0,N} = \left[B \ A\right] M \tag{44}$$

(43) can be written as

$$x(k+1) = X_{1,N}q(k) = X_{1,N}M^g v(k) \tag{45}$$

which is exactly (39). □

Now, using (44) one can immediately write

$$\left[B \ A\right] = X_{1,N}M^g \quad , \quad M = \begin{bmatrix} U_{0/N} \\ X_{0/N} \end{bmatrix} \tag{46}$$

which implies that the right-hand side of (46) is simply the minimization of the following least-square problem

$$\min_{\left[B \ A\right]} \quad ||X_{1,N} - [B \quad A]M||_F^2 \tag{47}$$

Similarly, the matrices C and D can be obtained from the output equation (2) with (41), which can be written as

$$y(k) = \left[D \ C\right] v(k) = \left[D \ C\right] Mq \tag{48}$$

Writing $Y_{o/l-1} = V_l x_0 + T_l U_{0/l-1}$ from (33) for $l = 1$, we get

$$Y_{0/0} \triangleq Y_{0/N} = CX_{0/N} + DU_{0/N} = \begin{bmatrix} D & C \end{bmatrix} M \tag{49}$$

$$\begin{bmatrix} D & C \end{bmatrix} = Y_{0/N} M^g \tag{50}$$

$$y(k) = \begin{bmatrix} D & C \end{bmatrix} Mq = Y_{0/N} q \tag{51}$$

and substituting q form (42) yields

$$y(k) = Y_{0/N} M^g v(k) \tag{52}$$

Consequently, the matrices C and D are captured by the product $Y_{0/N} M^g$. Using (50), one can interpret the solution as the minimization of the following least-square problem

$$\min_{\begin{bmatrix} D & C \end{bmatrix}} \quad ||Y_{0/N} - \begin{bmatrix} D & C \end{bmatrix} M||_F^2 \tag{53}$$

The above development clearly shows that if one applies [1] to obtain the nonnegative generalized inverse of the nonnegative data matrix M, then (46) and (50) guarantee the nonnegativity of state space matrices $\{A, B, C, D\}$. Otherwise, one can perform two separate least square problems defined by (47) and (53) with nonnegativity constraints.

The following section provides solution strategies for nonnegative least square problem.

4.2 Nonnegative Least-Square Procedure

Let us concentrate on the least-square problem (47), which can be written in a compact form

$$\min_{\tilde{x}} \quad ||\tilde{A}\tilde{X} - \tilde{B}||_F^2 \tag{54}$$

where $\tilde{A} = M^T, \tilde{X} = \begin{bmatrix} B & A \end{bmatrix}^T$, and $\tilde{B} = X_{1,N}^T$. Since the equation $\tilde{A}\tilde{X} = \tilde{B}$ represents an over-determined set of algebraic equations, $\tilde{B}$ is not in the range space of $\tilde{A}$ and the best approximate optimization solution is obtained by solving the unconstrained optimization problem (54). This leads to

$$\tilde{X} = (\tilde{A}^T \tilde{A})^{-1} \tilde{A}^T \tilde{B} \tag{55}$$

where the ith column of $\tilde{X}$ is given by $\tilde{x}_i = (\tilde{A}^T \tilde{A})^{-1} \tilde{A}^T b_i$. Thus the matrix algebraic equation $\tilde{A}\tilde{X} = \tilde{B}$ is the generalization of the over-determined algebraic equation $\tilde{A}\tilde{x} = \tilde{b}$, which admits the least-square solution.

However, the least-square problem (47) and (53) require the unknown parameter matrices $\{A, B, C, D\}$ to be nonnegative. In such cases, the problem must be modified to include nonnegativity constraints of the model parameters. The resulting problem is called the Nonnegative Least Square (NNLS), which can be formulated as follows:

1. NNLS problem for $\tilde{A}\tilde{x} = \tilde{b}$

$$\begin{aligned} \min \quad & f(\tilde{x}) = \frac{1}{2}||\tilde{A}\tilde{x} - \tilde{b}||_2^2 \\ \text{subject to} \quad & \tilde{x} \geq 0 \end{aligned} \tag{56}$$

2. NNLS problem for $\tilde{A}\tilde{X} = \tilde{B}$

$$\begin{aligned} \min \quad & f(\tilde{X}) = \frac{1}{2}||\tilde{A}\tilde{X} - \tilde{B}||_F^2 \\ \text{subject to} \quad & \tilde{X} \geq 0 \end{aligned} \tag{57}$$

There are several routines available for solving the NNLS problem. The algorithm in [2, 13] can be used to solve it, and its MATLAB code is available as "lsqnonneg".

A sequential coordinate-wise (SCW) algorithm was proposed in [5], which is easy to implement and can handle large scale problems. The basic idea stems from the following result, which is stated for case (1). The case (2) can equivalently be formulated and solved.

Lemma 3 *The NNLS problem*

$$\tilde{x}^* = arg \ \ min \frac{1}{2}||\tilde{A}\tilde{x} - \tilde{b}||_2^2 \quad \tilde{x} \geq 0 \tag{58}$$

is equivalent to the quadratic programming problem

$$\tilde{x}^* = arg \ \ \min_{\tilde{x}\geq 0}(\frac{1}{2}\tilde{x}^T P\tilde{x} + q^T\tilde{x}) \tag{59}$$

where $P = \tilde{A}^T\tilde{A}$ *is positive semidefinite symmetric matrix and* $q = -\tilde{A}^T\tilde{b}$.

5 Positive Subspace System Identification: Nonnegative Matrix Factorization Approach

In Sect. 3, it was shown that a decomposition of Hankel matrix allows to obtain state space realization. The same process can be adopted for data-based representation of Hankel matrix. However, the realization procedure does not guarantee the nonnegativity of $\{A, B, C, D\}$. In this section, we apply the iterative algorithm of Sect. 2 to obtain NMF of the Hankel matrix. It should be pointed out that one can employ a sequence of Alternate Least Square (ALS) with nonnegativity constraints to perform the task of NMF [13].

Consequently, our approach uses a sequence of NMF algorithms that combines with the initial step of specifying B, C matrices from realization algorihtm. From (19), consider a Hankel matrix of size $n \times n$, which is constructed from the impulse response of a positive system,

$$H(n,n) = \begin{bmatrix} h(1) & \dots & h(n) \\ \vdots & \ddots & \vdots \\ h(n) & \dots & h(2n-1) \end{bmatrix} \tag{60}$$

where $h(i) = CA^{i-1}B$. Then $H(n,n)$ can be factorized by nonnegative matrix factorization as follows,

$$H(n,n) = \begin{bmatrix} C \\ CA \\ \vdots \\ CA^{n-1} \end{bmatrix} \begin{bmatrix} B \ AB \ \dots \ A^{n-1}B \end{bmatrix} = V_n U_n \tag{61}$$

where V_n and U_n are the observability and controllability matrices of the system. Thus, the first row of V_n specifies C and the first column of V_n determines B.

The following strategy is used to perform the direct NMF algorithm.

Initialize V_n and U_n in (61) with the first n rows and the first n columns of H, respectively.

Concerning the observability matrix V_n, we have the following identity,

$$\begin{bmatrix} C \\ CA \\ \vdots \\ CA^{n-2} \end{bmatrix} A = \begin{bmatrix} CA \\ CA^2 \\ \vdots \\ CA^{n-1} \end{bmatrix} \rightarrow V_{n-1}A = \bar{V}_{n-1} \tag{62}$$

where V_{n-1} and $\bar{V}_{n-1}$ are the first $n-1$ rows and the last $n-1$ rows of the observability matrix V_n, respectively. Initialize A with the nonnegative least square solution of (62) or by generating the nonnegative generalized inverse of V_{n-1}, then NMF algorithm gives a positive solution A_1 of A from (62).

The observability matrix V_n can be factorized as follows,

$$V_n = \begin{bmatrix} C \\ CA \\ \vdots \\ CA^{n-1} \end{bmatrix} = \underbrace{\begin{bmatrix} C & & & \\ & C & & \\ & & \ddots & \\ & & & C \end{bmatrix}}_{\tilde{C}} \underbrace{\begin{bmatrix} I \\ A \\ \vdots \\ A^{n-1} \end{bmatrix}}_{\tilde{A}_o} \tag{63}$$

where $\tilde{C}$ is a diagonal block matrix, with C on its main diagonal. $\tilde{A}_o$ is also a block matrix, it contains different powers of matrix A, and the power ranges from 0 to $n-1$.

Initialize the factorization with $\tilde{C}$ and $\tilde{A}_o$ as block matrices. Then repeat (13) until the third block A^2 in $\tilde{A}_o$ converges to the square of the second block A. Thus, a second positive solution A_2 for matrix A is obtained. In other words, A_2 is the local best solution that the NMF algorithm can find near A_1 for (63).

Similarly, the controllability matrix U_n can be factorized as follows,

$$AU_{n-1} = \bar{U}_{n-1} \tag{64}$$

$$U_n = \begin{bmatrix} B & AB & \dots & A^{n-1}B \end{bmatrix} = \underbrace{\begin{bmatrix} I & A & \dots & A^{n-1} \end{bmatrix}}_{\tilde{A}_c} \underbrace{\begin{bmatrix} B & & & \\ & B & & \\ & & \ddots & \\ & & & B \end{bmatrix}}_{\tilde{B}} \tag{65}$$

U_{n-1} and $\bar{U}_{n-1}$ are the first $n-1$ columns and the last $n-1$ columns of the controllability matrix U_n, respectively. $\tilde{B}$ is a block diagonal matrix, with B on its main diagonal. $\tilde{A}_c$is a block row matrix, containing different powers of matrix A with the power ranging from 0 to $n-1$. The third positive solution A_3 of matrix A is obtained from the nonnegative matrix factorization of U_n in (65).

Once the matrix A is identified by finding the best approximation of H among A_1, A_2 and A_3, the state space representation $\{A, B, C\}$ is specified by this direct NMF algorithm. We also observe that the stability of the system remains with this direct NMF algorithm.

6 Data Driven Positive Stabilization

Sections 4 and 5 provided SSID methods for positive systems using the constraint NLS and NMF techniques, respectively. The subspace data-based identification of positive systems makes it possible to apply available positive control design for stabilization and performance. This approach resembles the classical indirect adaptive control e.g. self tuning regulator, in which the identification step is obtained first followed by regulation. However, it is also of practical interest to design control system directly from available data without intermediate system identification step. Effort in this direction was initiated by Willems et. al [30] and used for data-driven control design in [16–19]. Recently, the authors in [3] developed the theoretical foundation for data-driven control of linear systems. In this section, we achieve the same goal for stabilization of positive linear systems by state feedback as we explored in [24].

The fundamental result given by (37) and (38) allowed one to get a data-dependent state space representation of the open-loop positive system, as stated in Theorem 3.

Based on this theorem, we can also derive the data-based closed-loop representation with a state feedback control law $u = Kx$ for positive systems.

Theorem 4 *Let the condition (37) with $l = 1$ be satisfied. Then the system (1), (2) with positivity assumption in closed-loop with $u = Kx$ has the following equivalent data-based closed-loop representation*

$$x(k+1) = X_{1,N} G_k x(k) \tag{66}$$

where G_k is a $N \times n$ matrix satisfying

$$\begin{bmatrix} K \\ I \end{bmatrix} = \begin{bmatrix} U_{0/N} \\ X_{0/N} \end{bmatrix} G_k \tag{67}$$

where $u(k) = U_{0/N} G(k) x(k)$

Proof The proof is straightforward by writing

$$A + BK = \begin{bmatrix} B & A \end{bmatrix} \begin{bmatrix} K \\ I \end{bmatrix} = \begin{bmatrix} B & A \end{bmatrix} \begin{bmatrix} U_{0/N} \\ X_{0/N} \end{bmatrix} G_k = X_{1,N} G_k \tag{68}$$

where we intuitively assume the existence of feedback gain K such that $A + BK$ remains positive. The matrix G_k requires to satisfy $X_{0/N} G_k = I$ from (67). Its proper choice can ensure $X_{1,N} G_k$ to be an equivalent representation of the closed-loop system matrix $A + BK$ with $K = U_{0/N} G_k$. □

The above Theorem 4 on data-based closed-loop representation allows one to further develop a data-based approach in positive stabilization of the system (1), (2). We take advantage of the result of Theorem 2 to provide a solution using LMI.

Theorem 5 *Let the condition (37) with $l = 1$ be satisfied. Then there exists a state feedback control law $u = Kx$ such that the closed-loop system represented by (66) becomes positive stable if and only if there exists a diagonal positive definite matrix $Z \succ 0$ and a matrix Y such that*

$$\begin{bmatrix} -Z & Y^T X_{1,N}^T \\ X_{1,N} Y & -Z \end{bmatrix} \prec 0 \tag{69}$$

$$X_{1,N} Y \geq 0 \tag{70}$$

$$X_{0/N} Y = Z \tag{71}$$

where $K = U_{0/N} G_k$ and $G_k = Y Z^{-1}$

Proof It is clear that we need to find K such that $A + BK = X_{1,N} G_k$ as shown in (68). This requires to search a matrix G_k such that $X_{1,N} G_k$ satisfies the discrete Lyapunov condition

$$(X_{1,N}G_k)^T P(X_{1,N}G_k) - P \prec 0 \tag{72}$$

where G_k satisfies (67).

Applying Schur complement to the Lyapunov inequality and using congruent transformation $diag\{P^{-1}, P^{-1}\}$, we get (69) where $Z = P^{-1}$ and $Y = G_k Z$. The condition (70) guarantees the positivity structure of the closed-loop system matrix $A + BK$. □

7 Illustrative Examples

To evaluate the performance of NMF based SSID, we define the difference between the original Hankel matrix and the new Hankel matrix obtained from the positive system identification as follows,

$$H_{dif} = \frac{||H - H_{new}||_F}{||H||_F} \tag{73}$$

Simulation results show that positivity and stability remain with the Direct NMF algorithm.

Example 1: Stable SISO system (full order identification)
The following example shows the procedure of Direct NMF algorithm for positive system identification. Consider the original system $\Sigma_o = (A_o, B_o, C_o, 0)$ as follows,

$$A_o = \begin{bmatrix} 0.3298 & 0.2517 & 0.5478 \\ 0.0555 & 0.1834 & 0.2537 \\ 0.6003 & 0.06637 & 0.3671 \end{bmatrix}, B_o = \begin{bmatrix} 0.3196 \\ 0.5309 \\ 0.6544 \end{bmatrix}, C_o = \begin{bmatrix} 0.4076 & 0.8200 & 0.7184 \end{bmatrix}$$

where the Hankel matrix H is constructed from the impulse response data given by

$$H = \begin{bmatrix} 1.0357 & 0.8088 & 0.7731 \\ 0.8088 & 0.7731 & 0.7611 \\ 0.7731 & 0.7611 & 0.7484 \end{bmatrix}$$

Next, we apply NMF on the Hankel matrix by initializing $V_3 = H$ and $U_3 = H$. Then we have V_3 and U_3 as follows,

$$V_3 = \begin{bmatrix} 1.3046 & 0.6652 & 0.6042 \\ 0.6877 & 0.8325 & 0.8425 \\ 0.6106 & 0.8398 & 0.8586 \end{bmatrix}, U_3 = \begin{bmatrix} 0.5386 & 0.2837 & 0.2518 \\ 0.2744 & 0.3436 & 0.3466 \\ 0.2492 & 0.3477 & 0.3544 \end{bmatrix}$$

According to (61), we have,

$$B^T = \begin{bmatrix} 0.5386 & 0.2744 & 0.2492 \end{bmatrix}, C = \begin{bmatrix} 1.3046 & 0.6652 & 0.6042 \end{bmatrix}$$

Then we obtain a solution A_1 for A from (62)

$$A_1 = \begin{bmatrix} 0.3078 & 0.2836 & 0.2793 \\ 0.0008 & 0.0008 & 0.0009 \\ 0.4728 & 0.7646 & 0.7903 \end{bmatrix}$$

We can also provide two solutions for A from (63) and (65)

$$A_2 = A_1, \quad A_3 = \begin{bmatrix} 0.3078 & 0.2838 & 0.2795 \\ 0.0006 & 0.0003 & 0.0008 \\ 0.4726 & 0.7650 & 0.7904 \end{bmatrix}$$

Different solutions for A generate different Hankel matrices. The best solution results from A_2 obtained by (63) with a penalty term $A = A_2 + (A_2)^2 - A_2^2$, where A_2^2 is the third block of $\tilde{A}_o$ in (63), and it should be understood that the term $(A_2)^2 - A_2^2$ is not identically equal to 0 within the algorithm (Fig. 1).

Thus, the positive system obtained from the Direct NMF algorithm is represented as follows,

$$A = \begin{bmatrix} 0.3078 & 0.2836 & 0.2793 \\ 0.0008 & 0.0008 & 0.0009 \\ 0.4728 & 0.7647 & 0.7904 \end{bmatrix}, B = \begin{bmatrix} 0.5386 \\ 0.2744 \\ 0.2492 \end{bmatrix}, C = \begin{bmatrix} 1.3046 & 0.6652 & 0.6042 \end{bmatrix}$$

Example 2: Positive Stabilization

This example is constructed by obtaining the pair A, B of an unstable system from the NMF algorithm,

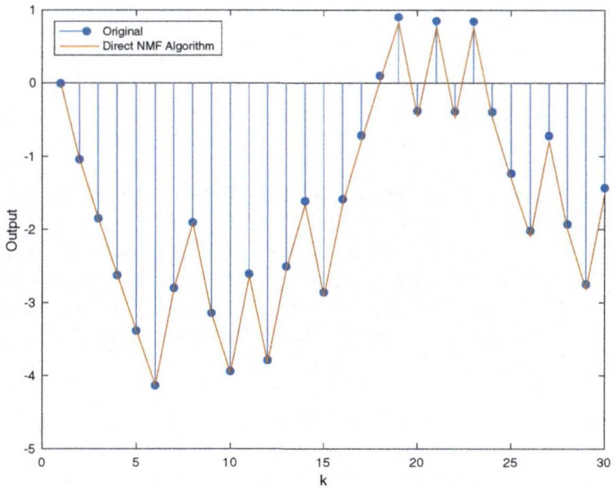

Fig. 1 Output of the original system and the system obtained from Direct NMF algorithm. Sampling period $k = 30$

$$A = \begin{bmatrix} 0.4393 & 0.4870 & 0.3407 & 0.5130 \\ 0.2403 & 0.0908 & 0.0417 & 0.0025 \\ 0.0447 & 0.4404 & 0.4266 & 0.4132 \\ 0.1221 & 0.2870 & 0.0569 & 0.5310 \end{bmatrix}, \quad B = \begin{bmatrix} 0.0636 & 0.0616 \\ 0.0032 & 0.0032 \\ 0.0292 & 0.0518 \\ 0.0190 & 0.0134 \end{bmatrix}$$

Using LMI of Theorem 5, one can obtain directly the state feedback gain K avoiding system identification step. However, here we use Theorem 2, since the parameters of the system are obtained by NMF based SSID. This results in

$$K = \begin{bmatrix} -0.3096 & -0.0270 & -0.9325 & -0.0145 \\ -0.0053 & -1.7566 & -0.0104 & -0.0667 \end{bmatrix}$$

which makes the closed-loop system matrix $A + BK$ stable with eigenvalues $\{0.9630, 0.0789, 0.1956 + 0.0667i, 0.1956 - 0.0667i\}$.

8 Conclusion

In this paper, we provided two main approaches for positive subspace system identification. The first approach is formulated as a constrained least square problem imposing nonnegativity in state space parameters. The second approach is based on nonnegative decomposition of Hankel matrix using NMF.

Using the proposed approach of SSID for positive system, the state space parameters can be obtained for subsequent control design. We also show that direct design of controller can be achieved from available data without intermediate system identification. The data-based closed-loop representation of positive system with state feedback is derived in terms of its open-loop representation. Then the stabilization is performed by solving data-dependent LMI.

References

1. Berman, A., & Plemmons, R. J. (1994). *Nonnegative matrices in the mathematical sciences*. SIAM.
2. Chen, D., & Plemmons, R. J. (2010). Nonnegativity constraints in numerical analysis. In *The birth of numerical analysis* (pp. 109–139). World Scientific.
3. De Persis, C., & Tesi, P. (2019). Formulas for data-driven control: Stabilization, optimality, and robustness. *IEEE Transactions on Automatic Control, 65*(3), 909–924.
4. Farina, L., & Rinaldi, S. (2000). *Positive linear systems: Theory and applications* (vol. 50). Wiley.
5. Franc, V., Hlaváč, V., & Navara, M. (2005). Sequential coordinate-wise algorithm for the nonnegative least squares problem. In *Computer analysis of images and patterns: 11th international conference, CAIP 2005, Versailles, France, September 5–8, 2005. Proceedings 11* (pp. 407–414). Springer.
6. Gillis, N. (2020). *Nonnegative matrix factorization*. SIAM.

7. Grussler, C., Umenberger, J., & Manchester, I. R. (2017). Identification of externally positive systems. In *2017 IEEE 56th annual conference on decision and control (CDC)* (pp. 6549–6554). IEEE.
8. HO, B., & Kálmán, R. E. (1966). Effective construction of linear state-variable models from input/output functions. *at-Automatisierungstechnik, 14*(1–12), 545–548 (1966).
9. Juang, J.-N. (1994). *Applied system identification*. Prentice-Hall, Inc.
10. Kaczorek, T. (2012). *Positive 1D and 2D systems*. Springer Science & Business Media.
11. Katayama, T. (2005). *Subspace methods for system identification* (Vol. 1). Springer.
12. Khosravi, M., & Smith, R. S. (2019). Kernel-based identification of positive systems. In *2019 IEEE 58th conference on decision and control (CDC)* (pp. 1740–1745). IEEE.
13. Lawson, C. L., & Hanson, R. J. (1995). *Solving least squares problems*. SIAM.
14. Lee, D. D., & Seung, H. S. (1999). Learning the parts of objects by non-negative matrix factorization. *Nature, 401*(6755), 788–791.
15. Lee, D. D., & Seung, H. S. (2001). Algorithms for non-negative matrix factorization. In *Advances in neural information processing systems* (pp. 556–562).
16. Markovsky, I., & Rapisarda, P. (2008). Data-driven simulation and control. *International Journal of Control, 81*(12), 1946–1959.
17. Maupong, T. M., & Rapisarda, P. (2017). Data-driven control: A behavioral approach. *Systems and Control Letters, 101*, 37–43.
18. Park, U. S., & Ikeda, M. (2009). Stability analysis and control design of lti discrete-time systems by the direct use of time series data. *Automatica, 45*(5), 1265–1271.
19. Proctor, J. L., Brunton, S. L., & Kutz, J. N. (2016). Dynamic mode decomposition with control. *SIAM Journal on Applied Dynamical Systems, 15*(1), 142–161.
20. Qin, S. J. (2006). An overview of subspace identification. *Computers and Chemical Engineering, 30*(10–12), 1502–1513.
21. Shafai, B., Chen, J., & Kothandaraman, M. (1997). Explicit formulas for stability radii of nonnegative and metzlerian matrices. *IEEE Transactions on Automatic Control, 42*(2), 265–270.
22. Shafai, B., Ghadami, R., & Oghbaee, A. (2013). Constrained stabilization with maximum stability radius for linear continuous-time systems. In *52nd IEEE conference on decision and control* (pp. 3415–3420). IEEE.
23. Shafai, B., & Hollot, C. (1991). Nonnegative stabilization of interval discrete systems. In *Control of uncertain dynamic systems* (pp. 471–490). CRC Press.
24. Shafai, B., Moradmand, A., & Siami, M. (2002). Data-driven positive stabilization of linear systems. In *2022 8th international conference on control, decision and information technologies (CoDIT)* (Vol. 1, pp. 1031–1036). IEEE.
25. Shafai, B., Naghnaeian, M., & Chen, J. (2019). Stability radius formulation of lσ-gain in positive stabilisation of regular and time-delay systems. *IET Control Theory and Applications, 13*(15), 2327–2335.
26. Shafai, B., Sadaka, H., & Ghadami, R. (2012). Robust stability and constrained stabilization of discrete-time delay systems. *IFAC Proceedings Volumes, 45*(14), 31–36.
27. Van Overschee, P., & De Moor, B. (2012). *Subspace identification for linear systems: Theory-implementation-applications*. Springer Science & Business Media.
28. Wang, Y., & Shafai, B. (2019). New initialization strategy for nonnegative matrix factorization. In *Proceedings of the international conference on image processing, computer vision, and pattern recognition (IPCV)* (pp. 72–77).
29. Wang, Y., & Shafai, B. (2021). Nonnegative matrix factorization approach for image reconstruction. In *2021 international conference on computational science and computational intelligence (CSCI)* (pp. 1639–1642). IEEE.
30. Willems, J. C., Rapisarda, P., Markovsky, I., & De Moor, B. L. (2005). A note on persistency of excitation. *Systems and Control Letters, 54*(4), 325–329.

Real-Time Ocean Prediction via a Grid of Autonomous Self-powered Swarm of Ocean Sensors

Prashant Chandra Pujari, Aneesh Jois, Jacob Lim, Stephanie Popielarz, Jianxi Wang, and Mohammad-Reza Alam

Abstract Ships are estimated to spend annually an extra \$25B worth of fuel because of encountering ocean waves en route. In terms of Greenhouse Gas (GHG) emissions, this extra fuel translates to more than 10% of total emissions by the world's transportation sector. Part of the reason that waves take this much energy to combat is that the details of incoming waves are usually not known a priori. Therefore, the effort at the bridge (pilothouse) is mainly focused on the ship's stability as it advances in waves, rather than optimizing fuel consumption in an unpredictable seaway. Our idea is to use a swarm of autonomous self-powered smart ocean sensors (hardware), combined with data-driven-based prediction algorithms already developed (software) in order to be able to predict ocean waves. Each unit of our ocean sensors can actively and autonomously position itself at a designated location for a given time period in order to measure wavesÃ¢â‚¬â„¢ height, frequency, and direction which are required for our prediction and smart routing/steering algorithm. This idea, when implemented, paves the path for a smart and efficient automatic steering in a seaway, that we estimated can easily save 15% of the cost of current method of advancing in waves. In this chapter, we provide a brief overview of state of the art technologies, challenges, current state of this project, and the future outlook.

1 Introduction

Transportation of 90% of the world's merchandise is done by ships. Because the typical ship fuel is highly polluting, the shipping industry accounts for about 3% of the world's CO_2 production, equal to the net footprint of Germany. With the current trends, if no action is taken, the CO_2 production by shipping industry can grow as much as 50% in the next 30 years [1].

In designing ships, calculations are first done for sailing in a calm water in the absence of waves because determining ship resistance in waves is a very complex

P. C. Pujari · A. Jois · J. Lim · S. Popielarz · J. Wang · M.-R. Alam (✉)
Department of Mechanical Engineering, University of California, Berkeley, CA 94720, USA
e-mail: reza.alam@berkeley.edu

M.-R. Alam and M. Fathi (eds.), *Integrated Systems: Data Driven Engineering*,
https://doi.org/10.1007/978-3-031-53652-6_18

problem. Then, at the next stage of the design, a so-called "weather margin" or "sea margin" is added to the ship's resistance as a safety factor because part of the journey of every ship will be in seaways [2]. The exact sea-margin is difficult to calculate and is chosen around 15%-20% above the fuel consumption of a ship in calm water. The number is usually assigned based on statistical data from similar ships.

Currently, ships sail in a wavy ocean with the main objective of maintaining stability. The roll motion of a ship is of greater concern, particularly when it is in synch with the incoming waves. In this case, waves can resonate the roll motion, potentially to the point of extreme healing that capsizes the ship. Another concern is the parametric roll, which is a roll motion caused by pitching through a nonlinear mechanism. Since details of encountering waves are not known a priori, the best course of action is to steer the ship safely based on what the helmsman *sees*. Particularly in higher sea states, this is limited to only a few waves ahead, giving little time to practice any maneuvers [3]. The focus of the steering is to pierce wave crests perpendicularly, in order to avoid any roll motion and potentially compromising the ship's safety. This is not the most efficient maneuver but is forced by the limited data available. An analogy is driving on a pitch-dark night on a winding road without a map. You do not know where the turns are, and how sharp they are, and this results in a sub-optimum driving both in terms of time and energy.

However, if we can have a detailed prediction of waves that a ship will encounter long enough in advance, much better steering can be practiced in terms of safety, speed, and efficiency [5]. Currently, we only have predictions of the average state of waves: we know areas with large wave activity, and areas that are calm, and we can roughly predict how this wave activity would propagate through the ocean. This is helpful in, when possible, steering ocean-going vessels away from the danger zone, usually sacrificing time and energy by taking a longer route. But if the phase-resolved (exact form of waves) can be predicted, then more efficient ship routes that may require specific maneuvers can be provided to ships [4]. Advanced maneuvers are now becoming possible due to advancements in the software (control strategies and computational algorithms), and the hardware (electronics, sensors, and actuators). This is similar to how self-driving cars are becoming a reality: they can now perform a large volume of calculations for pattern recognition and optimal decision-making in order to drive in a complex city environment that has a large number of unpredictable factors, many arising from interference from other human beings. On the positive side, as for now and in the open ocean, ships do not encounter each other very frequently, eliminating the need to include *nearby ship's human factor*, but fluid dynamics of waves make the problem a lot more complex than driving on a paved road.

In order to predict waves, there are a few outstanding challenges that need to be addressed:

1. Measurement: we need to have (relatively) accurate measurements of ocean waves (surface elevation), at least near the site where prediction is requested [6–8].
2. Reconstruction: it is to be noted that waves consist of a flow field below the surface and a change in the surface elevation. The former is, of course, more complex

as it extends in the three-dimensional water layer. The surface elevation that we see on the ocean surface and call it *waves* is only a facade of the entire complex mechanism called waves. A given surface elevation may evolve in an infinite number of ways depending on how the underlying current is. An analogy is a human who may look normal, but without knowing about their health condition (that more often than not cannot be seen from their face), it is difficult to predict how well they will be in the future. Going back to the ocean waves, since with today's technology we can only measure surface elevation, we therefore need to *infer* the underlying velocity field from the surface measurement. This is in general still an open problem, despite the great deal of research that has been devoted to the subject [9]. The bottleneck has to do with the nonlinearity of the equations, and the fact that real ocean waves are very nonlinear [10].
3. Prediction: let's assume waves are measured very accurately, and with reconstruction, we exactly know the underlying flow field, now we need to integrate the governing equation forward to be able to predict the future. Governing equations for waves are, nevertheless, expensive to integrate (time-wise). Integrating equations must be performed much faster than realtime in order to be able to provide us with a *prediction*. This is further highlighted for a ship problem where the size of a computer onboard is likely limited, both by size and its capability (cost).

Our proposed approach is to address challenge 1 via a grid of autonomous boat-shaped active sensors that can position themselves at a set of pre-determined locations. Ideally these sensors should talk to a satellite to send the data to a control center on land or on a ship. But with the lack of broadband satellite communication particularly in the open ocean and its extreme cost, it is decided that a mesh-network between units will relay the information from one unit to the other until it reaches a central command unit where all the information will be gathered and prediction is made. The area of satellite communication is also advancing fast, and once available, reliable, and cost-effective enough, the mesh-network will be replaced by direct communication with the satellite.

For challenges 2, and 3, advancement in machine learning has a promising outlook in accurately reconstructing and predicting the ocean waves [11–13]. Specifically, it has recently been shown [13], that a Convolutional Recurrent Neural Network (CRNN) can reliably predict the future of a given ocean state from a learning through a finite set of data. This includes both reconstruction and prediction from the measurement of surface elevation only. Once the training of the model is completed, the trained CRNN can provide prediction of about 5 min of the ocean ($\sim$ 5km in length) in advance. The study is performed on a long-crested polychromatic sea state, but the extension to directional seas follows the same procedure, though it will be computationally more expensive. For an area of about 5km x 5km, the extension of the methodology presented in [13] to a directional sea results in a 1-minute of prediction. Note that the estimation of prediction time in minutes is based on the computational power of regular laptops, and increases with more powerful computational resources. To keep this chapter within the word limit, for challenges 2, and 3 (reconstruction and prediction) we refer the reader to our manuscript [13] and references therein. Here, we focus our attention on the review of existing wave measurement techniques, and an update on our project of swarm of active ocean sensors from design to fabrication.

2 Review of Existing Wave Measurement Techniques

To be able to predict the future of a system, we need to be able to measure the current state of that system. In order to be able to predict the future of the ocean surface and waves, we need to have an accurate enough measurement of the current state of the ocean, at least in the vicinity of the point of interest. There exist several techniques for measuring ocean waves' characteristics in the ocean, each with its own pros and cons. Since accurate measurement of the surface is at the heart of this project, in this section we will briefly review the existing techniques.

2.1 Point Sensors

Point sensors are those that measure surface properties at a single point. An estimation of the two-dimensional surface elevation $\eta(x, y, t)$ can be found via a grid of such point sensors. The accuracy of this estimation depends on the distance between sensors, their configuration, the sampling rate, and whether they can also measure wave slope or not.

Point sensors are also divided into stationary sensors that are rigidly connected to the seabed or to a (relatively) fixed structure (like a floating offshore platform), and free drifters that can drift with waves, current, and wind. Most stationary sensors are used in shallow waters where it is practical and economical to tie them to the seabed. Here is a list of some of the existing technologies for point sensors:

- Wave Staff or "wave gauge" or "wave probe" works based on the electrical properties of water. They are usually resistive or capacitive type:
  - A resistive wave probe (aka twin-wire wave probe) consists of two parallel long wires at a very close distance. From the far, the two wires look like long rods. The probe is placed vertically in the water, in such a way that the water surface, at the maximum crest and minimum trough, is always between the top and the bottom of the two wires. A voltage is imposed between the two wires and current that goes through ocean water from one wire to the other is measured. The current is proportional to the height of the water (i.e. wave height), and with proper calibration, the time history of the wave height at the location of the wave probe can be recorded.

  - A capacitive wave probe consists of a single wire that goes into water. The capacitance of the wire and the ocean water changes depending on the height of the water, and that's the basis for measuring the surface height at each moment of time.

- Sonic and Optical wave sensors (aka vertical radars): they are placed right above the water surface, and send a signal (acoustic or optical) vertically to the surface. The time it takes for the signal to reflect back from the surface determines the wave height. If these sensors are equipped with Doppler shift measuring technology, they can also provide direct measurement of the surface velocity. Vertical radars usually work in the infrared frequency part of the spectrum.
- Pressure sensors: pressure sensors are placed on the seabed and can measure pressure fluctuations due to over-passing surface waves, from which surface elevation can be deduced. Pressure sensors are now sensitive enough that their application for the measurement of surface waves is not limited anymore to shallow waters.
- Buoys: Buoys are usually slackly moored to the seabed. They bob up and down with waves and their acceleration is measured by an on-board accelerometer (which upon twice integration gives surface elevation). The information can be communicated to land via satellite, or if they are close to the shore via a cable or wireless. Examples include:

  - Waverider from Datawell BV [14].

- Drifters: Drifters are free buoys. They float on the water surface and drift with currents, waves, and wind. They have instruments on-board and get power either from the internal battery for as long as it lasts, or equipped with solar panels with rechargeable internal batteries. Examples include:

  - Argo buoys is a group of about 4000 buoys that run on single-use batteries. An Argo float sends data once every ten days and can send such a signal for about 250 times, making its lifespan about 6.5 years [15].
  - Spotter from Sofar Ocean [16].

2.2 *Wide-Area Measurement Techniques*

Wide area measurements of ocean waves are done either by earth-based radars, or satellites. Both use electromagnetic waves to probe the ocean. One immediate advantage is that the sensor does not disturb the wave field (most point sensors do so one way or the other). The basic principle behind radar is that a signal is sent from an antenna, gets reflected back by the surface gets modulated in amplitude, frequency, and/or phase, and then arrives back at the sensor. The received signal undergoes signal processing techniques in order to extract as much information as possible. There are several challenges in using radar to measure surface waves. For example, ubiquitous capillary waves (short centimeter-scale waves) on the surface can significantly alter radar signals. These waves carry very small energy but can have a major effect on the back-scattering of the signal. Another major issue is the shadowing effect: water surface behind waves (e.g. when we have tall waves near the radar) cannot be seen [17]. In stormy seas, where the accuracy of measurements is more important, the shadowing effect is also more highlighted.

- Wave radar: These radars are installed on part of the land overlooking oceans, ships, or offshore structures. These radars may work with infra-red (micrometer wavelength), micro-wave (centimeter wavelength), or high-frequency (10-meter wavelength) parts of the electromagnetic wave spectrum [18].
- Satellite (microwave): About 40 years ago, a new possibility for the measurement of ocean surface waves became available: remote sensing by satellites which can be passive (called radiometers) or active. Satellites typically use electromagnetic waves in the microwave range which is a little distracted by cloud cover. Remote sensing by satellites uses the following technologies:

  - Altimeter: Radar altimeters are nadir-facing (looking right below) and active sensors that usually work around the frequency of 13.9 GHz and send about 1000 pulses per second. The resolution of radar altimeters is in the order of several kilometers. They are used mainly to measure the Significant Wave Height and average wave quantities only. Examples of Ocean altimeter satellites include TOPEX/Poseidon and Jason-series.
  - Synthetic Aperture Radar (SAR) are narrowband-coherent (phase-preserving), off-nadir (20–50 °C), and active sensors. They send a wave of 3cm(10 GHz)-25cm (1 GHz) at an oblique angle. The reason for the choice of frequency is that below 1 GHz we get a lot of contamination from cosmic microwaves, and above 10 GHz, atmospheric absorption is significant. The field of view of SAR is about 25 m, and it has a resolution of a few meters (closer to 5 m). SAR can provide information about not only the height but also the direction of waves as well [19].
  - Scatterometers are off-nadir active sensors (Ordinary non-SAR radar) that are used mainly for measuring near-surface winds with a resolution of 10km (they measure capillary waves amplitudes, which are strongly correlated with the wind speed). These sensors measure both wind speed and direction with about 10% accuracy.

3 Ocean Environment Challenges

The ocean presents a challenging environment characterized by salt, bio-organisms, and constant exposure to sunlight. These factors pose significant risks to vessels navigating its waters, leading to potential long-term damage. In this section, we will explore the impact of the demanding oceanic conditions on drones and outline the necessary preventive measures to mitigate these challenges.

- Corrosion: Sustained exposure to salt water can result in severe corrosion of metals such as aluminum and steel, compromising both the structural rigidity and efficacy of structural members and joints within the design. In the ocean, a maximum salinity of 40 parts per thousand can be expected, requiring very strong corrosion resistance for the vessel. As such, it is critical to ensure that the outer hull of the vessel is resistant to corrosion. For the purpose of resistance against salt water corrosion, composite materials, fiberglass, and plastics are all suitable for such environments [20].
- Sunlight and UV degradation: The intense sunlight that boats are exposed to in the ocean results in high levels of ultraviolet light affecting the surfaces of the boat. Many plastics degrade at an accelerated rate due to UV light exposure, and rubber seals can often become unfit for purpose. As such, appropriate material selection for the surfaces of the boat is essential to ensure the longevity of a vessel. Acrylic and high density polyethylene provide stronger UV resistance while also being resistant to corrosion, hence are strong candidates for use in the ocean drone.
- Biofouling: Biofouling is where pipes and underwater surfaces are "fouled" by organisms such as algae and barnacles. Biofouling is a multi-stage process. Within a few minutes of a vessel being submerged in water, organic carbon is absorbed onto the submerged surface of the vessel, increasing the roughness of the surface. Within a few hours, primary colonizers such as diatoms and bacteria settle onto the carbon-enriched surface. Over the course of the next few days, secondary colonizers such as protozoa and macroalgae spores also settle on the surface, forming a biofilm. Up to this stage, the biofouling is referred to as microfouling. Finally, over the course of a few weeks, tertiary colonizers such as macroalgae (e.g. seaweed) and large marine invertebrates bind to the submerged surface.
  Biofouling alters the profile of the boat in the water thereby reducing its efficiency. This has the adverse effect of reducing speed and increasing the energy consumption of the vessel [21]. The hull structure can be directly damaged and moving parts such as propellers can also be prevented from rotating due to the build-up of organic life on a vessel. Thus, when designing the hull, the team aimed to minimize the number of exposed joints and utilize active prevention strategies for biofouling build-up on external surfaces.

4 Design and Fabrication

The long-term ambition of this project is to create a drone that can autonomously navigate in the open ocean for about 20 years at a time. The drone will use a combination of sail power, solar, and hydrogeneration to sustain itself in the open ocean. Other design specifications are: each unit has to cost about $1k, the vessel must be able to travel on average (over all points of sail) at a speed at least half of the speed of wind, and the overall mass of the vessel must be less than 20kg. The main design and material considerations are briefly discussed in the following sub-sections.

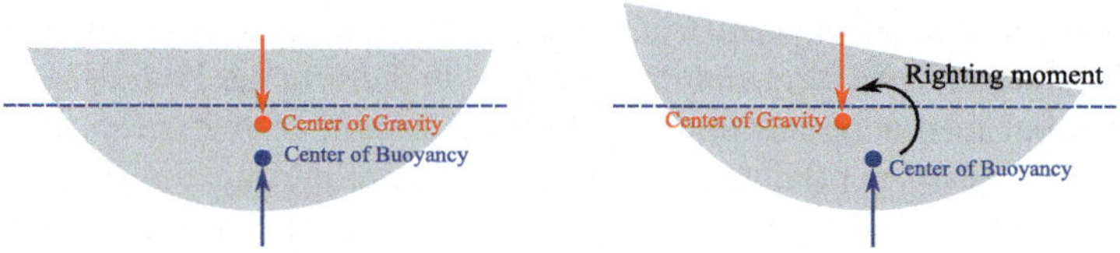

Fig. 1 Vessel Stability Diagram: left—at equilibrium, right—rolled vessel. If the center of gravity is above the center of buoyancy, for small roll angles, there is a *righting* moment that brings back the vessel to its equilibrium position. The vessel is said to be stable

4.1 Vessel's Stability

The vessel must be able to withstand waves and steep roll angles without capsizing out in the open ocean. The maximum angle to which a vessel can roll and its self-righting characteristics determine its stability. The basic stability of a vessel is based upon three key parameters: center of buoyancy, center of gravity, and the vessel's shape. The center of buoyancy is defined as the center of gravity of the volume of water that a hull displaces. The center of gravity is the point in a body where the gravitational force may be taken to act. The interaction between these two variables determines the stability of the vessel. This is illustrated in Fig. 1: the hull is considered stable if at equilibrium the center of gravity is above the center of buoyancy. Tilting of the hull results in the center of buoyancy and gravity no longer being aligned, this results in the formation of a righting moment which tries to correct the hull to an upright position. Past a certain angle, the buoyancy and gravitational forces create a moment that works in the same direction and can cause the hull to capsize.

For the prototype, we aimed to maximize stability while ensuring that the hull length was kept under 1m, and the sail length under 1.5 m. In order to maximize stability for such a small form factor, a catamaran hull design was selected. This consists of two hulls instead of the more traditional mono-hull design. Utilizing a twin hull design allows the boat to withstand greater gusts of wind before capsizing. Due to the inherent stability and the symmetrical geometry of a catamaran, it does not require a ballast [22]. This significantly reduces the weight of the hull and hence the overall cost.

4.2 Hull Design

The displacement of a hull is equivalent to the combined weight of the boat and payload. After evaluating the approximate weight of the electronics, hull, and sail for the vessel, a required displacement of 9L of seawater was determined. This required that each hull displace a minimum of 4.5 L of seawater. A midship section (waterline) of 0.14 m width, and 0.07m depth was determined to be suitable for the vessel. This enabled the maximum total weight of the vessel to be 9kg while still allowing for the vessel to reach the target speed of 4 knots in winds of 8 knots. The hull was designed to minimize sharp corner radii to allow for easier fiberglass layups. The final hull stayed within the 1m length limit.

4.3 Sail Design

When traveling at a constant speed, the hydrodynamic and aerodynamic forces are in static equilibrium. The aerodynamic force is generated by the rigid sail (which is at an angle of attack). The balancing hydrodynamic force is generated by the hull and keel of the vessel. In the case of a catamaran, only the hull provides the resistive hydrodynamic force.

Three different main sail options were explored for the ocean drone: a conventional cloth sail, a rigid sail actuated with an electric motor under the sail and a rigid sail actuated with a smaller control sail. The conventional cloth sail option was quickly eliminated due to cumbersome maintenance requirements, reduced aerodynamic efficiency, and increased control complexity when compared to a rigid sail. Using a motor to actuate the sail directly required a motor with high torque and would have a significantly higher power draw and weight penalty compared to the twin sail control system [23]. As a result, the twin sail system was selected. A schematic of the twin sail system is provided in Fig. 2a.

A NACA-0018 profile was selected for the main airfoil due to previous success in similarly sized vessels designed for research purposes at KTH Royal Institute of Technology [24, 25]. A chord length of 0.5 m and height of 1 m was used for the main sail. The small, control sail had a chord of 0.12 m and a height of 0.5 m. A servo is used to actuate the small sail, with wires being passed through the central shaft of the sail to the electronics box (a slip ring is used to facilitate 360 °C rotation of the sail). By adjusting the angle of the small sail, the incidence angle of the larger

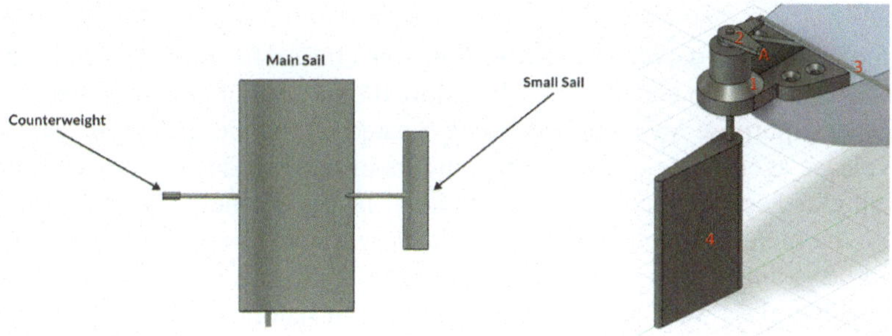

(a) (b)

Fig. 2 **a** Twin-sail configuration; **b** Left-side rudder system: 1. rudder mount, 2. servo horn, 3. connecting rod, and 4. the rudder sail to the incoming wind is adjusted. In order to facilitate this, a solid state wind vane was implemented into the design, providing information on the current wind direction relative to the heading of the vessel.

4.4 Rudder Design

The aim of the rudder system is to have a simple, low cost and lightweight design. The most straightforward method was to drive each rudder directly via a servo, forgoing any additional mechanical system. Though the mechanical design of such a system is simple, it is difficult to make sure two servos maintain synchronized in years of operation without adding additional cost. After some consideration, the sailboat employs a simplified push-pull rudder system, which is adapted from conventional rudder systems commonly found on full-size boats. Given that the sailboat is autonomous and considerably smaller in size, it is driven by a single servo, and both rudders turn synchronously. As illustrated in Fig. 2b, the system consists of the rudder mount (1), servo horn (2), connecting rod (3), and the rudder itself (4). The two servo horns are stacked and fixed together, oriented at 90 °C from each other. These servo horns are directly attached to the rudder and connected to the servo via joint A (Fig. 2b).

4.5 Fiberglass Layup of Hull and Sail

Once the physical design of the sailboat was completed, the manufacturing process began immediately. To make the boat lightweight and strong, fiberglass and rigid insulation foam were used for manufacturing the hull and sail. A female mold was created from the hull's CAD design for fiberglass layup, as shown in Fig. 3a. Due to the size constraints of the available 3D printer, the fiberglass was 3D printed in five separate sections. All sections were aligned and glued together to form a uniform mold.

A 5.85oz M32/5 woven fiberglass cloth was chosen for manufacturing the boat because of its strength and low cost. To ensure the successful release of the fiberglass part, general-purpose Vaseline was evenly coated on the mold surface, serving as a mold release agent. Due to the sharp contours of the hull design, it was difficult to use a single sheet of fiberglass for individual layers. To help the fiberglass better adapt to the mold's sharp contours, each layer was composed of two smaller sheets of fiberglass joining at the center-line. For the best balance of strength and cost, three layers of fiberglass were used to create the bottom section of the hull, as shown in Fig. 3b.

Once the bottom section had cured, excess material was carefully trimmed to achieve a smooth finish at the opening. To increase the hull's structural rigidity, four ribs made from thin layers of wood were placed evenly along the hull. Once the ribs were glued in place, a piece of rigid insulation foam was cut to shape and used to

Fig. 3 **a** Design of the hull mold, **b** Three layers of fiberglass are laid into the mold, **c** The final hulls, **d** Fabrication of the main sail, **e** The final assembly of the sailboat

enclose the hull. To achieve a watertight hull, three layers of fiberglass were laid on top of the foam. To ensure a perfect overlay over the edge due to the presence of a sharp contour, a specially designed cutting template was followed. The final products, fully enclosed hulls, are shown in Fig. 3c.

The manufacturing of the mainsail took a slightly different approach. Instead of using a female mold, a male mold was created out of rigid insulation foam. Since a hollow fiberglass may not provide the necessary structural rigidity, the foam mold itself served as an important structural insert for the fiberglass component. The mainsail had a relatively smooth surface, allowing a single sheet of fiberglass to be used for individual layers. During the layering process, both sides of the mainsail needed to be accessible, so the mainsail was suspended on the table, as shown in Fig. 3d. After two layers of fiberglass were applied and cured, a similar process to making the hull was followed. Excess material was trimmed off until a smooth edge was achieved, and the top and bottom cross-sections of the mainsail were capped with fiberglass.

5 Autonomous Navigation

A modern sailboat typically requires multiple crew members to navigate the open water. In contrast, an autonomous sailboat must have an equivalent control system that replicates the functions of human crew members [26]. Since the sailboat in this project is a scaled-down version of a manned sailboat, the vessel has fewer control surfaces, consisting of only one rigid mainsail and two synchronized rudders. To better understand the sailboat control system, the mainsail can be viewed as the throttle, controlling the sailboat's speed, while the rudder steers the sailboat. The rigid sail on the autonomous sailboat is a vertically placed, symmetrical airfoil. As air flows across the sail, a force is generated due to the pressure difference. The force that propels the sailboat forward is a combination of both the sail and the keel forces. The force produced by the sail is often not aligned with the boat's heading but is rather a force vector to the port or starboard of the sailboat. The underwater portion of the hull (or if a small keel is used) generates a counterforce on the opposite side of the sail force. Combining these two forces, the resulting force propels the boat in its intended direction.

However, the sail cannot propel the boat directly upwind. When the sailboat points upwind, the sail is unable to propel the boat at any angle, creating a range of sailboat headings called the "no-sail zone." On the open ocean, wind direction constantly changes, so it is highly likely that the autonomous sailboat will need to head in the direction of the no-sail zone during year-long missions. A special sailboat maneuver called "tacking" is used to overcome this problem [27]. When the sailboat needs to sail upwind, it travels in a direction just outside the no-sail zone and steers the boat into and through the no-sail zone to the other side, repeating the process after covering some distance. The tacking maneuver results in a zig-zag path, enabling the sailboat to progress upwind slowly [28].

Writing a complete control algorithm and developing compatible firmware to implement on the hardware is a complex and immense task. To expedite the development process, the Ardupilot software was chosen as the foundation for the control system. The Ardupilot library features numerous complete autonomous navigation systems, some of which are well-suited to the sailboat configuration of this project [29]. Since Ardupilot is also open-source, its versatility made it the ideal choice for implementation. To test the performance of Ardupilot's sailboat control system, a prototype was created for this purpose. The physical unit (sailboat) was equipped with necessary electronics and sensors, including a GPS, wind vane, onboard computer, telemetry radio, and batteries. The Ardupilot sailboat control system was rigorously tested in various wind conditions at Lake Merritt (Oakland-California). Through multiple attempts at servo input tuning based on the specific geometry of the sailboat, the test demonstrated that the control system's performance met the development standard.

6 Mesh Network

Data Transmission to a remote point in the ocean has challenges not only for navigation but also for reliable communication. The data thatÃ¢â‚¬â„¢s been gathered needs to be transmitted back to a server, while commands and waypoints need to be sent from the server to each drone. Traditional architectures rely on one transceiver, which connects the remotely located drone to the cloud, per sailboat. These communication devices are (relatively) expensive. Additionally, a regular service fee for every transceiver is also needed to maintain a connection.

A different architecture, a mesh network, where drones are connected to multiple other neighboring drones, provides a cost-effective and more reliable solution [30]. This architecture would allow multiple drones to share connections to the cloud, greatly reducing the cost per drone. These drones would also be able to connect to several transceivers, increasing reliability in the system by adding redundancy in the communication network [31]. The Xbee Digimesh is used for the mesh network. The solution abstracts and simplifies the setup and use of a mesh network at a reasonable cost. We employed XCTU, a configuration and management tool for the modules, to help configure and debug. Additionally, a Python library is available which makes development using Xbee much easier. One issue encountered was that the Xbee has a maximum payload size of 72 Bytes. We would need to send and receive rich sensor data and mission waypoints to and from the boat. To get around this, we employed Protobuf, which encodes the data and allows us to send data in a much smaller size. While the communication module can handle sending and receiving data, a controller is still needed to gather, process, and coordinate this data transfer.

To collect readings, transmit data, control the boat, and navigate, a computer would be needed to handle these tasks. The BeagleBone Blue, a single-board computer specially designed for robotics, was chosen. It has a 1GHz processor and 512MB of RAM, enough to power the complex and computationally intensive control algorithms. It is Linux-based, allowing a base for many possible applications. The Blue also offers built-in LiPo battery support and power management, eliminating the need for an additional battery management system. Additionally, the computer has built-in ports for motors and sensors, simplifying our system and connections. We connected an M8N GPS to the BeagleBone to receive geocoordinate data. A single-line schematic of the design is shown below in Fig. 4.

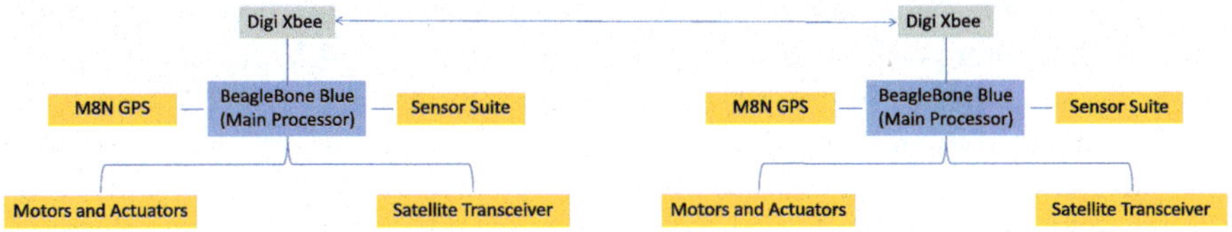

Fig. 4 Schematic of drone electronics

The goal of the project is to have drones that can be sent to mission waypoints via a mesh network, and each drone will autonomously navigate to the points. The mission waypoints will be the locations that the drone would need to collect data from and transmit these readings to a server where they can be analyzed. For the scope of our project, we tested whether mission waypoints can be sent to multiple drones at the same time and have the drones navigate to those waypoints without manual intervention. Because the sailboats were still being developed in parallel with the mesh communication, we built motor-powered boats in order to test the sending of waypoints and the autonomous navigation. These motorboats were made of plastic and wood and had two brushless motors mounted at the bottom that would allow the boat to go forward and turn left and right. Though not part of the final drone design, additional communication modules were added to the drone in order to get diagnostic information while also allowing the drone to be manually overridden in case the drone goes out of control. The tests were successful and we were able to send multiple waypoints to multiple points at the same time and the drones were able to navigate autonomously to these waypoints.

7 Energy Systems

The Drones are designed to be fully Energy Self-Sufficient. To enable autonomous navigation and mesh network capabilities, the sailboats must have a source of power. Approximately 50W of constant power is needed to power all the electronics on a single unit. The Beaglebone controller operates at 5 V with 2.5 A, the servo motor for the actuator operates at 6V with approximately 3A, and the rest of the sensors and XBee account for approximately 15 W. There is one main battery connected to all the electronic devices, including the controller. The battery chosen is a 12 V 500A H lithium polymer battery as it is lightweight, has long cycle life, and has high c-rates (fast charge and discharge) when compared to lead acid or lithium-ion. In the boat design, the battery is placed in the middle of the catamaran design along with the other electronic components. The sailboats are meant for long-term duration at sea and must be fully energy self-sufficient. Hydrogeneration and solar panels are used to obtain energy and charge the battery continuously in one integrated power system [32].

A low Kv motor (more efficient for power generation in lower shaft speeds) in the range of 300–500 is used and connected to a boost converter for the hydrogeneration system. The motor produces low voltages in the range of 2–4 V that are stepped up by the boost converter in order to be stored in the lithium polymer battery. Solar panels connected to an MPPT (Maximum Power Point Tracking) charge controller are connected to the same battery. The 24 V MPPT controller regulates the power generated by the solar panels as the sun's intensity varies throughout the day.

For the hydro-generator through a low Kv motor, the concept of tip speed ratio was used to calculate the nominal theoretical RPM. The tip speed ratio can be found by dividing the tip speed of the blade by the speed of the fluid. The tip speed ratio

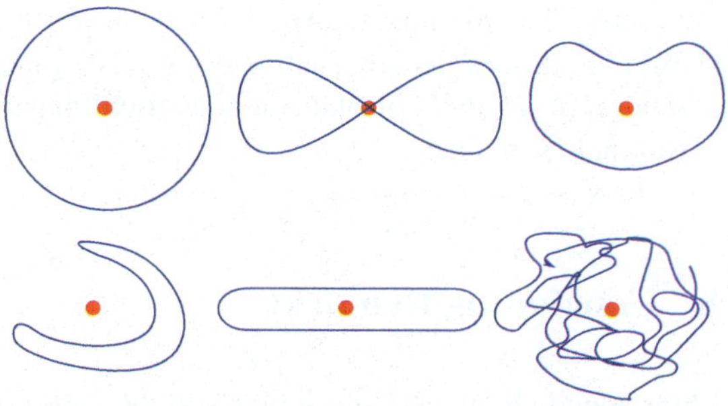

Fig. 5 Potential options for the path of the boat around the intended point of measurement, in order to maximize hydrodynamic energy generation. The optimum trajectory is a function of wind speed and direction, surface current, and waves

was found to be approximately 3. The nominal RPM was then calculated to be 700. An RPM of below 1000 is far too small for common brushed DC motors. Brushed motors are also not viable for the long term as maintenance is required to replace the brushes while the goal is for each drone to be able to last for over 20 years in the ocean with minimal maintenance. A brushless motor is best suited as it has an electronic controller and requires little maintenance, however, the drawback is the higher cost. A 300 Kv motor is used to reach approximately 2 to 3 V produced at 700 RPM, as the Kv times the voltage is equal to the RPM. This voltage will then be stepped up by a boost converter to reach 12 V to supply to the battery. The step-up of voltage will lower the current, however, we are focused on long-term smaller generation rather than short-term higher generation, as these drones are made for long-term duration.

When the unit (sailboat) is near the intended location, it can still undergo maneuvers in order to generate power via the hydro-generator. In other words, the unit can sail *around* the intended point as long as it stays within a given proximity of that point. The objective is to generate power and recharge onboard batteries. The optimum trajectory depends on the wind magnitude and direction, surface currents, waves, and their relative strength. The optimization may result in a simple curve (e.g. a circle around the intended location, as shown in figure below), or more complicated trajectories that may or may not encompass the target location, but the unit always stays within its permissible proximity (Fig. 5).

Optimization of power production must be done on the solar panels in order to gain maximum efficiency. A pair of 12 V 6 W solar panels were tested to study the effects of different shading from obstructions on the voltage and current outputs. It was found that the shading on the solar panel is directly proportional to the voltage; the more shade means the less voltage. It was also found that if the solar panels dropped below 11 V, there was no current being output. Additionally, a pair of 6V 6W solar panels were purchased that were placed in series to create 12 V 12 W to be able to power the 12 V Lithium Polymer battery and create a higher wattage. The effect of light intensity on the power output of the solar panels was also tested. It was found that an optimal angle of approximately 20 to 35 °C from the horizontal will provide the most current and voltage. There are a variety of methods by which solar panels can be integrated into the hull design [33]. A proposed design is to place the solar panels in a 27-degree pyramid shape within the optimal angle range on top of both hulls. The pyramid shape will account for the varying positions of the sun and stability in design. An alternative design is to place the solar panels on the rigid sail and neglect the angle of placement as it will have the most stability and take up the least amount of space.

8 Concluding Remarks

Ships spend about 20% more fuel to fight waves en route, equivalent to about $10B worth of fuel, and proportionally releasing pollutants and greenhouse gases to the atmosphere. The main reason for this large excess cost is that steering in waves is not very efficient due to the lack of knowledge (prediction) of waves the ship will encounter. To predict ocean waves, we need to first measure the current state of the ocean (at least in the vicinity of the vessel), then reconstruct the underlying flow field, and then perform fast prediction via efficient computational algorithms. Our idea is to obtain distributed measurements of waves via a swarm of active self-sufficient sensors that make a mesh-grid to transfer information to a central unit. For increased stability and control, a twin-sail catamaran hull is designed. The Ardupilot sailboat control system along with Xbee Digimesh is used for effective communication and navigation of the swarm. To monitor all these functions, the BeagleBone Blue, a special single-board computer is installed. The drones are made fully energy self-sufficient through solar panels and hydrogeneration. For the reconstruction and prediction using machine learning, the reader is referred to our paper [13].

Acknowledgements The authors would like to express their acknowledgment of the support received from the American Bureau of Shipping, and the National Science Foundation under the grant CPS-1932595.

References

1. A pathway to decarbonise the shipping sector by 2050. International Renewable Energy Agency (IRENA), Abu Dhabi (2021).
2. Wehausen, J. V. (1973). The wave resistance of ships. *Advances in Applied Mechanics, 13*, 93–245.
3. Onorato, M., Osborne, A. R., Serio, M., & Bertone, S. (2001). Freak waves in random oceanic sea states. *Physical Review Letters, 86*.
4. Wang, G., & Pan, Y. (2021). Phase-resolved ocean wave forecast with ensemble-based data assimilation. *Journal of Fluid Mechanics, 918*.
5. Köllisch, N., Behrendt, J., Klein, M., & Hoffmann, N. (2018). Nonlinear real time prediction of ocean surface waves. *Ocean Engineering, 157*, 387–400.
6. Blondel, E., Bonnefoy, F., & Ferrant, P. (2010). Deterministic non-linear wave prediction using probe data. *Ocean Engineering, 37*, 913–926.
7. Cousins, W., Onorato, M., Chabchoub, A., & Sapsis, T. P. (2019). Predicting ocean rogue waves from point measurements: An experimental study for unidirectional waves. *Physical Review E, 99*.

8. Qi, Y., Xiao, W., & Yue, D. K. P. (2016). Phase-resolved wave field simulation calibration of sea surface reconstruction using noncoherent marine radar. *Journal of Atmospheric and Oceanic Technology, 33*, 1135–1149.
9. Wu, J., Hao, X., & Shen, L. (2022). An improved adjoint-based ocean wave reconstruction and prediction method. *Flow, 2*, (2022)
10. Qi, Y., Wu, G., Liu, Y., Kim, M.-H., & Yue, D. K. P. (2018). Nonlinear phase-resolved reconstruction of irregular water waves. *Journal of Fluid Mechanics, 838*, 544–572.
11. Wu, G. (2004). Direct simulation and deterministic prediction of large-scale nonlinear ocean wave-field. Ph.D Thesis, Massachusetts Institute of Technology.
12. Qi, Y., Wu, G., Liu, Y., & Yue, D. K. P. (2018). Predictable zone for phase-resolved reconstruction and forecast of irregular waves. *Wave Motion, 77*, 195–213.
13. Mohaghegh F., Murthy, J., & Alam, M.-R. (). Rapid phase-resolved prediction of nonlinear dispersive waves using machine learning. *Applied Ocean Research, 117*, (2021)
14. Saetre, C., Tholo, H., & Hovdenes, J., et al. (2023). Directional wave measurements from navigational buoys. *Ocean Engineering, 268.*
15. Jayne, S. R., Roemmich, D., Zilberman, N., et al. (2017). The argo program: Present and future. *Oceanography, 30*, 18–28.
16. Houghton, I. A., Hegermiller, C., Teicheira, C., & Smit, P. B. (2022). Operational assimilation of spectral wave data from the sofar spotter network. *Geophysical Research Letters, 49.*
17. Belmont, M.R. (2023). Wave shadowing effects on phase resolved wave spectra estimated from RADAR backscatter: Part 1: An investigation of the errors due to wave shadowing in the spectral averaging approach. *Ocean Engineering, 267.*
18. Grønlie, Ø. (2004). Wave Radars – A comparison of different concepts and techniques. *Hydro International, 8.*
19. Li, X. M., & Huang, B. (2020) A global sea state dataset from spaceborne synthetic aperture radar wave mode data. *Science Data, 7.*
20. Ahmad, Z. (2006). *Principles of corrosion engineering and corrosion control.* Butterworth-Heinemann.
21. Munk, T., Kane, D., & Yebra, D. M. (2009). *Advances in marine antifouling coatings and technologies.* Woodhead Publishing.
22. Michel, W. H. (1961). The sea-going catamaran ship its features and its feasibility. *International Shipbuilding Progress, 8*, 390–401.
23. Ulysse, D., Niklas, R., & Jakob, K. (2019). *Energy efficient self-steering mechanism for an autonomous sailing vessel* (pp. 1–6). IEEE: OCEANS.
24. Tretow, C. (2017). *Design of a free-rotating wing sail for an autonomous sailboat.* Marina system: KTH.
25. Silva, M. F., Friebe, A., & Malheiro, B., et al. (2019). IGID wing sailboats: A state of the art survey. *Ocean Engineering, 187.*
26. An, Y., Yu, J., & Zhang, J. (2021). Autonomous sailboat design: A review from the performance perspective. *Ocean Engineering, 238.*
27. Angelou, M., & Spyrou, K. J. (2021). Course-keeping assessment of yachts' upwind sailing in waves. *Journal of Fluids and Structures, 101*, (2021)
28. Sauze, C., & Neal, M. J. (2008). *Design considerations for sailing robots performing long term autonomous oceanography* (pp. 21–29). Centrobot: IRSC.
29. dos Santos, D. H., & Garcia Goncalves, L. M. (2020). *Performance evaluation of propulsion control techniques for autonomous sailboat* (pp. 1–6). IEEE: LARS.
30. Akyildiz, I. F., Wang, X., & Wang, W. (2005). Wireless mesh networks: a survey. *Computer Networks, 47*, 445–487.
31. Sharma, A., Vanjani, P., & Paliwal, N., et al. (2020). Communication and networking technologies for UAVs: A survey. *Journal of Network and Computer Applications, 168.*
32. Zoolfakar, M. R., Shukur, M. S. S. A., & Suhainai, N. H. (2022). A portable hydro generator for fishing boats. Advanced Materials and Engineering Technologies. *Advanced Structured Materials, 162*, 93–105.
33. Friebe, A., Olsson, M., Gallic, M., Springett, J., Dahl, K., & Waller, M. (2017). *A marine research ASV utilizing wind and solar power* (pp. 1–7). IEEE: OCEANS.

Integrated System Approach for Peer-Peer Energy Trading

Hector K. Lopez and Ali Zilouchian

Abstract Peer-to-peer (P2P) energy trading platforms have gained significant attention in recent years, attracting engineers, utilities, and prosumers. Advancements in associated technologies have further accelerated the shift towards dynamic, consumer-centric, distributed electricity grid management. In this chapter, we present three integrated system approaches for P2P trading: A market solution based on blockchain and smart metering infrastructure, a trading mechanism based on the NRG-X-Change algorithm, and a fuzzy logic paradigm to enhance the robustness, simplicity, and flexibility of prosumer switching between P2P and Grid connections. We discuss the advantages and disadvantages of these approaches and provide simulation results showcasing their potential to maximize trading profits and optimize grid electricity usage. This study offers valuable insights into prosumers' decision-making processes and contributes to the knowledge base of P2P energy trading.

Keywords Peer-to-peer · Solar energy · Photo-Voltaic · Power demand · Micro-grid · Smart metering · Fuzzy systems · Decision making · Cost–benefit analysis · Energy management

1 Introduction

The Paris Agreement of 2015 started a global effort toward clean energy infrastructure [1]. Photo-voltaic (PV) systems are a pivotal innovation that has grown in popularity and affordability amongst residential power consumers [2, 3]. The reasons for adoption are government incentives and compensation for excess energy generation. But,

H. K. Lopez · A. Zilouchian
Department of Electrical Engineering and Computer Science, Florida Atlantic University, Boca Raton, FL 33431, USA
e-mail: hlopez5@fau.edu

A. Zilouchian (✉)
Director of Applied Engineering Research Center and Program Director at Keiser University, Fort Lauderdale, USA
e-mail: zilouchi@fau.edu

M.-R. Alam and M. Fathi (eds.), *Integrated Systems: Data Driven Engineering*,
https://doi.org/10.1007/978-3-031-53652-6_19

in 2023 the incentive mechanisms required reform due to claims of subsidization of wealthy homeowners [4]. In the United States excess generation is compensated at the retail electricity price. Utilities, however, want to use the wholesale cost because it doesn't include transmission costs [5]. Instead of raising rates for electricity prices or removing a profitable incentive for renewable energy researchers have proposed alternative schemes to the Net Metering (NEM) and Feed-in-Tarif (FiT) incentive programs [6, 7].

An energy infrastructure called Peer-to-Peer (P2P) energy trading is poised to disrupt current energy markets [6]. P2P promises that a central authority for financial transactions is not needed. Blockchain-based technology uses distributed ledgers that can confirm the validity of each transaction. Unlocking the possibility for a secure marketplace that can settle highspeed transactions [8]. These systems can remove significant administrative overhead from traditional energy trading markets but are complex to build in practice. They usually need a micro-grid, P2P metering, and a Distribution Service Operator (DSO) [6, 9]. Residents with a distributed energy resource (DER), also known as, prosumers would now have a natural incentive to generate excess energy, take ownership of the energy transition, benefit from new technologies to reduce their bills, and participate actively in the market [10, 11]. Additionally, consumer preferences would reflect broader trends favoring locally sourced and ethically produced products [12].

The success of P2P energy hinges on individuals' willingness to participate according to Kirchhof et al. [13]. Sousa et al. [14] consider that the lack of user engagement is one of the main threats to the concept of P2P. Likewise, Mengelkamp et al. [15] argue that the "socio-economic perspective of microgrid energy markets is mostly neglected in current research". An optimal fuzzy energy trading system by Shazad et al. considers user preferences [16]. Fuzzy Inference Systems (FIS) allow users to contribute their preferences and expertise, enhancing engagement and improving the overall efficiency of the P2P energy market [17]. Lopez et al. implemented a FIS control system that included user preferences to arbitrage between a simulated P2P market (NRG-X-Change) [18] and the grid (NEM) [19]. The results of the simulation showed greater profits by switching between both sources. More importantly, the simulation provided clarity into the robustness of the FIS and sensitivity of the user-defined inputs [18].

2 Related Works

The related works around the intersection of P2P energy markets and FIS may have begun by Liu et al. in 2015 [20] and Rubio et al. in 2016 [21]. They highlighted the need for scalability and personalization in decision-making for prosumers. While Alam et al. 2017 [22] and Shahzad et al. 2021 [16] proposed methods with better scalability by adding additional trading factors, but the research lacked practical simulations of prosumer impact and profitability. K.G.H. Kong et al. 2022 [23] used fuzzy optimization for multi-objective optimization in P2P systems, allowing for trade-offs between conflicting objectives, but focused on climate policies over personal trading preferences. Lopez et al. 2022 [17], presented a novel approach by breaking down the FIS control system for prosumers into three separate systems: one for producers, one for consumers, and one for combined prosumers. This simplified the rules and made it easier for prosumers to understand how to set membership functions for personalized trading factors.

3 Organization and Contributions

This chapter is structured into three major sections to provide an understanding of Peer-to-Peer (P2P) energy markets and their ability to be used as an alternative to changing net metering incentives for prosumers. P2P is a possible alternative if it has the correct infrastructure to support it. Energy trading is made possible due to a distribution infrastructure amongst prosumers on the P2P network. Additionally, the use of a trading algorithm is popular in P2P networks because it allows for real-time transactional payments instead of the bidding and settlements in traditional energy trading markets. Blockchain technology is a popular solution that allows secure transactions on a distributed ledger. P2P markets are usually built using blockchain, resulting in a fully decentralized system. This chapter will not dive into the application of a blockchain-based P2P market or the infrastructure requirements of the prosumer network. The contributions will focus on comparing existing payment schemes given real-time payment algorithms amongst prosumers in a P2P simulation. Additionally, the contributions will focus on extending the comparison to include additional trading factors such as a prosumer's preference to participate in either payment scheme.

In the first section, we provide an overview of net metering (NEM) as a mechanism that prosumers depend on in today's grid. Secondly, a review of P2P markets is presented to provide clarity on infrastructure requirements along with an example of a payment mechanism called NRG-X-Change. Thirdly, we present a simulation that attempts to use a fuzzy logic controller to switch between profits from a grid connection and a P2P market connection. Finally, the last section concludes the chapter by summarizing the main findings and insights derived from the simulations.

4 Net Metering Overview

NEM has emerged as a global policy tool to promote renewable energy adoption, by providing a financial incentive for excess energy generated by prosumers [24]. Smart metering devices capture the bi-directional flow of energy from prosumers. The NEM policies often focus on payment based on the kilo-watt-hours that flow back into the bulk electric grid. The history of NEM in the United States demonstrates its significance in promoting the adoption of Distributed Energy Resources (DERs) such as residential PV systems. Since its inception in the early 1980s, NEM has gained widespread acceptance, with 41 states implementing mandatory net metering policies by 2022 [25]. Utilities have lobbied to reform the NEM payment for excess energy from retail pricing to the much lower rate of wholesale pricing, also known as full-avoided cost. By lowering the payment to prosumers, the utilities can cover the costs of maintaining and operating the distribution of electricity to the prosumers [26].

The successful transition to NEM 3.0 in California went into effect on April 14, 2023, resulting in a 75% reduction of incentives for PG&E, SCE, and SDG&E customers [27]. Different states have begun to propose incentives at varying levels of success [28]. In Florida, similar legislation for NEM 3.0 was vetoed by the governor, due to concerns of adverse economic impact after the pandemic [29]. The trends point to continued reform for NEM that will either push new kinds of incentives or decrease the adoption of renewables by prosumers.

In the context of Florida, NEM has played a significant role in the state's renewable energy landscape. Florida ranks fifth in the nation in terms of total solar power generating capacity, with small-scale solar installations contributing significantly to renewable-sourced net generation [30, 31]. Major utilities in Florida, such as Florida Power and Light (FPL), Tampa Electric Company (TECO), Gulf Power, and Duke Energy, have adopted NEM policies that compensate prosumers at retail rates. However, limitations are imposed on Distributed Generation (DG) systems to prevent them from exceeding 115% of the home's annual consumption [32]. These limitations are further defined by generation tiers based on system size. For instance, Tier 1 includes systems with a capacity of 10 kW or less, Tier 2 covers systems above 10 kW up to 100 kW, and Tier 3 encompasses systems above 100 kW up to 2,000 kW [31]. Despite the popularity of NEM in Florida, ensuring fairness and a sustainable energy transition remains a key consideration for the state's policymakers and utilities.

5 Peer-to-Peer Markets

When energy is generated into electricity it is fed into a transmission infrastructure so that it can be consumed by a load instantaneously. The current wholesale electricity market is made up of future scheduled transactions of electricity so that physical constraints of moving the energy are considered, as well as emergency responses [33]. The wholesale electricity market is based on a centralized structure where participants are limited and have a minimum capacity barrier to participate in trading. Independent System Operators (ISOs) collect bids from large-scale power generation participants and then depending on the market-clearance price, dispatch the electricity [34]. The central model is shifting toward due to the integration of Distributed Generation (DG) [35]. Smart metering infrastructure has enabled the bi-directional flow of energy from consumers who have adopted DERs [36]. Due to the small scale and volatility of prosumer-level generation payment for excess energy is provided as a Power Purchase Agreement (PPA), or in NEM. In both cases, the method allows for the reduced

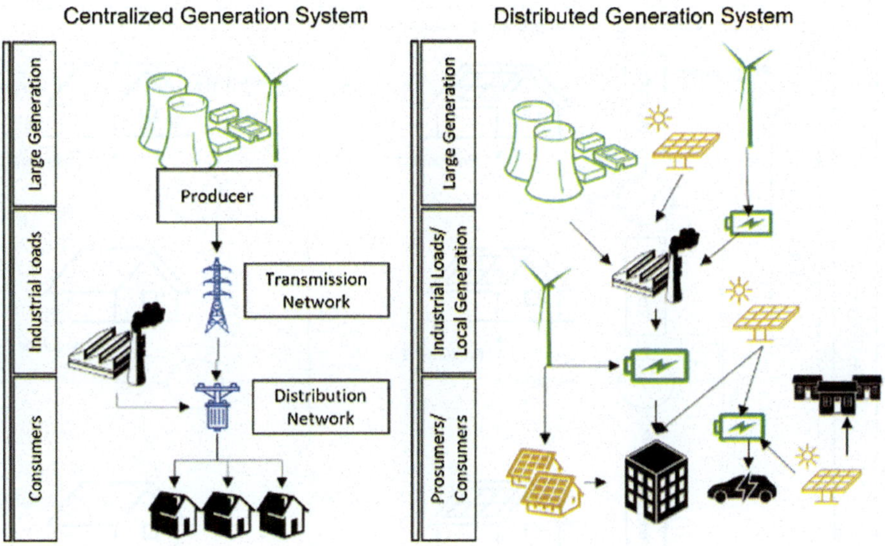

Fig. 1 The centralized generation system versus distributed generation system highlighting the bi-directional flow of electricity between producers and consumers

risk of price fluctuations [34]. The centralized and distributed generation systems shown in Fig. 1 do not show the management systems, communication layers, and maintenance programs that add additional costs and complexity to managing either infrastructure. Costs cannot be ignored when trying to decide between these systems since sustainability, safety, and reliability are all constrained by the economics of implementing either solution.

5.1 Peer-to-Peer Market Infrastructure

To set up a micro-grid P2P market the following is required: a sufficient number of prosumers connected to a micro-grid, the ability to island or synchronize with the traditional grid, an information system that supports the market and pricing mechanisms, and finally an energy management trading system that enforces regulation [34, 37] (Fig. 2).

Blockchain technology like Ethereum is a possible solution for P2P trading because of its capability to resist cyber-security attacks [34, 38]. Whole-sale market transactions run on 15-min cycles, allowing blockchain block transactions to operate with enough time to resolve consensus algorithms. Finally, smart contracts built on blockchain can include rich information that can facilitate energy management, pricing mechanisms, and regulation [39, 40].

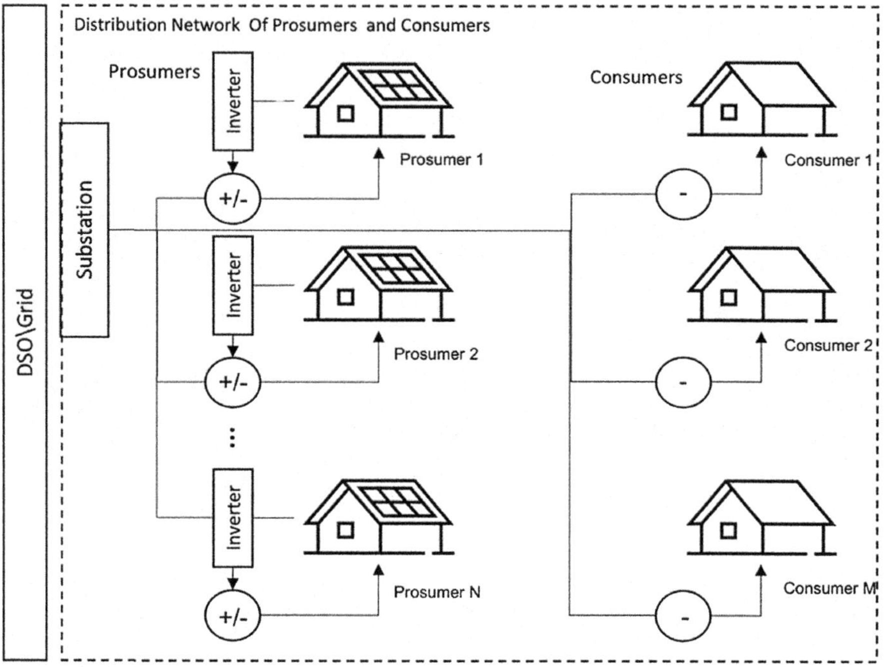

Fig. 2 Distribution network of prosumers with PV installed. The network facilitates an energy trading market that is local and can be an alternative to the connected grid connection (P = Prosumers, C = Consumers, M = Meter, DSO = Distribution Service Operator)

Using blockchain for P2P also has challenges. The transaction speeds are limited up to 30 transactions per second, using a popular platform like Ethereum. Limiting the ability to scale past a certain number of participants. The current regulations do not allow P2P trading to interact with the wholesale market. Despite the challenges, a micro-grid with a blockchain-based P2P market can be leveraged to perform real-time demand response, optimal power flow, ancillary service compensation, energy storage, and even electric vehicle integrations [34].

5.2 Peer-to-Peer Market Pricing Mechanism

Various P2P market pricing mechanisms exist such as uniform, discriminatory, constrained optimized, continuous double auction, formula pricing, and negotiation [40, 41]. The NRG-X-Change algorithm by Mihaylov et al. is a specific formula-based pricing mechanism that is highlighted here due to its simplicity and effectiveness. The simplicity stems from removing the number of operations required to settle the payment of a prosumer. Instead of using an auction or a bidding process between multiple buyers and sellers, the algorithm applies a dynamic value to a prosumer's real-time contribution based on the calculation of the overall demand on the network. Reducing time to settle the value and simplifying the process. Additionally, the system is effective because it rewards prosumers who change their behavior to balance the demand and the generation on the network, adding to the stability of the grid. The NRG-X-Change algorithm relies on smart metering infrastructure for real-time monitoring of energy generation and consumption. It determines a fair price for the energy being traded based on the total local demand and capacity among participants [19, 42]. The variable x, is calculated as the difference between a specific prosumer's load $P_{n_{\text{load}}}(\tau)$ and generation $P_{n_{\text{gen}}}(\tau)$, at some time τ. The prosumer's payment $P_{n_{\text{pay}}}(\tau)$ is determined by multiplying the net energy with a time-dependent retail price, denoted as $q(\tau)$.

$$P_{n_{\text{pay}}}(\tau) = x(\tau) \cdot q(\tau)$$

NRG-X-Change algorithm employs a payment function that considers both individual contributions and the overall state of the network to balance supply and demand within the network. The maximum retail price, $q_{\max}$, is based on the prosumer's net energy x and the network's total generation t_p and total consumption t_c, the payment function can be expressed as:

$$g\left(x, t_p, t_c\right) = \frac{x \cdot q_{\max}}{e^{\left(t_p - t_c\right)^2/a}}$$

The constant a is used to adjust the payment distribution. The variable x represents the net energy produced by the prosumer, t_p signifies the total generation of the network, and t_c denotes the total consumption of the network. In Fig. 1 a curve shows the value of the payment function $g(.)$ on the y-axis and the value of the total generation on the x-axis to demonstrate the relationship between payment price and alignment with overall generation. The curve incentivizes prosumers to generate more when demand is high and less when demand is low, attempting to balance the grid through economic incentives (Fig. 3).

To calculate a cost function that is used when consumers on the network need to purchase electricity the following can be used:

$$h\left(y, t_p, t_c\right) = \frac{y \cdot r_{\max} \cdot t_c}{t_c + t_p}$$

In this equation, y represents the energy consumed by consumers, t_p signifies the total generation of the network, t_c denotes the total consumption of the network, and $r_{\max}$ represents the maximum cost of energy delivered by the Distribution Service Operator (DSO) when prosumers' energy supply is low.

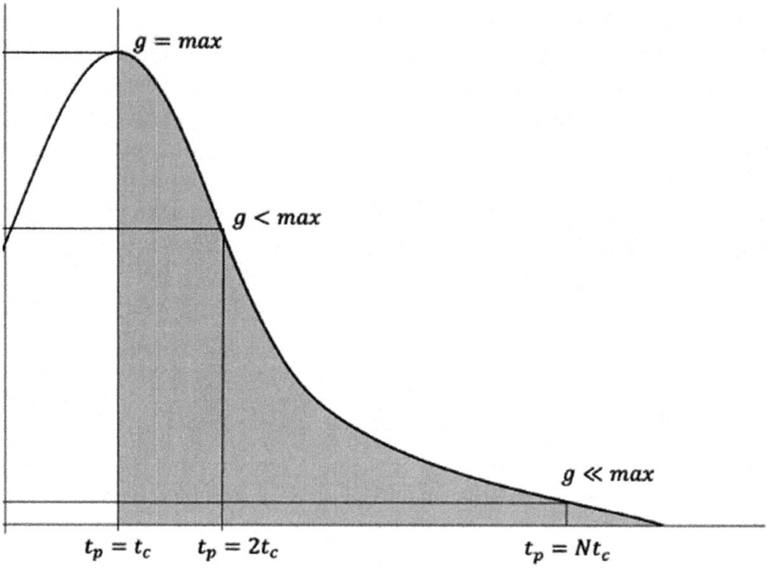

Fig. 3 Payment function for NRG-X-Change, showing the limits of payment function, $g(.)$, as total production, t_p, reaches total consumption, t_c, times number of prosumers N

5.3 *Performance of a Prosumer Market*

A three-prosumer P2P market can demonstrate the effectiveness of the NRG-X-Change payment mechanism. In Lopez et al. a simulation was created that represented a 3-prosumer network with real-world conditions [17]. A reference dataset from the government resource EIA.gov provides the monthly real-time price, residential prosumer load, and PV solar generation over the same 14-month period, in the same region of Florida. With this dataset and an assumed statistical variance from typical home size, one can instantiate many possible prosumer consumption and generation profiles. The size of the home would determine how much generation a typical Florida Tier-1 PV system would generate. The result is N number of prosumers that can be included in a simulated P2P network and paid based on the NRG-X-Change algorithm. For a small network of, $N = 3$. The results of the simulation show that prosumers are paid less when the total consumption is low (Prosumer #1, and #2), also the cost for consuming is high when the production is low (Prosumer #3). At $N = 100$ the simulation resulted in prosumers having a greater than usual consumption resulting in negative revenue throughout the simulation. This was because the variation between the prosumers was not large enough to make a difference (Table 1 and Fig. 4).

Table 1 Total results by prosumer

#	Total demand	Total generation	Total consumption	Total net energy	Revenue ($)
1	11,578.96	13,201.34	1184.77	−2807.14	−11.455765
2	12,930.85	14,372.98	1279.87	−2722.00	−12.385499
3	13,269.68	13,152.68	1491.75	−1374.75	−79.031829

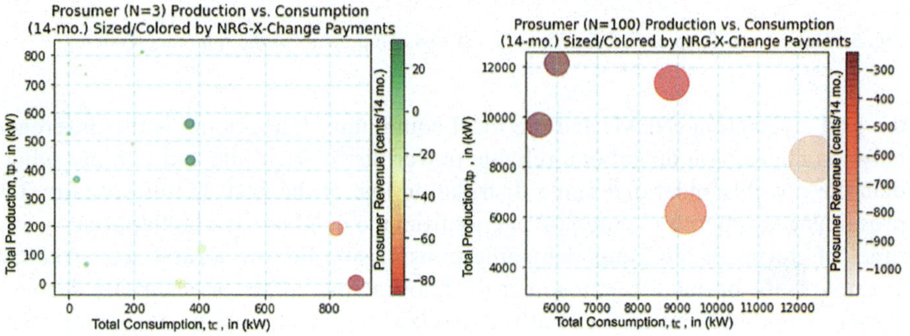

Fig. 4 Production versus consumption in a 3 prosumer P2P market (left) and 100 prosumers (right) with NRG-X-change payment mechanism over 14-mo. simulation

6 A FIS to Optimize Prosumer Profits and Preferences

While economic requirements are often emphasized in P2P market literature, the significance of personalization for prosumers is frequently overlooked. However, personal preferences are crucial for creating a user-centric energy system. Shazad et al. proposed an FIS solution to incorporate personal trading factors into P2P markets [16]. Fuzzy logic systems were pioneered by Mamdani to improve the robustness of traditional control systems [Mamdani]. An expert in the system provides intuitive inputs, rules, membership functions, and thresholds to create a controlled output response [43]. The fuzziness of the inputs makes it easier for an expert to encode many possible inputs that result in a desired output. In general, the crisp inputs are converted to fuzzified inputs and passed through a rule-base to find the fuzzy conclusions. The overall inference system uses the experience and heuristic information from the expert that created the rule base to achieve decision-making and form a fuzzy output, finally, the output is then defuzied to create a crisp output [44] (Fig. 5).

A FIS is successful in reducing the complexity of a control system and including human ambiguity in personal preferences. FIS enables a more tailored and individualized trading experience that can include ecological and situational trading factors as well as profit optimization [16, 23].

An example of using a FIS is when including preferences that are unique to a prosumer. Consider a residence that is in a rural community and choosing between

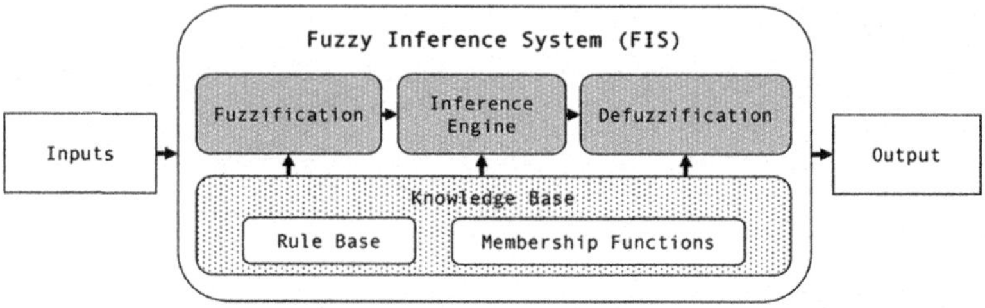

Fig. 5 The general form of FIS showing the various components of the inference system

the grid connection (NEM) and the local community P2P pricing is the difference between choosing a neighbor's reliable power distribution line that is more robust than the possibly older and aging distribution line of the grid. In this scenario, the prosumer may choose to purchase energy from the neighbor at a higher price during times of storms or blackouts damaging transmission lines in a rural area. Another example is the option for a prosumer to choose to pay more for electricity during a certain period because they can afford and desire to contribute to supporting clean energy providers. In this case, the prosumer will select to purchase power from neighbors rather than the grid so that they can propagate the adoption of solar PV. Finally, another example of prosumer choice can be the option to select the cheapest price for consumed energy and maximize profit while away on vacation but then choose to maximize reliable grid connection when at home. These examples show the possibility for prosumers to accommodate a lifestyle into the decision-making of an energy trading platform.

7 P2P Market or Grid NEM Selection Using FIS Controller

In this section, a method to use FIS is presented to help prosumers choose between P2P or a grid connection. The method involves breaking up the knowledge base of the prosumer FIS control system into three separate FIS systems: a producer FIS control system, a consumer FIS control system, and a combined prosumer FIS control system. By dividing the knowledge base in this way, we aim to simplify the rules and make it easier for prosumers to understand and use the FIS control system to include preferential trading factors. To model a prosumer, we use an aggregate dataset from EIA.gov to determine the prosumer's load and generation curve. Then a statistical variance is added to the generation and load to create unique prosumer data. In the simulation, the EIA.gov data set is collected for 14 months between 10/1/2019 and 7/1/2020. The sales and purchase price for P2P electricity is assumed to have more volatility and is calculated as a normal random variable with a mean at 75% of the retail price of electricity sold, and a normal distribution of 0.02 \$/kWh of the retail price of electricity sold (Fig. 6).

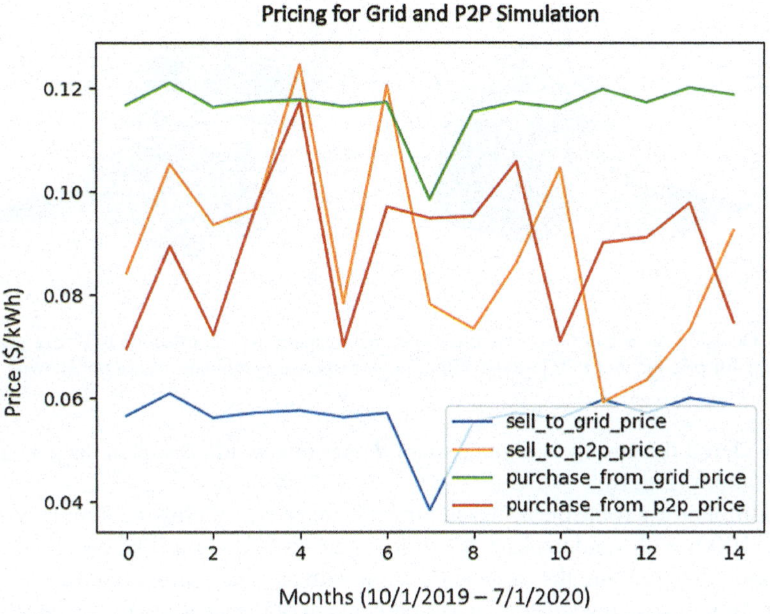

Fig. 6 Pricing for the grid and P2P purchase and sales price for electricity

The advantages of a FIS are the low computational requirement and the flexibility for changes in the control scheme based on unique user preferences. Fuzziness occurs in the uncertainty of linguistic ideas without clear boundaries. Most of the factors linked with energy trading can be optimally expressed using fuzzy logic theory. Moreover, fuzzy logic is beneficial because there is no necessity for mathematical modeling compared to more complex solutions to price and preference optimization [45].

The personal preference inputs chosen are arbitrary, but the pricing inputs are taken from the same P2P market dataset used in Lopez et al. [17] and feed into a rule-base that has been crafted to favor P2P market participation when profit is 'medium' and below. The inputs are the pricing for buying and selling to either the Grid or P2P market as well as the two personal preferences called 'reliability preference' and 'p2p interest'. The output is a switch that selects either the P2P distribution line or the Grid distribution line. The following decision surfaces demonstrate the response based on the inputs (Fig. 7).

A simulation is created that runs through the demand and generation profiles of a prosumer from the P2P network presented in Lopez et al. [17] that leverages the NRG-X-Change payment mechanism. The month-to-month revenue of one prosumer is recorded. The following figure shows the prosumer fully connected to either the P2P distribution line or the Grid distribution line, and finally a scenario where the FIS controller switched from either source based on the inputs. A sensitivity analysis of the inputs indicates the most important input to the FIS controller. The P2P pricing

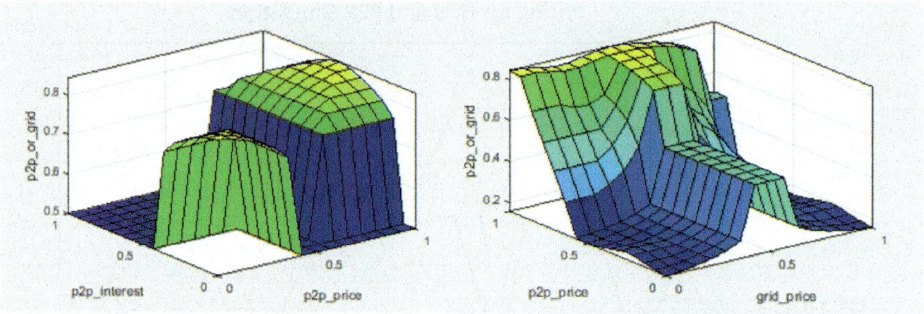

Fig. 7 Decision surface of fuzzy inference system to respond with connection to P2P or grid based on inputs for pricing, and P2P interest. Other input response combinations are not shown

is considered the strongest signal, small deviations would result in large responses in the output (Fig. 8).

Additionally, a robustness measurement is taken by running the simulation many times (1000) with random user preference curves and compiling the profit of the prosumer, then running the same simulation with reduced rules in the rule base. The distribution of total revenue drops as a percentage of the rules are randomly removed from the rule base. Below 75% of the rules in the original rule base of 30 rules resulted in a breakdown of the FIS to provide prosumer profit (Fig. 9).

The simulation shows that the total profit for a prosumer who is only connected to the grid is $3.36, while a prosumer who is only connected to P2P has a total loss of $63.81. However, using the FIS controller to switch between P2P and grid

Fig. 8 First order sensitivity analysis of the FIS controller inputs

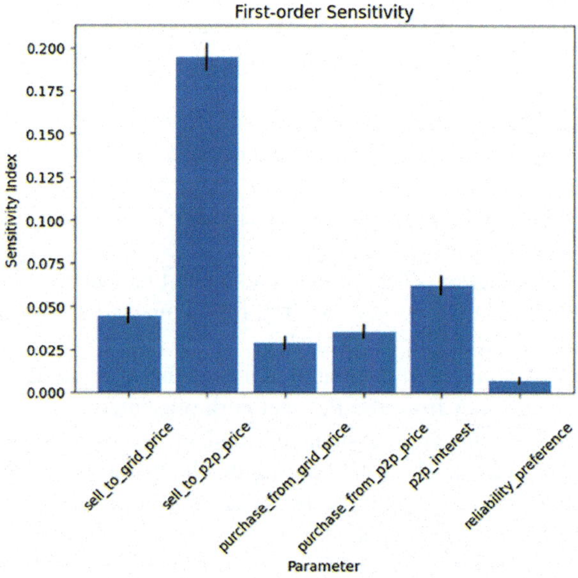

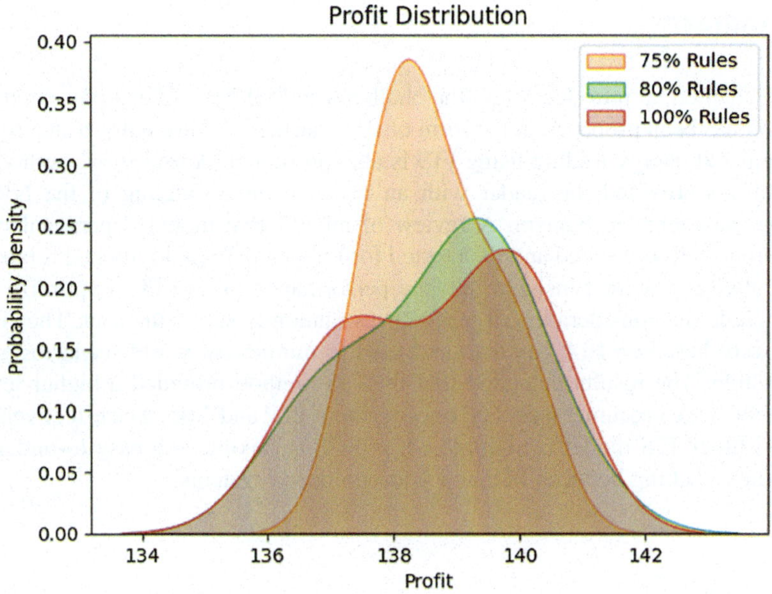

Fig. 9 A probability density versus profit plot of the simulation for the FIS output

connections results in a total profit of $21.97, indicating that the controller can optimize the choice for the prosumer. The FIS controller is shown to be a more profitable choice than being exclusively connected to either the grid or the P2P network (Fig. 10).

Fig. 10 Prosumer fuzzy inference systems (FIS) with a model that selects the FIS based on inputs and current month as grouped by seasonal FIS to improve the accuracy of rules

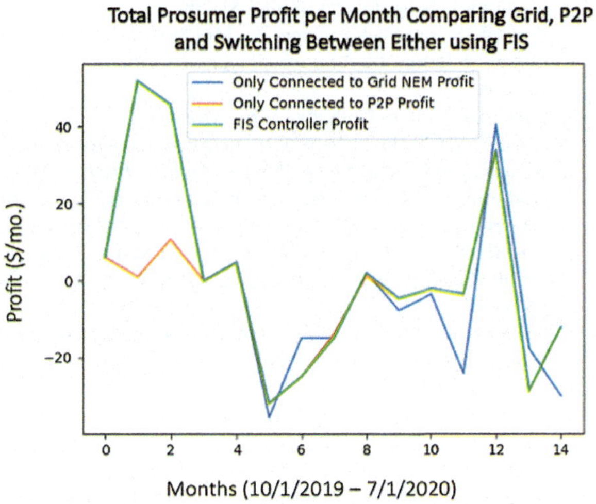

8 Summary

In this chapter, an introduction to the challenges of NEM has been presented and a solution has been proposed in the form of P2P markets. Additionally, a step towards personalized energy trading using FIS is also presented. A review of methods and simulations provided the reader with an intuitive understanding of the NRG-X-Change payment mechanism. A review of an FIS that included personal preference inputs was presented and then tested for First-Order sensitivity and robustness. A simulation provided insights into the performance of an FIS to optimize profit and include the consideration of preferences randomly set by the user. The simulations were based on EIA.gov data and tried to mimic real-world scenarios as best as possible. The results indicated that the FIS method provided a higher level of profit when the prosumer switched between both P2P and Grid. The effectiveness of a multi-block FIS model is highlighted, providing flexible and easy-to-understand rules for switching between P2P and grid-connected systems.

References

1. U. E. I. Administration. (2008). eia.gov. Retrieved July 17, 2023, from www.eia.gov.
2. Christian, B., Christian, B., Meiss, J., Goldschmidh, J. C., & Riede, M. (2013). A top-down analysis: determining photovoltaics R&D investments from patent analysis and R&D headcount. *Energy Policy, 62*(0301–4215), 1570–1580.
3. Uribe-Perez, N., Hernandez-Callejo, L., Vega, D., & Angula, I. (2016) State of the art and trends review of smart metering in electricity grids. *Applied Sciences, 6*(10.3390), 68–92.
4. Enrique, J., Durán, E., Sidrach-de-Cardona, M., & Andújar, J. (2007). Theoretical assessment of the maximum power point tracking efficiency of photovoltaic facilities with different converter topologies. *Solar Energy, 81*(1), 31–38.
5. Pociask, S., Pusok, K., Sigaud, L., McPherson-Smith, O. (2019) The Unintended Consequences of Net Metering. In: *A nonprofit educational and research organization.* .American Consumer Institute
6. Tushar, W., Yuen, C., Saha, T. K., Morstyn, T., Chapman, A. C., Alam, M. J. E., Hanif, S., & Poor, H. V. (2021). Peer-to-peer energy systems for connected communities: A review of recent advances and emerging challenges. *Applied Energy, 282*(0306–2619), p. Part A (2021).
7. Cummings, J. (2009). SIRE: Database of state incentives for renewables and efficiency (Website Review). *Reference Reviews, 23,* 44–45. https://doi.org/10.1108/09504120910969131
8. Mishra, S., Crasta, C. J., Bordin, C., & Mateo-Forné, J. (2022). Smart contract formation enabling energy-as-a-service in a virtual power plant. *International Journal of Energy Research, 46*(3), 3272–3294.
9. Hashemipour, N., Crespo del Granado, P., & Aghaei, J. (2021). Dynamic allocation of peer-to-peer clusters in virtual local electricity markets: a marketplace for EV flexibility. *Energy, 236*(121428).
10. Klein, L., Matos, L., & Allegretti, G. (2020).A pragmatic approach towards end-user engagement in the context of peer-to-peer energy sharing. *Energy, 205*(118002).
11. Ableitner, L., Tiefenbeck, V., Meeuw, A., Wörner, A., Fleisch, E., & Wortmann, F. (2020). User behavior in a real-world peer-to-peer electricity market. *Applied Energy, 270,* 115061. https://doi.org/10.1016/j.apenergy.2020.115061

12. Hu, W., Batte, M. T., Woods, T., & Ernst, S. (2011). Consumer preferences for local production and other value-added label claims for a processed food product. *European Review of Agricultural Economics, 39*(3), 489–510. https://doi.org/10.1093/erae/jbr039
13. Kirchhoff, H., & Strunz, K. (2019). Key drivers for successful development of peer-to-peer microgrids for swarm electrification. *Applied Energy, 244*, 46–62. https://doi.org/10.1016/j.apenergy.2019.03.016
14. Sousa, T., Soares, T., Pinson, P., Moret, F., Baroche, T., & Sorin, E. (2019). Peer-to-peer and community-based markets: A comprehensive review. *Renewable and Sustainable Energy Reviews, 104*, 367–378. https://doi.org/10.1016/j.rser.2019.01.036
15. Mengelkamp, E., Gärttner, J., Rock, K., Kessler, S., Orsini, L., & Weinhardt, C. (2018). Designing microgrid energy markets. *Applied Energy, 210*, 870–880. https://doi.org/10.1016/j.apenergy.2017.06.054
16. Shahzad, K., Iqbal, S., & Mukhtar, H. (2021). Optimal fuzzy energy trading system in a fog-enabled smart grid. *Energies, 14*(4), 881. https://doi.org/10.3390/en14040881
17. Lopez, H. K., & Zilouchian, A. (2022). "Peer-to-peer energy trading for photo-voltaic prosumers. *Energy,* 125563. https://doi.org/10.1016/j.energy.2022.125563.
18. Mihail, M., Sergio, J., Kristof, V. M., Narcís, A., & Ann, N. (2014). NRG-X-change a novel mechanism for the trading of renewable energy in smart grids. In: *Proceedings of the 3rd International Conference on Smart Grids and Green IT Systems.*
19. Mihaylov, M., Jurado, S., Avellana, N., Moffaert, K. V., Abril, I. M. d., & Nowé, A. (2014). NRGcoin: Virtual currency for the trading of renewable energy in smart grids. In: *11th International Conference on the European Energy Market (EEM14)* (pp. 1–6).
20. Liu, N., Yu, X., Fan, W., Hu, C., Rui, T., Chen, Q., & Zhang, J. (2018). Online energy sharing for nano grid clusters: A Lyapunov optimization approach. *IEEE Transactions on Smart Grid, 9*(5), 46244636.
21. Rubio-Solis, A., Panoutsos, G., & Thornton, S. (2016). A data-driven fuzzy modelling framework for the classification of imbalanced data. In: *Proceeding of 2016 IEEE 8th International Conference on Intelligent Systems* (pp. 302–307).
22. Muhammad Raisul, A., St-Hilaire, M., & Kunz, T. (2017). An optimal P2P energy trading model for smart homes in the smart grid. *Energy Efficiency, 10*(6), 1475–1493.
23. Kong, K. G. H., et al. (2022). Fuzzy optimization for peer-to-peer (P2P) multi-period renewable energy trading planning. *Journal of Cleaner Production, 368*, 133122. https://doi.org/10.1016/j.jclepro.2022.133122
24. Smith, K. M., Koski, C., & Siddiki, S. (2021). Regulating net metering in the United States: A landscape overview of states' net metering policies and outcomes. *The Electricity Journal, 34*(2), 106901. https://doi.org/10.1016/j.tej.2020.106901
25. Haas, R., et al. (2008). *Promoting electricity from renewable energy sources—Lessons learned from the EU, United States, and Japan* (pp. 419–468) . https://doi.org/10.1016/b978-008047172-3.50016-7.
26. Revesz, R. L., & Unel, B. (2016). Managing the future of the electricity grid: Distributed generation and net metering. *SSRN Electronic Journal.* https://doi.org/10.2139/ssrn.2734911
27. Thoubboron, K. (2023). NEM 3.0 in California: What you need to know, EnergySage. EnergySage Blog. https://news.energysage.com/net-metering-3-0/.
28. Sensiba, J. (2022). *Solar net metering is under threat all over the US.* CleanTechnica. Retrieved July 17, 2023, from https://cleantechnica.com/2022/08/03/solar-tariffs-under-threat-all-over-the-us/.
29. House Bill 741. (2022). The Florida Senate. Flsenate.gov, 2022. https://www.flsenate.gov/Session/Bill/2022/741.
30. U. E. I. Administration. Florida State Energy Profile. US Energy Information Administration. Retrieved July 17, 2023, from https://www.eia.gov/state/print.php?sid=FL#34.
31. EnergySage. (2009). How does Florida Power and Light net metering work?" EnergySage, Inc. Retrieved July 17, 2023, from https://www.energysage.com/local-data/net-metering/fpl/.
32. F. P. S. Commission. (2023). Find Your Utility. Florida Public Service Commission, 2021. Retrieved July 17, 2023, from http://www.psc.state.fl.us/ConsumerAssistance/FindYourUtility.

33. Silva, P. D., & Karnouskos, S. (2013). The impact of smart grid prosumer grouping on forecasting accuracy and its benefits for local electricity market trading. *IEEE Transactions on Smart Grid, 5*(402e10).
34. Thukral, M. K. (2021). Emergence of blockchain-technology application in peer-to-peer electrical-energy trading: A review. *Clean Energy, 5*(1), 104–123. https://doi.org/10.1093/ce/zkaa033
35. Soto, E., Bosman, L., Wollega, E., & Leon-Salas, W. (2020). Peer-to-peer energy trading: A review of the literature. *Applied Energy,* 116268.
36. Ekanayake, J. B., Jenkins, N., Liyanage, K., Wu, J., & Yokoyama, A. (2012). *Smart grid: Technology and applications*. John Wiley & Sons.
37. Parag, Y., & Sovacool, B. (2016). Electricity Market design for the prosumer era. *Nature Energy, 1*. https://doi.org/10.1038/nenergy.2016.32.
38. Doan, H. T., Cho, J., & Kim, D. (2021). Peer-to-peer energy trading in smart grid through blockchain: A double auction-based game theoretic approach. *IEEE Access, 9*, 49206–49218. https://doi.org/10.1109/access.2021.3068730
39. Xiong, X., Qing, G., & Li, H. (2022). Blockchain-based P2P power trading mechanism for PV prosumer. *Energy Reports, 8*, 300–310. https://doi.org/10.1016/j.egyr.2022.02.130
40. Kim, H. J., Chung, Y. S., Kim, S. J., Kim, H. T., Jin, Y. G., & Yoon, Y. T. (2023). Pricing mechanisms for peer-to-peer energy trading: Towards an integrated understanding of energy and network service pricing mechanisms, *183*, 113435–113435 (2023). https://doi.org/10.1016/j.rser.2023.113435.
41. Muhsen, H., Allahham, A., Al-Halhouli, A., Al-Mahmodi, M., Alkhraibat, A., & Hamdan, M. (2022). Business Model of Peer-to-Peer Energy Trading: A Review of Literature. *Sustainability, 14*(3), 1616.
42. Mihaylov, M., Jurado, S., Avellana, N., Razo-Zapata, I., Van Moffaert, K., Arco García, L., Bezunartea, M., Grau, I., Canadas, A., & Nowe, A. (2015). SCANERGY: A scalable and modular system for energy trading between prosumers. In: *2015 International Conference on Autonomous Agents and Multiagent Systems*.
43. Garcia-Trivino, P., Torreglosa, J. P., Fernandez-Ramirez, L. M., & Jurado, F. (2018). Decentralized fuzzy logic control of microgrid for electric vehicle charging station. *IEEE Journal of Emerging and Selected Topics in Power Electronics, 6*(2), 726–737. https://doi.org/10.1109/jestpe.2018.2796029
44. Yin, H., Zhou, W., Li, M., Ma, C., & Zhao, C. (2016). An adaptive fuzzy logic-based energy management strategy on battery/ultracapacitor hybrid electric vehicles. *IEEE Transactions on Transportation Electrification, 2*(3), 300–311. https://doi.org/10.1109/tte.2016.2552721
45. Ross, T. J. (1994). Fuzzy Logic with Engineering Applications.

Innovation and Design Model in Marine Decarbonization Activities

Zuhal ER

Abstract Although there is no exact numerical data about electric marine vehicles in the world, it is observed that the trend continues. According to some sources, there are about 2000 electric vessels worldwide as of 2020, the value of this vessel market is estimated at $ 2.6 billion, and it is expected to reach $ 15.6 billion by 2027 with an annual growth rate of 25.2%. In this study, which aims to present the current analyzes and make new design boat models, there are two basic approaches as quantitative (market reviews and new boat design) and qualitative (computation) methods. Additionally, in this study's target, the place of this boat design in sectoral application was examined and electric boat design (14 m long; 60 kW) was also examined. In this study, 14.95 m long, approximately 0.02 drag coefficient, approximately 78 m2 total wet surface area, 16 knots (8.23 m/s) speed, 3 h cruising time for this boat, its power is 58 kW and its battery capacity is approximately 174 kWh and each battery's capacity is 4 kWh, and the total number of batteries is calculated as approximately 42. Furthermore, it was requested to introduce new terminology by the naming of the designed boat.

Keywords Electric boat · Electric ship market · Marine e-mobility · Boat design · Python

1 Introduction

Depending on their functions, structural features, areas of use or the type of cargo they carry, ships can be of different types such as cargo ships (container, liquid or dry cargo), tankers, yachts, river ships. The most common type of ship in the world fleet are container ships. Of these, small containers can vary in length from around 50 to 150 m, ideal for regional and coastal lines over short distances and in smaller ports. It is difficult to have an exact number of how many container ships there

Z. ER (✉)
Faculty of Science and Arts-Physics Engineering Department, Istanbull Technical University, Maritime Faculty, Istanbul, Turkey
e-mail: erzuh@itu.edu.tr

M.-R. Alam and M. Fathi (eds.), *Integrated Systems: Data Driven Engineering*,
https://doi.org/10.1007/978-3-031-53652-6_20

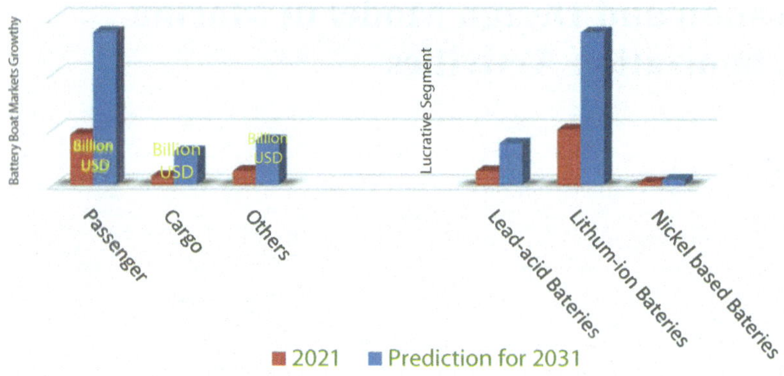

Fig. 1 Electric boat markets perspectives

are worldwide, but according to the International Maritime Organization (IMO) and similar organizations, it is estimated that there are an average of more than 60,000 container ships worldwide. On the other hand, as can be seen in Fig. 1, which I created from literature reviews, according to the world electric boat market analysis, cargo boats are predicted to be the most profitable segment in terms of returns. [1–13, 16, 17, 30–35].

Table 1 typically gives the required power ranges along with the dimensions of the small river type and container ships, but river type ships have different transport methods and capacities than container ships [11–13, 18].

On the other hand, depending on the types of ships and their intended use, AC or DC types of electrical equipment used in ships may be preferred [14, 16, 19, 23–28]. In Table 2, which I summarized from literature examinations, the AC and DC electrical equipment and ship types used in small container ships and river type ships are presented.

Table 2 also presents an overall analysis, taking into account that while sampling equipment selections are made where AC or DC power is preferred depending on vessel types and intended use, these choices may actually vary depending on the ship design, intended use, technical requirements and energy efficiency goals. Thus, it is considered to be examined in the context of the possibility and consistency of the results of the study. In this study, I created a Python code sample to calculate the total DC power consumption of a ship containing such as control panels, sensors, computers and communication equipment. Lithium-Ion (Li-I) batteries are a technology often used in electric vessels. Therefore, what are the World Shipping

Table 1 Comparison of container and river ships

Type of vessels	Size of the vessel	Required power range
Small container ships	50 to 80 m	1,000 to 4,000 horsepower
River type ships	40 to 70 m	500 to 2,000 horsepower

Table 2 Electrical equipment used on ships by ship types (Example: Length assumed 50 m)

Vessels	Equipment type	AC (Alternating Current)	DC (Direct Current)
Container small ships	Lighting, cranes, Ref. sys., others	For example, Total Power: 250 kW: 50 kW, 100 kW, 30 kW, 70 kW	Population system, for example: 500 kW DC LEDs
River type ships	Lighting, cranes, Ref. Sys.,control panels, sensors	For example, total AC power requirement (example: 200 kW: 20 kW, 50 kW, 30 kW, 40 kW, 60 kW	Electric drive system, for example: 400 kW

requirements for such an electric vessel, and are their advantages important? The question comes to mind at this point. The World Maritime Organization (IMO) has a plan called "Strategy 2050", the main objective of which is to reduce greenhouse gas emissions in the maritime sector by about half of current levels by 2050. In addition, IMO presents a range of studies and directions relating to the development of electrically propulsion and managed ships, supporting the maritime industry's efforts to reduce carbon emissions and achieve sustainability targets. At this point, let us emphasize that the urgency of current maritime electric efficiency incentives is due to a number of factors [1, 2, 9, 12, 18–20, 27–35].

There is no exact numerical data on electric marine vehicles in the world. However, it is stated that the use of electric marine vehicles is increasing and this trend is expected to continue in the coming years. According to some sources, as of 2020, there are about 2000 electric marine vehicles worldwide. These vehicles include various types such as yachts, boats and ferries [21, 22, 25]. At this point, this study aimed to provide the objectives of making new design model design by examining the maritime market.

2 Methodology

As a result of the information presented above and various observations, two main objectives: market analysis and new design model design have emerged as the objectives of the study. In this study, two basic approaches were used to achieve these goals, quantitative and qualitative methods. The quantitative information in the study was carried out by examining the relationships between variables such as market and global requirements for electric and green ships. Qualitative methods are to present a new design model that will be included in the analysis of the study. In this context, the motivational elements of the study arose from environmental requirements, increased market share and excitement to introduce new technologically feasible models. In this study, the steps of the methodology applied to analyze the electric ship market are presented in Fig. 2 and each step is tried to be explained briefly below.

Fig. 2 Methodology of the study

In the first step of the methodology of this study for electric solar boat design, the purpose of the study, the research questions and the importance of the design are defined. In addition, a basis was established by examining the literature review and current solar boat designs. In this context, first of all, market research was conducted in the methodology. This has been carried out in the form of various market reports and analyzes in the sector, as well as the examination of academic studies, and the outputs of the information obtained are presented in this study with references at the relevant places. In order to understand the market and statistics, it was necessary to examine the different segmentation of hand-powered marine vehicles. Thus, how the design is planned, how the modeling and analysis will be done, and how the data will be collected have been clarified. As a result of market segmentation and customer analysis, design parameters were determined and the performance and efficiency of the design were analyzed. The results obtained are compared with the targets set at the beginning, and the results obtained are interpreted and the potential effects of the design and future development areas are discussed.

Information on market research has been presented in detail in the previous section to emphasize the purpose of the study. Up-to-date and reliable data on the electric ship market, such as sectoral reports, have been collected. In addition, the past, current and future size of the electric ship market has been examined. Different segments of the market (for example, passenger ships, cargo ships, other sea vehicles) were defined and analyzed. In addition, the future growth and development expectations of the market are mentioned with the analysis of developments and innovations in electric ship technologies.

Another important step of the methodology was SWOT. SWOT analysis can be an important guide for successfully making strategic decisions. In this context. The SWOT for electric vessels is presented in Table 3.

Table 3 Overview of electric vessels with SWOT analysis

Strong	Weakness
Eco-friendly and sustainable alternative	High start-up cost
Low energy costs	Limited range and charging infrastructure
Reduced carbon emissions, environmental	The need to develop battery technology
Quiet operation and vibration-free experience	Limited applicability on large ships
Low maintenance and long service life	Power management and energy storage challenges
Government incentives and support programs	Challenges of integration by existing infrastructure
High energy efficiency and performance	Performance degradation at slow speeds
Opportunities	Threats
Environmental regulations and objectives	Weak use in conventional fuel-powered vessels
Increasing environmental awareness...	High costs and economic constraints
Emergence of new market segments	Increased competition and market saturation
Technological battery developments	Uncertainties and regulations
Expansion of charging infrastructure	Battery technology risks and safety
Innovation and R&D investments	Limited range and charging infrastructure

It was necessary to understand the market and statistics, as well as to examine the different segmentations of electric vessels for the new design model. Related parameters and their considered categorization have exhibited in Fig. 3 in this study.

On the other hand, the market research that can be done to understand the electric ship market is examined in this study with technological and market size parameters

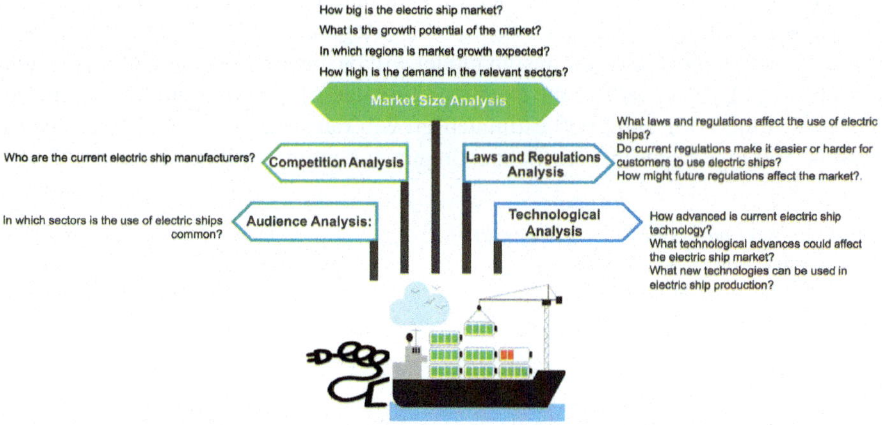

Fig. 3 Questions and classifications used when conducting Ship Market Research

[1, 2, 7, 10, 28, 30–35]. In this case, this study shows that strategies to stand out from the competition can be determined with the design of a new electric marine vehicle, as well as the feasibility of the idea of the new design model. Furthermore, "electromobility", also known as "E-mobility", is the general name given to electrically powered cars and includes fully electric vehicles, hybrid electric vehicles and vehicles that use hydrogen fuel cell technology [1, 2, 7, 16, 23]. Innovative model design has been made according to the usage areas, battery capacities, regions and power sources of the electric marine vehicles described above, which have different segmentations, so that the segment in which the design will be included is exhibited. In this direction, in this study, the segments were based on Segmentation by Power Source and the new design of the electric boat was first studied by determining the power consumption according to the purpose and making other calculations. The Python codes have also used for some calculations in this study.

3 Results and Discussions

With the examinations made, in 2020, the value of the electric ship market is estimated at $2.6 billion, and with an annual growth rate of 25.2%, it is expected to reach $15.6 billion by 2027, and the main reason for this is the environmental regulations of governments, as well as the lower operating costs of electric ships [1, 16, 22, 25]. The fastest growth in the electric watercraft market is expected to be in the Asia–Pacific region [2, 7, 10, 20, 29–35]. Different sources can give different estimates of the growth rate of the electric ship market. In addition, when we look at the issues emphasizing the increase in market share, formations are seen in ports and financing factors. It can be said that Environmental Ship Index, Green Award, Clean Transport Index, GG Emission values, which express the environmental performance of a single ship according to e-indices, are effective in the increase in the green port fees market. In this study, the segments of the electric ship market have been evaluated such as Area of Use, Battery Capacity, Power Sources, Regional, and calculations have been made by deciding that the new design model will be in the class of small ships, which is widely used. As a result of the examined literature, Fig. 4 provides a proportional analysis of the 2021 and 2030 estimated growth rates for hybrid and fully electric vessels, and Fig. 5 presents segment estimates for energy storage systems [28–35].

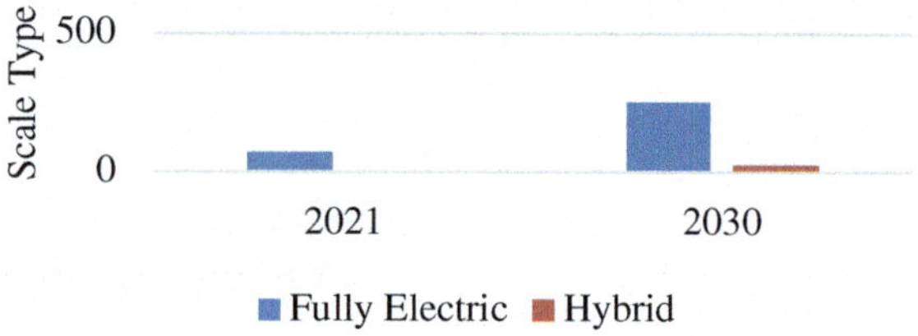

Fig. 4 By type, projection of the electric ship market of the all-electric segment from 2021 to 2030

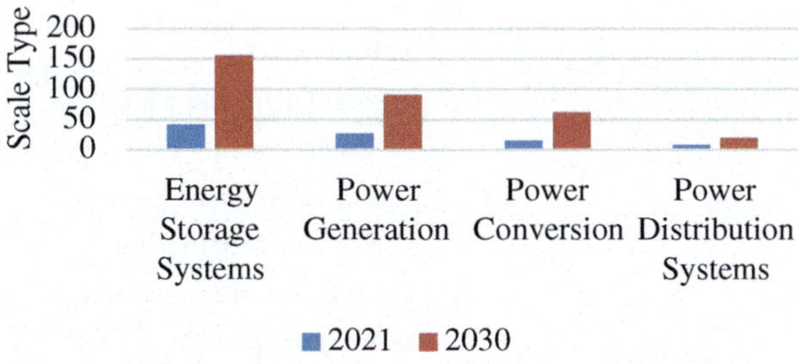

Fig. 5 Energy storage systems segment forecasts by systems in electric ship markets

As a result of all the above considerations, the fact that the most common ship type in maritime is the container and the rapid increase in the electric ship market has demonstrated the importance of new design ships. For this reason, as mentioned in the methodology section, the purpose and requirements of the electric boat, which is the first step for the new design model, were determined in the study.

The design of a bot is based on many factors and is a complex process. In this study, a Python code (Appendix 1) sample has been created to calculate the total DC power consumption of the boat. In this code, a function named calculate_dc_power_consumption has defined as below:

```
def calculate_dc_power_consumption(panels, sensors, computers,
communication):
```

This function includes the parameters as given below.

```
panels_power = float(input("Enter the power consumption (watts) of
the control panels: "))
sensors_power = float(input("Enter the power consumption (watts)
of the sensors: "))
computers_power = float(input("Enter the power consumption (watts)
of computers: "))
communication_power = float(input("Enter the power consumption
(watts) of communication equipment: "))
```

Within the function, it sums the power consumption of the given components and assigns them to the total_dc_power_consumption variable. Finally, the function returns the total_dc_power_consumption value. Prints the result to the screen using the print function.

As a result of the needs analysis conducted to determine the factors such as the purpose of use of the boat, passenger/cargo capacity, cruising time, speed requirements, in this study, an innovative model of an innovative boat for a 14-seater, 14 m long, 4 m wide, 60kWs boat with a speed of 16 knots, the estimated weight, coefficient, estimated resistance and electric motor power are calculated as shown in the Table 4. Schematic diagram of the innovated is given in Fig. 6.

In addition, a Python code was written in this study, and thus the horsepower required by a motor running at 16 knots is calculated in kW and Hp. This code first performs resistance calculations, then calculates weight and motor power. In the design in this study, the propulsion system was connected to the power supply

Table 4 New innovative design model properties

Parameters	Value
Boat length and width	14 m—4 m
Displacement—Estimated total weight	128.8 ton-129.85 ton
Estimated load	1050 kg
Estimated coefficients	0.99
Cp Predictive resistance	0.8
V (Speed)	16 knot (8,24 m/sn)
S (Underwater section)	56 m2
ρ (Density of sea water)	1025 kg/m3
Cd (Coefficient of resistance)	0.6
Resistance	118,361 N
Electric motor power	60 kW
HP motor power	1773 BG

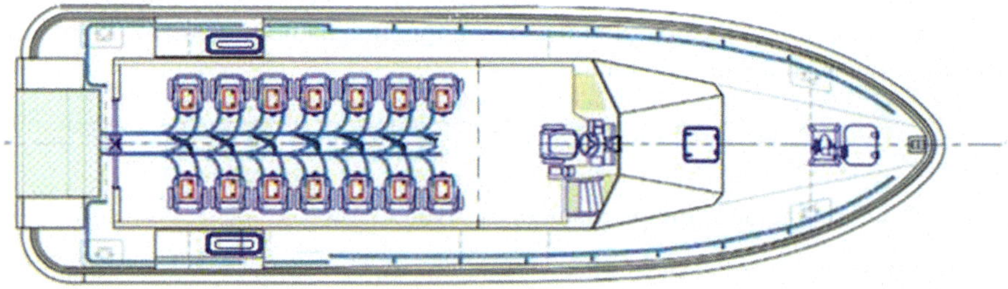

Fig. 6 New design electric battery boat schematic diagram

and moved the boat. The power supply is designed as a battery pack consisting of lithium-ion batteries. The electric motor moves with the electrical energy it receives from the battery pack. In the study, 30% of the battery for a boat with a speed of 16 knots was accepted as the operating time required to maintain speed. The charging time of ship batteries will vary depending on the capacity of the batteries used and the characteristics of the charging system. However, to make an approximate assessment, another Python code (Appendix 2) is also written that takes input from the user, both to calculate the battery capacity and to calculate the number of batteries required.

```
# Prompting the user to enter power consumption (watts), operating
time (hours) and charging power (kW)
power_consumption = float(input("Enter the power consumption
(watts): "))
operational_hours = float(input("Enter the operating time (hours):
"))
charging_power = float(input("Enter charging power (kW): "))
```

is made in the form of. This code, calculate_charging_time function calculates the charging time by taking the total capacity and charging power. The function assigns the charging time to the charging_time variable by dividing the total capacity by the charging power and returns this value. calculate_battery_capacity function calculates the battery capacity by taking the power consumption and operating time. The function calculates the battery capacity by multiplying the power consumption by the operating time in the battery_capacity variable and returns this value. calculate_required_batteries function calculates the required number of batteries, taking the total capacity and battery capacity. The function calculates the required number of batteries by dividing the total capacity by the battery capacity to the required_batteries variable and returns this value. Asking the user for power consumption, operating time, charging power and total capacity, it assigns these values to the appropriate variables. Using the calculate_battery_capacity function, it calculates the battery capacity and assigns it to the battery_capacity variable. Calculates the charging time using the calculate_charging_time function and assigns it to the charge_time variable. The user is prompted to enter the total capacity. Using the calculate_required_batteries function, it calculates the required number of batteries and assigns it to the required_batteries variable. To print the results on the screen, it uses the print function to print the battery capacity, the number of batteries required and the charging time on the screen.

In this way, both the ease of calculation and the flexibility of doing the work with Python code is the original part of this work that offers originality and novelty because it is a new design model that includes analyzes. Discharged batteries can be easily replaced with charged batteries upon arrival at port. In the boat, where renewable energy is used for support purposes, there are spare batteries to be used in emergency situations with controller that monitors and manages battery status. The coding parts of this study have been supporter that will find a place in these subjects. This is the feature of this study and it is shown in Fig. 7.

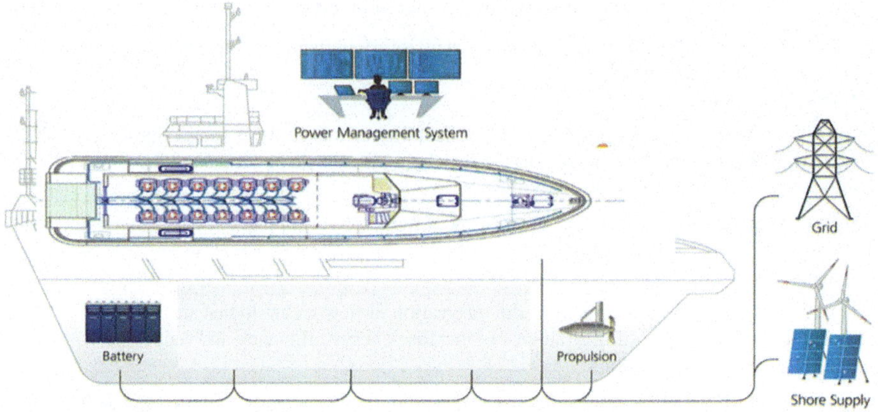

Fig. 7 Battery servicing boat and port opportunities schematic illustration

When developing, designing and creating an electromotor complex for high-power vessels with an installed capacity of a single electric propulsion of more than 5 MW, there are problems with the feasibility of such systems on the existing element base; Most of the designed/manufactured vessels have a single electric propulsion installed capacity of 2 ÷ 3 MW as shown in the Fig. 8.

Comparisons of this study with innovative electric watercraft, battery supply boat studies have been made and it has been seen that it is consistent with both literature studies and those in practice. Some literature benchmark compatibilities were exhibited in the Table 5.

At this point, the answer to the question "What is the total share of this new design?" will explain the place of this new innovative design model in the maritime sector. As mentioned above, in the designed electric vessel, the transfer of energy from

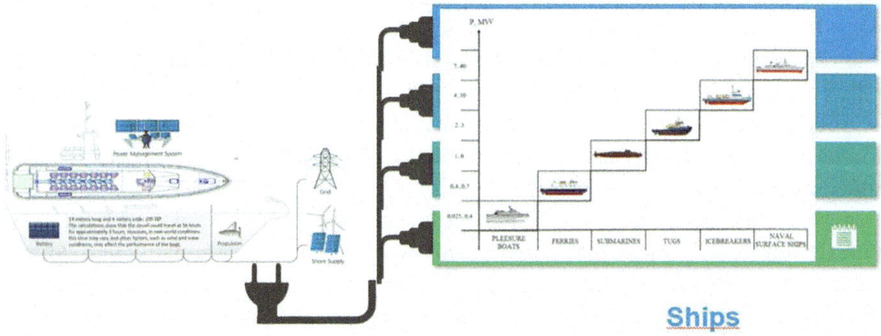

Fig. 8 Installed capacity of a single electric propulsion of vessels and new design model schematic illustration

Table 5 Some examples of literature benchmark compatibilities

DOI	This study is compatible with doi in left column	Year
https://doi.org/10.1016/j.ijepes.2021.107484. [31]	In the context of power availability in the all-electric marine industry	2022
https://doi.org/10.3390/en14196246 [32]	In the context of the applicability of this study, it shows that it enables power sharing between all-electric batteries and facilitates soft mode switching	2021
https://doi.org/10.1109/jsyst.2021.3111685 [33]	In the context of stability of marine hybrid power systems using dc power architecture	2021
https://doi.org/10.1109/tec.2020.2997307 [34]	In the context of the integration of new technologies and innovative operating strategies to increase energy efficiency and reduce gas emissions	2020
https://doi.org/10.1109/tia.2020.3034290 [35]	In the context of inspection and maintenance of ship microgrids on purely electric ships (AES) and forecasting electricity prices by coding in different ports	2020

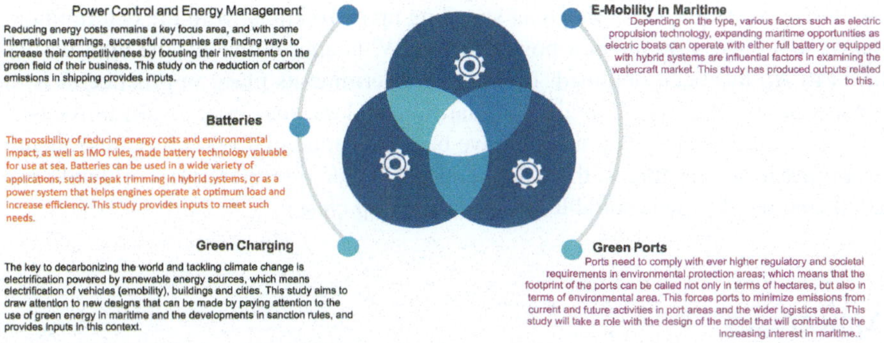

Fig. 9 Output evaluations of the study by review statistical analyses and the designed model

the power supply to the engines has done directly, without transmission, differential or other mechanical components. This increases efficiency and reduces maintenance costs. In addition, the electric boat is quiet and low emission, and therefore environmentally friendly. With this study, the requirements and technological examinations in maritime affairs have been made and a new innovative design model has been made, and the electric marine vehicle related to the application with green ports has been revealed. The new design can be shown in the figure of the opportunities for the Battery Supply Electric Boat. Finally, the association of all these aspects with each other is also presented in the Fig. 9.

This study is a model study, but there are things to be done for the production of this model. As part of the design process, it can include measures such as drag reduction, aerodynamic design, energy recovery and optimized cruise profiles, with the energy efficiency analysis of the boat. By creating a computational prototype of the designed electric boat, performance, safety, energy efficiency and other features were verified from other online opportunities. From the point of view of this study, when previous studies were examined, it was found that it was possible to design such a boat, and many electric boats had these characteristics. Nonetheless, for practical use, it is desirable to introduce the boat designed in this study as a "battery service (refueling) boat" which is offering new terminology in the sector.

4 Conclusion

While the urgency of current marine electricity efficiency incentives stems from a number of factors, there are effective developments such as IMO requirements. In this study, the rapid growth statistics of the electric ship market in maritime are presented and the determinations for the design are presented. As a result, with the calculations made in this study, if the engine power is 80.5 HP (approximately 60 kW), assuming 90% propulsion efficiency, the calculation of the battery capacity corresponding to the engine power required for this boat speeding up to 16 knots, with the determination that it will have a propulsion power of 54 kW, is approximately the end that will be 1146 Ah has been obtained. In order to determine its place in practice, it is the most common ship type container in maritime and the statistical results of the rapid increase in the electric ship market have been compiled in this study. Consequently, according to some sources, there are about 2000 electric watercraft worldwide as of 2020, and the design is suitable for the small ship class.

Appendixes

Appendix-1

```
def calculate_dc_power_consumption (panels, sensors, computers,
communications):
# Calculate total DC power consumption for control panels, sensors,
computers and communication equipment
total_dc_power_consumption = panels + sensors + computers + commu-
nication
Back to total_dc_power_consumption
# Get power consumption (watts) of control panel, sensor, computer
and communication equipment from the user
panels_power = float(input("Enter the power consumption (watts) of
the control panels: "))
sensors_power = float(input("Enter the power consumption (watts)
of the sensors: "))
computers_power = float(input("Enter the power consumption (watts)
of computers: "))
communication_power = float(input("Enter the power consumption
(watts) of communication equipment: "))
# Call function to calculate total DC power consumption
total_dc_power = calculate_dc_power_consumption(panels_power,
sensors_power, computers_power, communication_power)
# Print the result to the screen
print("Total DC power consumption: {:.2f} watts".format(total_dc_
power))
```

Appendix-2

```
import math
boat_length, drag_coefficient, wetted_surface_area, boat_speed,
cruise_time, power, battery_capacity = 14.95, 0.02, 78, 8.23, 3, 58,
174
battery_count = math.ceil(0.5 * drag_coefficient * wetted_surface_
area * (boat_speed ** 3) * cruise_time * 3600 / (battery_capacity *
1000))
print("Number of batteries required for the boat:", battery_count)
```

References

1. Tupper, E. C. (2013). Chapter 15 ship types. In B. Levenspiel (Ed.), (s. 15–1 - 15–10). https://doi.org/10.1016/b978-0-08-098237-3.00015-1.
2. Zhang, Y., and Cramer, A. M. (2017). "Market-based control of electric ship power systems," *Electric Ship Technologies Symposium 14–17 August 2017 (s. 1–8).*, Arlington, VA, USA, IEEE. https://doi.org/10.1109/ests.2017.8069309.
3. Longobardi, G., Efthymiou, L., & Arnold, M. (2018). "GaN power devices for Electric Vehicles State-of-the-art and future perspective," *IEEE International Conference on Electrical Systems for Aircraft, Railway, Ship Propulsion and Road Vehicles & International Transportation Electrification Conference (ESARS-ITEC) 07–09 November 2018 (s. 1–6)*, Nottingham, UK, IEEE. https://doi.org/10.1109/esars-itec.2018.8607788.
4. Kobzar, T., Savinok, O., & Marinicheva, K. (2022). Demands and challenges for construction of marine infrastructures in China. *Frontiers of Structural and Civil Engineering. Springer, 16*(5), 551–563. https://doi.org/10.1007/s11709-022-0839-8.
5. Olsen, A. A. (2023). *Merchant Ship Types* (p. 9781032378756). Routledge.
6. Brigham, L. W. (2001). The Northern Sea Route, 1999–2000. *Polar Record, 37*(203), 329–336. https://doi.org/10.1017/S003224740001706X.
7. United Nations, Review of Maritime Transport 2021, New York, USA, 2021. ISBN: 978–92–1–113026–3.
8. Bucci, V., Marino, A., Bosich, D., & Sulligoi, G. (2015) Inland waterway gas-fuelled vessels: An innovative proposal of a hybrid ship for the European network. *International Conference on Electrical Systems for Aircraft, Railway, Ship Propulsion and Road Vehicles 2015, Aachen,* Germany, IEEE, 2015. https://doi.org/10.1109/esars.2015.7101474.
9. Ivce, R., Tomic, L., Zekic, A., and Miskovic, D. (2022). Impact of the COVID-19 on Container Ships Efficiency. *Journal of Maritime Sciences (JMS)*, UDC: 629.544:616.98:578.834, https://doi.org/10.56080/jms221109.
10. Silva C. A.,Guedes Soares C. (2014). Sizing a fleet of container ships for a given market. *Traffic and Transportation, 26*: 4, 333–344. https://hrcak.srce.hr/file/192222.
11. Yu, Z. Z. (2021). Design and implementation of charging and discharging device for electric inland river cruise ship. *International Journal of Applied Engineering Research, 10*(1), 22–30. https://doi.org/10.1088/1742-6596/1750/1/012061.
12. Mykhailova, I., Navrozova, Y., & Steba, A. (2022). Features of the organization of cruise transportation of passengers on yachts. *Technical Sciences, 45*(2), 276–282. https://doi.org/10.31498/2225-6733.45.2022.276272
13. Grida, M., & Lee, C. Y. (2018). An empirical model for estimating berth and sailing times of mega container ships. *Maritime Policy & Management, 45*(6), 789–805. https://doi.org/10.1080/03088839.2018.1452304
14. Perabo, F., Park, D., Karbalaye Zadeh, M., Smogeli, O., & Jamt, L. (2020). Digital twin modelling of ship power and propulsion systems: application of the Open Simulation Platform (OSP). *International Symposium on Industrial Electronics, 17–19 June 2020, pp. 234–240.* Delft, Netherlands, IEEE, 2020. https://doi.org/10.1109/isie45063.2020.9152218
15. Baldi, F., Nguyen, T. V., & Ahlgren, F. (2016). The application of process integration to the optimization of cruise ship energy systems: a case study. *Energy Conversion and Management*, http://lnu.diva-portal.org/smash/get/diva2:957811/FULLTEXT01.pdf
16. Huberts, J. (2018). Battery boat: Concept study in bulk electric energy sea transport. *Master of Science at the Delft University of Technology*. available at http://repository.tudelft.nl/
17. Jin, Z., Wang, X., Sun, J., & Xu, Q. (2022). Optimization of fleet structure and investment evaluation from the cargo owner's fleet perspective. *Maritime Business Review, 7*:3, 2022 pp. 239–254 Emerald Publishing Limited 2397–3757 https://doi.org/10.1108/MABR-07-2021-0054
18. Wang, W., Liu, Y., Zhen, L., & Wang, H. (2022). How to deploy electric ships for green shipping. *Journal of Marine Science and Engineering, 10*(11), 1611. https://doi.org/10.3390/jmse10111611

19. Bennabi, N., Charpentier, J. F., Menana, H., Billard, J. Y., & Genet, P. (2016). Hybrid propulsion systems for small ships: Context and challenges, *International Conference on Electrical Machines 4–7 September. 2016, pp. 456–461*, Lausanne, Switzerland, IEEE, 2016. https://doi.org/10.1109/icelmach.2016.7732943
20. Mjelde, A., Endresen, O., Bjorshol, E., Gierloff, C. W., Husby, E., Solheim, J., Mjosa, N., & Eide, M. S. (2019). Differentiating on port fees to accelerate the green maritime transition. *Marine Pollution Bulletin, Elseiver, 149*, 110561. https://doi.org/10.1016/j.marpolbul.2019.110561
21. Mella, G., & Pentucci, P. P. (2022). EU policies on sustainable transport in Cross-Border Maritime areas connecting european coasts. *European Journal of Sustainable Development*, pp 1–26. https://doi.org/10.30687/978-88-6969-617-6/001
22. Nuchturee, C., Li, T., & Xia, H. (2020). Energy efficiency of integrated electric propulsion for ships—A review. *Renewable and Sustainable Energy Reviews, 134*, 110145. https://doi.org/10.1016/j.rser.2020.110145
23. Nuchturee, C., Li, T., (2018). A review of energy efficient methods for all-electric ships. *IOP Conference Series: Earth and Environmental Science 188*(2018) 012050. https://doi.org/10.1088/1755-1315/188/1/012050
24. Bolbot, V., Theotokatos, G., Boulougouris, E., & Vassalos, D. (2019). Comparison of diesel-electric with hybrid-electric propulsion system safety using system-theoretic process analysis. *RINA-Power & Propulsion Alternatives for Ships 2019 Papers, GBR*. ISBN 9781909024878. https://doi.org/10.3940/rina.ppa.2019.08
25. McCoy, T. J., & Amy, J. (2009). The state-of-the-art of integrated electric power and propulsion systems and technologies on ships. *IEEE Electric Ship Technologies Symposium Baltimore, 20–22 April 2009 MD, USA,* 2009. Available: IEEE Xplore, https://ieeexplore.ieee.org/document/4906534 https://doi.org/10.1109/ests.2009.4906534
26. Pense, C., & Akinoglu, B. G. (2020). Using renewable energy on electric vessels in coastal regions. *2nd International Conference on Photovoltaic Science and Technologies (PVCon) 2020*. Ankara, Turkey https://doi.org/10.1109/pvcon51547.2020.9757767
27. Chalfant, J., Kite-Powell, H., Bonfiglio, L., & Chryssostomidis, C. (2022). Decarbonization of the cargo shipping fleet. *Journal of Ship Production and Design, 38*(1), 1–18. https://doi.org/10.5957/jspd
28. Prousalidis, J., Spathis, D., Spathis, D., Papaspilitopoulos, V., Georgiou, V., Kourmpelis, T., Kanellos, F. D., & Kanellos, F. (2019). Facing safety challenges towards smart ships and ports. *Proceedings of Marine Electrical and Control Systems Safety Conference (MECSS), 2–3 July 2019*. https://doi.org/10.24868/issn.2515-8198.2019.005
29. Kalgora, B., & Christian, T. M. (2016). The financial and economic crisis, its impacts on the shipping industry, lessons to learn: The Container-Ships market analysis. *Open Journal of Social Sciences, 4*, 38–44. https://doi.org/10.4236/jss.2016.41005
30. Minami, S., Tsukuda, K., Koizumi, K., & Ikeda, H. (2015). Remarkable CO2 reduction of the Fixed-Point fishing Plug-in hybrid boat. *Journal of Asian Denmark Vehicles*. https://doi.org/10.4130/jaev.13.1719
31. Allied Market Research. (2023). Electric boat market. Retrieved from https://www.alliedmarketresearch.com/electric-boat-market-A08766
32. Milousi, C., et al. (2019). [Report Brochure]. Retrieved from https://www.researchandmarkets.com/reports/5317213/global-electric-ships-market-size-share-value
33. International Transport Forum. (2018). [Report]. In toward zero carbon shipping: ambitious decarbonization agenda for the maritime transport sector. Paris, France: OECD Publishing. https://doi.org/10.1787/9789282108114-en
34. Wartsila. (2021). [Report]. Retrieved from https://www.wartsila.com/insights/article/green-shipping-is-the-new-gold
35. Chris Busch, John Laitner, Rob McCulloch, Ivana Stosic. (2012). Gearing up: smart standards create good jobs building cleaner cars. Blue Green Alliance/ACEEE, Washington, D.C., [Report]. Retrieved from https://www.bluegreenalliance.org/wp-content/uploads/2016/08/062812-GearingUp_Final.pdf

36. Doerry, N. H., & Fireman, H. (2006). Designing All Electric Ships. *Proceedings of the Ninth International Marine Design Conference, 16–19 May* 2006, Ann Arbor, MI, USA http://doerry.org/norbert/papers/060325imdcdesigningallelectricwarships1-presentation-final.pdf
37. Esteve-Pérez, J., & Gutiérrez-Romero, J. E. (2015). Renewable Energy Supply to Ships at port. In Sixth International Workshop on Marine Technology, Martech, Cartagena (pp. 169–172). ISBN: 978–84–608–1708–6
38. Faber, J., et al. (2021). Fourth IMO GHG Study 2020 executive summary. *International Maritime Organization.* Retrieved from https://www.imo.org/en/MediaCentre/PressBriefings/pages/19-GHG-Fourth-IMO-GHG-Study-Executive-Summary.aspx
39. Power Calculation. (2021). Calculation of battery pack capacity, c-rate, run-time, charge and discharge current. Retrieved from https://power-calculation.com/battery-storage-calculator.php.
40. IEA. (2021). Global EV Data Explorer. Retrieved from https://www.iea.org/articles/global-ev-data-explorer
41. Mordorintelligence. (2021). Electric boat and ship market - growth, trends, covid-19 impact, and forecast (2022—2027). Retrieved May 20, 2021, from https://www.mordorintelligence.com/industry-reports/electric-boat-and-ship-market
42. Hand Marcus. (2021). With sustainability demand growing maritime is moving towards a global green ship finance market with a major demand expected for retrofit funding, according to KfW IPEX Bank.. Retrieved June 17, 2021, from https://www.seatrade-maritime.com/finance-insurance/moving-towards-global-green-ship-finance-market
43. Wartsila. (2021). Green shipping is the new gold. Retrieved August 4, 2021, from https://www.wartsila.com/insights/article/green-shipping-is-the-new-gold
44. Sidun Fang; Yu Wang; Bin Gou; Yan Xu. Toward Future Green Maritime Transportation: An Overview of Seaport Microgrids and All-Electric Ships. *IEEE Transactions on Vehicular Technology. 69*:1, January 2020 https://doi.org/10.1109/TVT.2019.2950538
45. Z. Er. D. Ates. Evaluation of electric ships' markets, and considered new design marine cabin combined with solar power. *Journal of Maritime Transport and Logistics*, 04(01): 01–10 (2023) https://doi.org/10.52602/mtl.1036256

The manufacturer's authorised representative in the EU is Springer Nature Customer Service Centre GmbH, Europaplatz 3, 69115 Heidelberg, Germany. If you have any concerns regarding our products, please contact ProductSafety@springernature.com

Printed and bound by CPI Group (UK) Ltd, Croydon, CR0 4YY
01/10/2026
02225870-0011